T0186626

Air Pollution and Turbulence

Modeling and Applications

Air Pollution and Turbulence

Modeling and Applications

Edited by
Davidson Moreira and
Marco Vilhena

CRC Press
Taylor & Francis Group
Boca Raton London New York

CRC Press is an imprint of the
Taylor & Francis Group, an **informa** business

CRC Press
Taylor & Francis Group
6000 Broken Sound Parkway NW, Suite 300
Boca Raton, FL 33487-2742

First issued in paperback 2019

© 2010 by Taylor and Francis Group, LLC
CRC Press is an imprint of Taylor & Francis Group, an Informa business

No claim to original U.S. Government works

ISBN-13: 978-1-4398-1144-3 (hbk)
ISBN-13: 978-0-367-38481-4 (pbk)

This book contains information obtained from authentic and highly regarded sources. Reasonable efforts have been made to publish reliable data and information, but the author and publisher cannot assume responsibility for the validity of all materials or the consequences of their use. The authors and publishers have attempted to trace the copyright holders of all material reproduced in this publication and apologize to copyright holders if permission to publish in this form has not been obtained. If any copyright material has not been acknowledged please write and let us know so we may rectify in any future reprint.

Except as permitted under U.S. Copyright Law, no part of this book may be reprinted, reproduced, transmitted, or utilized in any form by any electronic, mechanical, or other means, now known or hereafter invented, including photocopying, microfilming, and recording, or in any information storage or retrieval system, without written permission from the publishers.

For permission to photocopy or use material electronically from this work, please access www.copyright.com (http://www.copyright.com/) or contact the Copyright Clearance Center, Inc. (CCC), 222 Rosewood Drive, Danvers, MA 01923, 978-750-8400. CCC is a not-for-profit organization that provides licenses and registration for a variety of users. For organizations that have been granted a photocopy license by the CCC, a separate system of payment has been arranged.

Trademark Notice: Product or corporate names may be trademarks or registered trademarks, and are used only for identification and explanation without intent to infringe.

Library of Congress Cataloging-in-Publication Data

Air pollution and turbulence : modeling and applications / edited by Davidson Moreira and Marco Vilhena.
    p. cm.
"A CRC title."
Includes bibliographical references and index.
ISBN 978-1-4398-1144-3 (alk. paper)
    1. Air--Pollution--Simulation methods. 2. Atmospheric turbulence--Simulation methods. I. Moreira, Davidson. II. Vilhena, Marco.

TD890.A364 2010
628.5'3011--dc22
                                2009039105

**Visit the Taylor & Francis Web site at
http://www.taylorandfrancis.com**

**and the CRC Press Web site at
http://www.crcpress.com**

Dedication

We thank God for the opportunity given to mankind to uncover the beauty and mystery of the masterpiece of creation: Nature.

To my daughter, Evelyn
To my wife, Márcia
I give my gratitude for
her loving patience
and support during this
episode of my life,

In memoriam to my father, Paulo
To my mother, Ieda
To my sister, Tânia
To my wife, Sônia
With all my love and gratitude,

Davidson Martins Moreira

Marco Túllio M. B. de Vilhena

Contents

Foreword

Air pollution is inherently linked to human activities and was already mentioned as a nuisance in antic Roman texts and in the middle ages. The industrial revolution in the nineteenth century worsened its effects and increasingly turned it to a nonlocal problem. Parallel developments in physical sciences provided new tools to address this problem. Understanding the rate and patterns of atmospheric dispersion is crucial for environmental planning (location of industrial plants) and for forecasting high pollution episodes (above legislation thresholds inducing detrimental effects on human health, ecosystems, and/or materials). Last but not least, local emissions are transported by air motions to create regional environmental problems, and, finally, the accumulation of pollutants in the global atmosphere yield and interfere with climate change processes. Consequently, there is a strong need for developing ever-better models and assessment tools for air pollution concentration, dispersion, and effects. These tools can span from simple analytical models for monitoring and predicting short-range effects to regional or global three-dimensional models assimilating a wide range of physical and chemical in situ and satellite observations. The breadth of the different mathematical, physical, chemical, and biological processes and issues has generated a lot of basic and applied research that should also take into account the needs of environmental managers, physicians, and also of process engineers and lawyers. No book can tackle all these issues in a balanced way; therefore, this book mainly addresses issues of atmospheric dispersion modelling and their effects on building surfaces.

To assess spatial and temporal distributions of pollutants and chemical species in the air and their deposition on the Earth's surface, atmospheric dispersion and chemical transport models are used at different scales, addressing different applications from emergency preparedness, ecotoxicology, and air pollution effects on human health to global atmospheric chemical composition and climate change. During the last two decades, several basic aspects of air pollution modeling have been substantially developed, thanks to advances in computer technologies and numerical mathematics, as well as in the physics of atmospheric turbulence and the atmospheric boundary layer (ABL).

Most air quality modeling systems consist of a meteorological model coupled offline or online to emission and air pollution models, and, sometimes, also to a population-exposure model. The meteorological model calculates three-dimensional fields of wind, temperature, relative humidity, pressure, and, in some cases, turbulent diffusivity, clouds, and precipitation. The emissions model estimates the amount and chemical composition of primary pollutants based on process information (e.g., traffic intensity) and day-specific meteorology (e.g., temperature for biogenic emissions). The outputs of these emission and meteorological models are then inputs to the air pollution model, which calculates concentrations and deposition rates of gases and aerosols as a function of space and time. There are various mathematical models that can be used to simulate meteorology and air pollution in a mesoscale domain (San Jose et al., 2008). Although models differ in their treatment of different mechanisms and feedbacks, they all employ a similar framework and consist of the same major modules:

- *Transport and diffusion*—calculating three-dimensional motion of gases and aerosols in a gridded model domain
- *Gas-phase chemistry*—calculating changes in gaseous concentrations due to chemical transformations
- *Aerosol*—calculating size distribution and chemical composition of aerosols accounting for chemical and physical transformations
- *Cloud/fog meteorology*—calculating physical characteristics of clouds and fog based on the information from the meteorological model (or from observations)
- *Cloud/fog chemistry*—calculating changes in chemical concentrations in clouds/fog water
- *Wet deposition*—calculating the rates of deposition due to precipitation (and, possibly, cloud impaction and fog settling) and the corresponding changes in chemical concentrations
- *Dry deposition*—calculating the rates of dry deposition for gases and aerosols and the corresponding changes in their concentrations

Consequently, the quality of the air pollution forecasts using such systems critically depends on the adequacy in mapping emissions, representing meteorological fields, and modeling the transport, dispersion, and transformation of chemicals/pollutants. Various scientific developments now allow models to reasonably predict simple flow situations within a factor of 2 or so.

What is more challenging is to predict episodes of high pollutant concentrations, which may cause dramatic impacts on human health. Such situations, moreover, are often induced by special situations, such as complex terrains, low winds, and very stable stratification causing shallow ABLs with low level of turbulent mixing. These situations create problems for current methods and models to realistically reproduce meteorological input fields.

The key physical mechanisms controlling concentrations of pollutants in the atmosphere are advection, turbulent diffusion, wet and dry deposition, and gravitational settling. Their representation requires 3D fields of the wind velocity and direction, static stability (lapse rate), the ABL height (often called "mixing height"), basic characteristics of turbulence (eddy diffusivities and velocity variances across the atmosphere, and turbulent fluxes of momentum, buoyancy, and scalars at the surface and at the ABL outer boundary), and precipitation. Additionally, boundary conditions described by the basic physical and geometric characteristics of the surface (in particular, the roughness lengths for momentum and scalars, and the displacement heights) are very critical.

Most of the emissions are situated and most of the pollutants are dispersed within the ABL, whose upper boundary (the layer at which the intensity of turbulence strongly drops down) serves as a kind of a semi-impervious lid. Hence the mechanisms controlling concentrations strongly depend on the ABL turbulence, and, first of all, on the ABL height. The temporal and spatial variations in the ABL height and the entrainment processes at the ABL upper boundary lead to the penetration of pollutants from the ABL to the free troposphere, and, vice versa, to the intrusion of some chemical compounds (e.g., ozone) from the upper atmospheric layers down to the surface. Physical processes controlling the ABL height and the turbulent entrainment (e.g., Zilitinkevich, 1991; Zilitinkevich et al., 2007a; and references therein) are, therefore, of crucial importance for the air-pollution applications.

Furthermore, some physical processes at the ABL upper boundary, crucially important for air pollution modelling, are still insufficiently understood, e.g., turbulent entrainment in rapidly deepening convective ABLs and nonsteady interactions between the stable ABLs and the free flow. The latter are comparatively simple at mid-latitudes where nocturnal stable ABLs develop on the background of almost neutrally stratified *residual layers*, whereas at high latitudes, long-lived stable ABLs develop against very stable stratification typical of the free troposphere, yielding the formation of strong *capping inversions* and making the theory much more complicated (e.g., Zilitinkevich and Esau, 2007).

For short-range dispersion of simple cases or targeted plumes, one classical modeling approach is based on using the so-called statistical technique or the eddy diffusivity concept. Several chapters in this book address new developments with these techniques. Therefore, new developments in turbulence theory and ABLs will have a direct impact on these techniques as well. For instance, one typical long-lasting issue has been the turbulence closure for very stable stratification (including the turbulent diffusion formulations), whereby the energetics of turbulence is modeled using solely the turbulent kinetic energy budget equation, leading to a cut off in turbulence at "supercritical" stratification, though observations showed the presence of turbulence in typical atmospheric and oceanic sheared flows. The problem was treated heuristically by prescribing a "minimal diffusivity"—just to avoid the total decay of turbulence. New insight might come from recent work based on the concept of total turbulent energy and applicable to "supercritical" flows with *no cut off* (Mauritsen et al., 2007; Zilitinkevich et al., 2007b; Canuto et al., 2008). Another area of potential development is the generalization of the Monin–Obukhov similarity theory, taking into account the nonlocal effect of free-flow stability on stably stratified ABLs and also nonlocal mixing due to large-scale, organized eddies in the shear-free convection (Zilitinkevich et al., 2006; Zilitinkevich and Esau, 2007). Further work is also needed to extend the ABL theory to the sheared convection and to ABLs over complex and sloping terrains.

During the last decade, meso-scale modeling of pollution dispersion and air quality employing the integrated modelling approach together with advances in ABL physics reported above have been developed in both research and operational modes (see an overview of European models in COST-WMO, 2007).

Short-term pollution episodes occurring during adverse meteorological conditions and causing strong short-term exceedances of air quality standards in ambient air are presently one of the major concerns for the protection of human health, ecosystems, and building materials, especially in cities. Reliable urban-scale forecasts of meteorological fields are, therefore, of primary importance for urban emergency management systems, addressing accidental or terrorist releases, and fires, of chemical, radioactive, or biological substances.

The urban environment presents challenges to atmospheric scientists—theoreticians, experimentalists, and modellers—because of very high roughness elements penetrating deeply into the ABL (thus requiring the revision of such classical concepts as the *surface layer, roughness length,* and *displacement height;* see Zilitinkevich et al., 2008), the heterogeneous distribution of surface features, and the strong spatial and temporal variabilities of surface fluxes of heat, moisture, momentum, and pollutants. Additionally, the structure of the conurbation may enhance vertical motions, changing the residence times of atmospheric compounds (Hidalgo et al., 2008) and triggering local meteorological circulations (e.g., caused by "heat islands"), and the production of condensation nuclei, thus affecting cloud formation, precipitation patterns, and the radiation balance. The increased relevance of urban meteorology is reflected in the number of experimental campaigns performed in urban areas in Europe and America during the last decade, e.g., BUBBLE (Rotach et al., 2005), ESCOMPTE (Mestayer et al., 2005), CAPITOUL (Masson et al., 2009), and MILAGRO (Molina et al., 2007).

The incorporation of urban effects into air pollution models is generally carried out through the "urbanization" of meso-meteorological or numerical weather prediction (NWP) models (which act as driver models), or using special urban meteo-pre-processors to improve non-urbanized NWP input data (COST-715, 2005).

The persistently increasing resolution in NWP models allows to reproduce more realistically urban air flows and air pollution, and triggers interest in further experimental and theoretical studies in urban meteorology. Recent works performed by a consortium of an European project, EMS-FUMAPEX 2005, on integrated systems for forecasting urban meteorology and air pollution, and by the U.S. EPA and NCAR communities employing the models MM5 (Dupont et al., 2004; Taha, 2008) and WRF (Chen et al., 2006), as well as other relevant works (see COST-728, 2009), have disclosed many options for the urbanization of NWP and meso-meteorological models.

It goes without saying that no single book could cover the entire range of problems listed above. The scope of this book does not intend such a grand task. It rather reflects and summarizes some recent developments relevant to the key issues in modeling atmospheric turbulence and air pollution. Chapter 1 deals with the modelling of deposition, transformation and remobilization of soot and diesel particulates on building surfaces, damage to facades and decoration by air pollution, and the human health aspect of air pollution (Brimblecombe and Grossi, 2005). Chapter 2 describes observational studies of convective ABLs over pastures and forests in Amazonia (Fisch et al., 2004). Chapter 3 discusses the theoretical studies of turbulence and turbulent diffusion in convective ABLs (Degrazia and Anfossi, 1998; Goulart et al., 2003). Chapter 4 describes the parameterization of convective turbulence and clouds in atmospheric models based on the combination of the eddy-diffusivity and mass-flux approaches (Soares et al., 2004; Siebesma et al., 2006). Chapter 5 contains a general discussion of analytical solutions to the advection-diffusion equations (Tirabassi, 1989, 2003). Chapter 6 describes analytical models for air pollution including those for low wind conditions (Sharan et al., 1996; Sharan and Modani, 2005). Chapter 7 deals with the analytical solutions to the advection-diffusion equations using the generalized integral Laplace transform technique (GILTT) and the decomposition method (Moreira et al., 2005, 2006, 2009). Chapter 8 describes the Lagrangian stochastic dispersion models with applications for airborne dispersion in the ABL (Anfossi et al., 1997, 2006). Chapter 9 deals with the large eddy simulation (LES) of dispersion within ABLs using the Lagrangian and the Eulerian approaches (Rizza et al., 2003, 2006). Chapter 10 describes the modelling of photochemical air pollution for better air quality management (Borrego et al., 2000; Monteiro et al., 2005). Finally, Chapter 11 describes the analysis of the transport of a trace gas (CO_2) at the global scale and overviews the inverse-problem techniques for deducing emissions from known concentrations (Enting, 2002, 2008).

The book is of interest for the entire boundary-layer meteorology and atmospheric turbulence communities, including both students and researchers, especially those interested in the nature, theory, and modeling of air pollution. For a deeper acquaintance with these fields, we recommend the following monographs and collections of papers on boundary-layer meteorology: Sorbjan (1989), Zilitinkevich (1991), Garratt (1992), Kraus and Businger (1994), Holtslag and Duynkerke (1998), Kantha and Clyson (2000), Baklanov and Grisogono (2007); turbulent diffusion: (Pasquill and Smith, 1983; Arya, 1999); and air pollution (Seinfeld and Pandis, 2006; Jacobson, 2005).

A. A. Baklanov
S. M. Joffre
S. S. Zilitinkevich

REFERENCES

Anfossi, D., Ferrero, E., Tinarelli, G., and Alessandrini, S., 1997: A simplified version of the correct boundary conditions for skewed turbulence in Lagrangian particle models. *Atmospheric Environment*, **31**, 301–308.

Anfossi, D., Alessandrini, S., Trini Castelli, S., Ferrero, E., Oettl, D., and Degrazia, G., 2006: Tracer dispersion simulation in low wind speed conditions with a new 2-D Langevin equation system. *Atmospheric Environment*, **40**, 7234–7245.

Arya, S.P., 1999: *Air Pollution Meteorology and Dispersion*. Oxford University Press, New York, 310 pp.

Baklanov, A. and Grisogono B. (Eds.) 2007: *Atmospheric Boundary Layers: Nature, Theory and Application to Environmental Modelling and Security*. Springer, Berlin, Germany, 248 pp., ISBN: 978-0-387-74318-9.

Borrego, C., Gomes, P., Barros, N., and Miranda, A.I., 2000: Importance of handling organic atmospheric pollutants for assessing air quality. *Journal of Chromatography A*, **889**, 271–279.

Brimblecombe, P. and Grossi, C.M., 2005: Aesthetic thresholds and blackening of stone buildings. *Science of the Total Environment*, **349**, 175–189.

Canuto, V.M., Chen, Y., Howard, A.M., and Esau, I.N., 2008: Stably stratified flows: A model with no Ri(cr). *Journal of the Atmospheric Sciences*, **65**, 2437–2447.

Chen, F., Tewari, M., Kusaka, H., and Warner, T.L., 2006: Current status of urban modelling in the community. Weather Research and Forecast (WRF) model. Sixth AMS Symposium on the Urban Environment, Atlanta, GA, January 2006.

COST-715: Fisher, B., Joffre, S., Kukkonen, J., Piringer, M., Rotach, M., and Schatzmann, (Eds.), 2005: COST 715 Final Report: *Meteorology Applied to Urban Air Pollution Problems*. Demetra Ltd Publ., Bulgaria, 276 pp., ISBN 954-9526-30-5.

COST-728: Baklanov, A., Grimmond, S., Mahura, A., and Athanassiadou, M. (Eds.), 2009: *Meteorological and Air Quality Models for Urban Areas*. Springer, Berlin, Germany, 2009, 140 pp.

COST-WMO, Baklanov, A., Fay, B., Kaminski, J., and Sokhi, R. 2007: *Overview of Existing Integrated (Off-line and On-line) Mesoscale Meteorological and Chemical Transport Modelling Systems in Europe*, WMO GAW Report No. 177, Joint Report of COST Action 728 and GURME, 107 pp, available from: http://www.cost728.org.

Degrazia, G.A. and Anfossi, D., 1998: Estimation of the Kolmogorov constant from classical statistical diffusion theory. *Atmospheric Environment*, **32**, 3611–3614.

Dupont, S., Otte, T.L., and Ching, S., 2004: Simulation of meteorological fields within and above urban and rural canopies with a mesoscale model (MM5). *Boundary-Layer Meteorology*, **113**, 111–158.

EMS-FUMAPEX, 2005: Urban Meteorology and Atmospheric Pollution, Baklanov, A., Joffre, S., and Galmarini, S. (Eds.), Special Issue, *Atmospheric Chemistry and Physics*, http://www.atmos-chem-phys.net/special issue24.html.

Enting, I.G., 2002: *Inverse Problems in Atmospheric Constituent Transport*. Cambridge University Press, Cambridge, U.K.

Enting, I.G., 2008: Assessing the information content in environmental modelling: A carbon cycle perspective. *Entropy*, **10**, 556–575.

Fisch, G, Tota, J., Machado, L.A.T. et al. 2004: The convective boundary layer over pasture and forest in Amazonia. *Theoretical and Applied Climatology*, **78**, 47–59.

Garratt, J.R., 1992: *The Atmospheric Boundary Layer*. Cambridge University Press, Cambridge, U.K., 316 pp.

Goulart, A., Degrazia, G., Rizza, U., and Anfossi, D., 2003: A theoretical model for the study of convective turbulence decay and comparison with large-eddy simulation data. *Boundary-Layer Meteorology*, **107**, 143–155.

Hidalgo, J., Masson, V., Baklanov, A., Pigeon, G., and Gimeno, L., 2008: Advances in urban climate modelling. Trends and directions in climate research. *Annals of the New York Academy of Sciences*, **1146**, 354–374. doi: 10.1196/annals.1446.015.

Holtslag, A.A.M. and Duynkerke, P.G. (Eds.), 1998: *Clear and Cloudy Boundary Layers*. Royal Netherlands Academy of Arts and Sciences, Amsterdam, the Netherlands, 372 pp.

Jacobson, M.Z., 2005: *Fundamentals of Atmospheric Modelling*, 2nd edn. Cambridge University Press, New York, 813 pp.

Kantha, L.H. and Clyson, C.A., 2000: *Small Scale Processes in Geophysical Fluid Flows*. Academic Press, San Diego, CA, 888 pp.

Kraus, E.B. and Businger, J.A., 1994: *Atmosphere-Ocean Interaction*. Oxford University Press/Clarendon Press, New York/Oxford, 352 pp.

Masson, V., Gomes, L., Pigeon, G. et al. 2009: The Canopy and Aerosol Particles Interactions in TOulouse Urban Layer (CAPITOUL) experiment. *Meteorology and Atmospheric Physics*, 102, 135–157.

Mauritsen, T., Svensson, G., Zilitinkevich, S.S., Esau, I., Enger, L., and Grisogono, B., 2007: A total turbulent energy closure model for neutrally and stably stratified atmospheric boundary layers. *Journal of the Atmospheric Sciences*, **64**, 4117–4130.

Mestayer, P.G., Durand, P., Augustin, P. et al. 2005: The urban boundary layer field experiment over Marseille, UBL/CLU-ESCOMPTE: Experimental set-up and first results. *Boundary-Layer Meteorology*, **114**, 315–365.

Molina, L.T., Kolb, C.E., de Foy, B. et al. 2007: Air quality in North America's most populous city—overview of the MCMA-2003 campaign. *Atmospheric Chemistry and Physics*, **7**, 2447–2473.

Monteiro, A., Vautard, R., Lopes, M., Miranda, A.I., and Borrego, C., 2005: Air pollution forecast in Portugal: A demand from the new Air Quality Framework Directive. *Internation Journal of Environment and Pollution*, **25**(2), 4–15.

Moreira, D.M., Vilhena, M.T., Tirabassi, T., Buske, D., and Cotta, R.M., 2005: Near source atmospheric pollutant dispersion using the new GILTT method. *Atmospheric Environment*, **39**(34), 6290–6295.

Moreira, D.M., Vilhena, M.T., Buske, D., and Tirabassi, T., 2006. The GILTT solution of the advection–diffusion equation for an inhomogeneous and nonstationary PBL. *Atmospheric Environment*, **40**, 3186–3194.

Moreira, D.M., Vilhena, M.T., Buske, D., and Tirabassi, T., 2009: The state-of-art of the GILTT method to simulate pollutant dispersion in the atmosphere. *Atmospheric Research*, **92**, 1–17.

Pasquill, F. and Smith, F.B., 1983: *Atmospheric Diffusion*. Ellis Horwood, Chichester, U.K., 437 pp.

Rizza, U., Gioia, G., Mangia, C., and Marra, G.P., 2003: Development of a grid-dispersion model in a large eddy simulation generated planetary boundary layer. *Nuovo Cimento Sezione C*, **26**, 297–309.

Rizza, U., Mangia, C., Carvalho, J.C., and Anfossi, D., 2006: Estimation of the Lagrangian velocity structure function constant C_0 by large eddy simulation. *Boundary-Layer Meteorology*, **120**, 25–37.

Rotach, M.W., Vogt, R., Bernhofer, D. et al. 2005: BUBBLE—An urban boundary layer meteorology project. *Theoretical and Applied Climatology*, **81**, 231–261.

San Jose, R, Baklanov A., Sokhi R.S., Karatzas, K., and Perez, J.L., 2008: Air quality modeling. In: *Ecological Models*, Vol. 1, *Encyclopaedia of Ecology*, 5 Vols., Elsevier, Oxford, U.K., pp. 111–123.

Seinfeld, J.H. and Pandis, S.N., 2006: *Atmospheric Chemistry and Physics—From Air Pollution to Climate Change*, 2nd edn. John Wiley & Sons, New York, 1232 pp., ISBN 9780471720188.

Sharan, M. and Modani, M., 2005: An analytical study for the dispersion of pollutants in a finite layer under low wind conditions. *Pure and Applied Geophysics*, **162**, 1861–1892.

Sharan, M., Singh, M.P., and Yadav, A.K., 1996: A mathematical model for the atmospheric dispersion in low winds with eddy diffusivities as linear functions of downwind distance. *Atmospheric Environment*, **30**, 1137–1145.

Siebesma, A.P., Soares, P.M.M., and Teixeira, J., 2006: A combined eddy diffusivity mass-flux approach for the convective boundary layer. *Journal of Atmospheric Science*, **64**, 1230–1248.

Soares, P.M.M., Miranda, P.M.A., Siebesma, A.P., and Teixeira J., 2004: An eddy-diffusivity/mass-flux turbulence closure for dry and shallow cumulus convection. *The Quarterly Journal of the Royal Meteorological Society*, **130**, 3365–3384.

Sorbjan, Z., 1989: *Structure of the Atmospheric Boundary Layer*. Prentice Hall, Englewood Cliffs, NJ, 317 pp.

Taha, H., 2008: Sensitivity of the urbanized MM5 (uMM5) to perturbations in surface properties in Houston Texas. *Boundary-Layer Meteorology*, **127**, 193–218.

Tirabassi, T., 1989: Analytical air pollution and diffusion models. *Water, Air, and Soil Pollution*, **47**, 19–24.

Tirabassi, T., 2003: Operational advanced air pollution modeling. *Pure and Applied Geophysics*, **160**(1–2), 5–16.

Zilitinkevich, S.S., 1991: *Turbulent Penetrative Convection*, Avebury Technical, Aldershot, 180 pp.

Zilitinkevich, S. and Esau, I., 2007: Similarity theory and calculation of turbulent fluxes at the surface for the stably stratified atmospheric boundary layers. *Boundary-Layer Meteorology*, **125**, 193–296.

Zilitinkevich, S.S., Hunt, J.C.R., Grachev, A.A. et al. 2006: The influence of large convective eddies on the surface layer turbulence. *The Quarterly Journal of the Royal Meteorological Society*, **132**, 1423–1456.

Zilitinkevich, S., Esau, I., and Baklanov, A., 2007a: Further comments on the equilibrium height of neutral and stable planetary boundary layers. *The Quarterly Journal of the Royal Meteorological Society*, **133**, 265–271.

Zilitinkevich, S.S., Elperin, T., Kleeorin, N., and Rogachevskii, I., 2007b: Energy- and flux-budget (EFB) turbulence closure model for the stably stratified flows. Part I: Steady-state, homogeneous regimes. *Boundary-Layer Meteorology*, **125**, 167–192.

Zilitinkevich, S.S., Mammarella, I., Baklanov, A.A., and Joffre, S.M., 2008: The effect of stratification on the aerodynamic roughness length and displacement height. *Boundary-Layer Meteorolology*, **129**, 179–190.

Preface

In this book, we aim to put together important topics covering the theoretical aspects of air pollution modeling and applications. This was possible thanks to the dedication of the researchers who kindly accepted to write chapters for this book. We are grateful to all of them for helping us to accomplish our objective of publishing this book. We hope that this book will play an important role in helping researchers and graduate students with their investigations. We also hope that this book can raise the interest of people working with the different topics of this research field. Finally, we would like to express our gratitude to our universities for supporting us in this enterprise and also to the National Council for Scientific and Technological Development (CNPq) for financing part of this work. Furthermore, we would like to express our special thanks to Professors Sergej S. Zilitinkevich, Alexander Baklanov, and Sylvain M. Joffre for the generous contribution to write the Foreword of this book. Finally, we would like to thank CRC Press (Taylor & Francis Group) for offering us the possibility of presenting these contributions.

Davidson Martins Moreira
Marco Túllio M. B. de Vilhena

Editors

Davidson Martins Moreira is a graduate in physics with a doctorate in mechanical engineering at the Federal University of Rio Grande do Sul (UFRGS). He is a researcher at the National Council for Scientific and Technological Development (CNPq), Brasilia, Brazil. Currently, he works as a liaison officer for the International Atomic Energy Agency (IAEA) in Brazil. He has supervised degree and postgraduate theses and was a member of the examining board for awarding PhD and master's degrees. He has taught in seminars and tutorials both in universities and in courses organized by other public and private entities, besides acting as a scientific adviser and as a referee for a number of international scientific journals.

His research activities have increasingly dealt with the phenomenological and theoretical aspects of atmospheric transport and diffusion. During the last years, he has worked on developing mathematical air pollution models, which have in common the utilization of analytical solutions of the advection–diffusion equation. At present, he is working on turbulence in the atmospheric boundary layer with the analysis of the turbulent field during the transitions, focussing specifically on the residual layer. He is also conducting research on the mathematical description of the turbulent transport of atmospheric contaminants.

Marco Túllio M. B. de Vilhena is a senior professor of mathematics and mechanical engineering at the undergraduate and graduate levels and has supervised approximately 80 master's and doctoral dissertations. He has significant professional experience in teaching in seminars and tutorials both in universities and in courses organized by other public and private entities, besides acting as a scientific adviser and as a referee for a number of international scientific journals. He is also a researcher at the National Council for Scientific and Technological Development (CNPq), Brasilia, Brazil.

His research interests include the modeling of the dispersion of pollutants in the atmospheric boundary layer in all stability conditions, the dispersion simulation of radioactive pollutants in the atmosphere using generalized integral transform, and the transport theory of neutral particles, reactor physics, radiative transfer in the atmosphere, and physical medical applications.

Contributors

Domenico Anfossi
National Research Council
Institute of Atmospheric Sciences and
Climate
Torino, Italy

Carlos Borrego
Center for Environmental and Marine
Studies
Department of Environment and
Planning
University of Aveiro
Aveiro, Portugal

Peter Brimblecombe
School of Environmental Sciences
University of East Anglia
Norwich, United Kingdom

Daniela Buske
Departamento de Matemática e
Estatística
Instituto de Física e Matemática
Federal University of Pelotas
Pelotas, Brazil

Gervásio Annes Degrazia
Department of Physics
University of Santa Maria
Santa Maria, Brazil

Ian G. Enting
ARC Centre of Excellence for
Mathematics and Statistics
University of Melbourne
Melbourne, Victoria, Australia

Joana Ferreira
Center for Environmental and Marine
Studies
Department of Environment and
Planning
University of Aveiro
Aveiro, Portugal

Gilberto Fisch
Atmospheric Science Division
Institute of Aeronautics and Space
São José dos Campos, Brazil

Giulia Gioia
National Research Council
Institute of Atmospheric Sciences and
Climate
Lecce, Italy

Antonio Gledson Oliveira Goulart
Centro de Ciências Exatas e
Technológicas
Universidade Federal do Pampa
Campus Bagé, Brazil

Carlota M. Grossi
School of Environmental Sciences
University of East Anglia
Norwich, United Kingdom

Pramod Kumar
Centre for Atmospheric Sciences
Indian Institute of Technology Delhi
New Delhi, India

Guglielmo Lacorata
National Research Council
Institute of Atmospheric Sciences and
Climate
Lecce, Italy

Cristina Mangia
National Research Council
Institute of Atmospheric Sciences and
    Climate
Lecce, Italy

Gian Paolo Marra
National Research Council
Institute of Atmospheric Sciences and
    Climate
Lecce, Italy

Ana Isabel Miranda
Center for Environmental and Marine
    Studies
Department of Environment and
    Planning
University of Aveiro
Aveiro, Portugal

Pedro M. A. Miranda
Centro de Geofisica da Universidade de
    Lisboa
Instituto Geofisica do Infante Dom Luiz
University of Lisbon
Lisbon, Portugal

Davidson Martins Moreira
Department of Mechanical Engineering
Federal University of Rio Grande
    do Sul
Porto Alegre, Brazil

Umberto Rizza
National Research Council
Institute of Atmospheric Sciences and
    Climate
Lecce, Italy

and

Departamento de Física
Universidade Federal de Santa Maria
Santa Maria, Brazil

Debora Regina Roberti
Department of Physics
University of Santa Maria
Santa Maria, Brazil

Maithili Sharan
Centre for Atmospheric Sciences
Indian Institute of Technology Delhi
New Delhi, India

Pedro M. M. Soares
Centro de Geofisica da Universidade de
    Lisboa
Instituto Geofisica do Infante Dom Luiz
University of Lisbon
Lisbon, Portugal

and

Department of Civil Engineering
Instituto Superior de Engenharia de
    Lisbon
Lisbon, Portugal

João Teixeira
Jet Propulsion Laboratory
California Institute of Technology
Pasadena, California

Tiziano Tirabassi
National Research Council
Institute of Atmospheric Sciences and
    Climate
Bologna, Italy

Silvia Trini Castelli
National Research Council
Institute of Atmospheric Sciences and
    Climate
Torino, Italy

Marco Túllio M. B. de Vilhena
Instituto de Matemática—Departmento
    de Matemática Pura e Aplicada
Universidade Federal do Rio Grande
    do Sul
Porto Alegre, Brazil

1 Deposition, Transformation, and Remobilization of Soot and Diesel Particulates on Building Surfaces

Peter Brimblecombe and Carlota M. Grossi

CONTENTS

1.1 INTRODUCTION

The late twentieth century saw a growing awareness that particles in the atmosphere have a significant effect on urban health. This came as a surprise because of the large decreases in air pollution that typified urban areas since the 1960s and 1970s. These improvements had often come about through a declining use of coal as a source of energy in cities. Although the improvements in traditional air pollutants such as sulfur dioxide and smoke had been widespread these pollutants had been replaced by photochemical oxidants in smog: ozone, nitrogen oxides, and more recently fine particles. Much of the change in the nature of air pollution was the result of the extensive use of the automobile, which released volatile organic compounds into urban air. More recently the popularity of diesel engines has increased the emissions of fine particles in cities. While these changes have raised concerns in terms of human health there have been parallel worries about the damage it is causing to the architecture of cities.

1.2 POLLUTION AND ARCHITECTURE

In the early twentieth century the dominant impact of air pollution on stone was the sulfation of the surfaces through the deposition of sulfur dioxide and its oxidation to sulfuric acid. The concomitant deposition of soot onto the surfaces led to thick disfiguring black gypsum crusts. Blackened buildings typified the coal-burning cities of *fin de siècle* Europe and led not only to a century of scientific concern and a range of interventions, but also influenced the nature of modern architecture. Architects were forced to abandon detailed moldings in soft light-colored stone and looked to produce buildings with simpler lines and darker colors constructed in more resistant materials (Bowler and Brimblecombe, 2000).

The cleaner air found in cities of the second half of the twentieth century gave some respite to the formation of black crusts on the surfaces of buildings. Although there was a public debate over the role of acid rain in causing damage, in reality the corrosion rates of metals and building stones in cities such as London declined through the last half of the century. These improving conditions allowed a developing civic desire for clean buildings. People were suddenly confronted by architecture that was very bright and unfamiliar. The buildings were now clean and may have looked closer to the architect's original intent, but this change was not free of criticism from those who feared that history had been scraped away (Andrew, 1992; Ball et al., 2000; Grossi and Brimblecombe, 2004a). Somehow our most loved and valuable buildings had lost their patina.

Despite the overall improvements in air quality, diesel-derived particulate matter was increasing in concentration in heavily trafficked areas. Contemporary particulate material is finer than that from coal smoke and blacker and richer in organic matter. Even more problematic for those charged with caring for the urban fabric was a population accustomed to lower soiling rates. The last years of the twentieth century saw black crusts on buildings reemerge as an air pollution issue with some buildings grossly disfigured (Figure 1.1).

1.3 CHEMISTRY

The traditional view of the way in which stones and metals are degraded is from acid produced by the oxidation of sulfur dioxide. In the case of limestone and other calcareous stones this leads to a transformation of the carbonate to a sulfate. Calcium sulfate or gypsum is more soluble in water such that the carbonate readily dissolves from the building. Gypsum also has a larger molecular volume than calcium carbonate so the increasing volume of the mineral on the outside of the stone imposes mechanical stress and disrupts the surface.

In the contemporary atmosphere, sulfur dioxide concentrations are much lower than in the past. However, ozone and metal ions could act as oxidants or oxidation

FIGURE 1.1 (See color insert following page 234.) Palacio del Marqués de Sta. Cruz (Oviedo-Spain), a building with rain streaking and biological and pollutant staining showing the forms that can disfigure architecture. (Photo courtesy of Carlota Grossi.)

catalysts increasing the effectiveness of even low sulfur dioxide concentrations in damaging building materials (Johansson et al., 1986). The higher concentrations of oxides of nitrogen in the air can increase the amount of nitric acid deposited on urban surfaces. It is extremely corrosive and traces of nitrate are found in rainwater that drains from contemporary buildings. Furthermore, there is a noticeable change in the microflora on buildings (see Warscheid and Braams, 2000 for a review of causes). Sulfur dioxide is phytotoxic, so in atmospheres with less sulfur, plants (especially lichen) grow more effectively on buildings. When combined with the greater rate of delivery of nitrate to buildings surfaces, which are usually poor in this nutrient, an increasing rate of biological attack is expected. Where lichens are growing an oxalate layer often forms, although this can have a protective role.

The impacts of air pollution on architecture in the twenty-first century derive from a range of novel interactions. The increase in the number of diesel particles are accompanied by associated organic compounds (Hermosin et al., 2004), such as polycyclic aromatic molecules or organic acids. These organic compounds can act as photosensitizers inducing oxidation processes or polymerization. The polymers formed may create a kind of adhesive, and thus replace the calcareous cements that have characterized the outer layers of buildings in the past.

1.4   OLDER CRUSTS

If we look at the thick deposits found on monuments such as the Tower of London, we find layered structures that reflect both the change in deposition with time and the changes brought about by physical, chemical, and biological processes. Deeper in the deposits there are traces from wood and coal burning in the past (Del Monte et al., 2001). Older crusts also tend to be thicker with dendritic aluminosilicate and iron-containing particles. At the Tower, younger crusts are thinner and tabular. Their structure is clear under microscopy (Sabbioni et al., 2004), with the presence of coal and wood smoke in oldest layers.

These changes seem to relate to changes in the nature of urban pollution. This may also be true of the type of carbon present. Particulate carbon in the modern atmosphere tends to be associated with significant fractions of organic matter. In the past there was a smaller amount of organic material. Today those cities with a large amount of pollution generated from two-stroke motor vehicles (motor cycles especially) have large amounts; this leads to contemporary thin crusts observed with high organic carbon/elemental carbon ratios (OC/EC). In Florence, it varies between 1.5 and 2.2, while in older crusts the OC/EC ratio is smaller, such as those of the cathedral of Milan varying from 0.1 to 0.7 (Bonazza et al., 2005).

1.5   TRANSFORMATIONS

We can also see transformations taking place in the crusts. These can be considered in terms of a simple model (Figure 1.2). The concentrations of insoluble components, such as elemental carbon and oxalate that are largely immobile, are correlated in the crust at the Tower of London. By contrast, soluble aerial components are not well

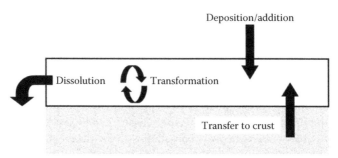

FIGURE 1.2   Proposed model showing the fluxes within the crust. It illustrates a dynamic system that involves atmospheric deposition, transfer from the stone substrate, chemical transformation, dissolution, migration, and loss of soluble compounds. (From Bonazza, A. et al., *Environ. Sci. Technol.*, 41, 4199, 2007. With permission.)

correlated with these insoluble materials. Nevertheless, they correlate well among themselves, so chloride and formate concentrations are related because they are both soluble and removed from the crust in similar ways by rainfall. There is also evidence of biological transformations within the crust such as the production of oxalate or even acetate (Bonazza et al., 2007).

1.6 DARKENING OF BUILDINGS

Declining concentrations of corrosive components in air has made the accumulation of black particles more important and raised the relevance of aesthetic considerations. Today diesel soot has become the major source of elemental carbon in urban air. Nevertheless, in some locations decreasing atmospheric soot concentrations have meant that in recent years there has been less darkening and in some cases rainfall removal has made buildings much cleaner (Davidson et al., 2000). Biological activity, perhaps supported by an ongoing increase in organic pollution, may increasingly contribute to stone blackening (Viles and Gorbushina, 2003).

Particle deposition gradually darkens the surfaces of buildings over time. As might be imagined this change in color can be represented as an exponential decrease in the lightness of the surface (Figure 1.3). The process is doubtless dependent on the concentration of particles and a range of transfer processes. These are highly variable, but as the timescales of these changes are shorter than the darkening it is possible to consider average conditions and pollutant loads. Although the darkening of buildings appears as an exponential of the process, slight difficulties arise with the boundary conditions. This requires knowledge of the reflectance of the building stone and the final color after it is covered with urban soot (Brimblecombe and Grossi, 2004).

The blackening process is a consequence of the increasing accumulation of black carbon on the surfaces, which can be measured as a reflectance change. Although deposited carbon might be assumed to be chemically inert, it can be slowly oxidized

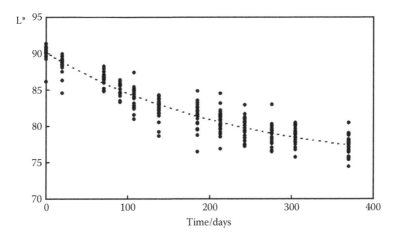

FIGURE 1.3 Reflectance or lightness change of a whitish limestone during 1 year exposure under rain-sheltered locations in an urban environment. (From Esbert, R.M. et al., *Atmospheric Environment*, 35, 441, 2001.)

in modern crusts. It is increasingly recognized that carbon can be converted to humic-like substances (HULIS) in the atmosphere (Graber and Rudich, 2005). Such oxidative processes on soot is likely to be promoted by traces of the organic compounds associated with the diesels. The products exhibit yellowish or brownish colors rather than black. This is increasingly noticed at sites such as the Tower of London (see Figure 1.4), where there are already observations of a warmer tones to the buildings (Grossi et al., 2006, 2007).

In spite of the subtle change in tone, it is elemental carbon in crusts that prove the major control on appearance. At the Tower of London (Figure 1.5) reflectance

FIGURE 1.4 (See color insert following page 234.)   Carved stone at Tower of London exhibiting the warmer tones found today. (Photo courtesy of Carlota Grossi.)

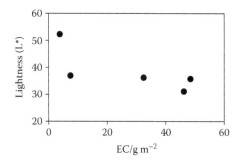

FIGURE 1.5   Measured lightness (L*) versus elemental carbon area concentrations in crusts from the Tower of London. (From Bonazza, A. et al., *Environ. Sci. Technol.*, 41, 4199, 2007. With permission.)

of the surfaces was related to the area concentrations of EC (Bonazza et al., 2007). These tended to reach a minimum reflectivity of around 35% and appeared very black when EC concentrations in the crust reached more than $10\,g\ m^{-2}$. Such high loadings of EC were probably reached during the last century, when the airborne load of EC was higher than at present (Brimblecombe and Grossi, 2005).

1.7 APPEARANCE OF BUILDINGS

The changing color and blackening are key drivers of concern about appearance. This can be especially troublesome in the case of historic buildings and provides a challenge for management in some cases. Take for example the White Tower of the Tower of London, where the name implies a specific color, so while managing this important building one has to consider that visitors may be disappointed to find that the tower does not appear white (Grossi and Brimblecombe, 2005). More generally the color of buildings affects our appreciation by creating a sense of care or civic pride as distinct from neglect.

The issues above suggest that considerations of the darkening process need to go beyond measurements of reflectance and examine aesthetic aspects. Public sensitivity to blackening has increased in recent decades in line with improved air quality and the acid rain debate. Public perception of darkening can be investigated using *in situ* questionnaires at historic buildings, which can exhibit varying degrees of blackening (Brimblecombe and Grossi, 2005). This technique can also be used to assess the success of treatments and choice of replacement stone (Valach et al., 2006).

Such studies of perception have been used to establish aesthetic thresholds by comparing the proportion of the public finding a building discolored to the perception of its lightness or reflectance. The "perceived lightness" of a façade can be assessed by inviting visitors to compare the tone of the building with a grey scale (Brimblecombe and Grossi, 2005). Most respondents experience little difficulty in deciding the tone of the building even though they are offered a grey scale and building façades are typically colored. It proved relatively simple to relate the perceived lightness of the building to their views on whether the building is "dirty" or needs cleaning. As would be imagined where the building is perceived to be light in color respondents also believe it to be clean (see Figure 1.6a). Where reflectance or lightness was low, respondents usually believed the building was dirty and correlated with a view that it was in need of cleaning (Figure 1.6b).

Aggregating data from a range of buildings enabled us to establish the average perceived lightness. We were also able to assess approximate concentrations of EC at the survey sites and thus relate perceived lightness of the façade to amount air pollution (Figure 1.7). A range of approaches allowed us to assess the perceived lightness that was likely to be acceptable to the public (see Brimblecombe and Grossi, 2005 for details). This band of acceptability is also shown on the diagram so it would appear that at $10\,\mu g\ m^{-3}$ the façades built from light-colored stone are typically found to be unacceptably dark. However, by the time the EC concentrations drop to $2–3\,\mu g\ m^{-3}$ the buildings have a more acceptable appearance.

These ideas offer the possibility for setting acceptable concentrations of EC in the atmosphere on the basis of building aesthetics. Nevertheless a more detailed analysis

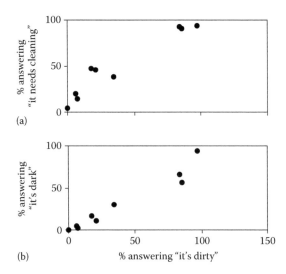

(a)

(b)

FIGURE 1.6 (a) Relationship between the frequency respondents found a building light or dark to the frequency of answering "it is dirty." (b) Relationship between answers that the building is "dirty" and "it needs cleaning." (From Grossi, C.M. and Brimblecombe, P., Aesthetics and perception of soiling. In C. Saiz-Jimenez (ed.), *Air Pollution and Cultural Heritage*, Balkema, Rotterdam, the Netherlands, pp. 199–208, 2004a. With permission.)

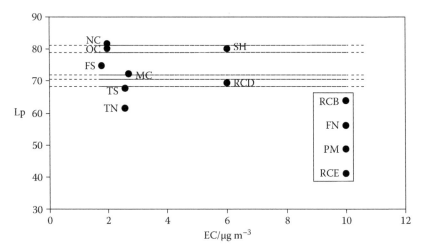

FIGURE 1.7 Mean values of perceived lightness (Lp) at the historic building sites (as dots) as a function of elemental carbon concentrations. Lines represent the different suggested aesthetic thresholds (see Brimblecombe and Grossi, 2005 for details). Abbreviations refer to buildings: FN, FS Florence Cathedral; MC, Cathedral of Milan; NC, Norwich Cathedral; OC, Oviedo Cathedral; PM, Palacio Marques Sta Cruz, Oviedo; RCD, RCB, RCE, Cathedral of St John Norwich; SH, Surrey House Norwich; TSE, TN, Tower of London. (From Brimblecombe, P. and Grossi, C.M., *Science of the Total Environment*, 349, 175, 2005. With permission.)

would be needed to assess whether a value of 2–3 µg m⁻³ would be appropriate as an air pollution standard (Brimblecombe and Grossi, 2005). In the near future we expect lower concentrations of EC in many European cities where regulatory controls are increasingly in place. This means that it should be possible to meet the 2–3 µg m⁻³ level except for historic buildings close to very busy roads.

1.8 BLACKENING PATTERNS

The analysis above treats buildings as though they take on a homogeneous dark color over time. This is not true and despite the fact that respondents to our questionnaires found little difficulty in assessing the tone of a building by comparison with a grey scale, the black discoloration takes distinct forms and patterns. One respondent to our survey thought the building might be darker or lighter, but should not appear as it does, with long black streaks washed by rainwater (see Figure 1.1).

A desktop exercise allowed us to explore public response to blackening patterns on buildings, using a methodology derived from the psychology of art. Exercises used a range of simulated soiling patterns around a simple architectural element, a pedimented window (Grossi and Brimblecombe, 2004b). Respondents were asked to arrange the images from the "most to the least acceptable" pattern. The order hinted at the importance of certain soiling features in driving the ranking. Figure 1.8 shows the results of this exercise and a surprising amount of agreement on what constitutes an acceptable pattern. Simple forms that shadow the architecture seemed more acceptable. This has earlier been noted by Andrew (1992) who wrote

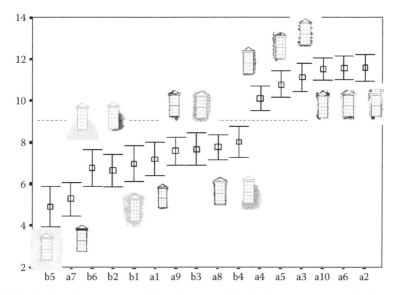

FIGURE 1.8 Ranking of acceptability of simulated blackening patterns, from 1 which is the most acceptable pattern to 16 the least acceptable. (From Grossi, C.M. and Brimblecombe, P., *Environ. Sci. Technol.*, 38, 3971, 2004b. With permission.)

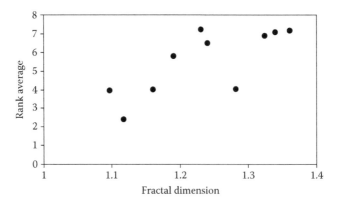

FIGURE 1.9   Averaged ranking of some of the images of pedimented windows (in Figure 1.6) as a function of the fractal dimension of the blackening patterns. (From Grossi, C.M. and Brimblecombe, P., *Environ. Sci. Technol.*, 38, 3971, 2004b. With permission.)

that "light or moderate blackening around architectural details could improve the visual appearance of the building by increasing contrast and enhancing shadowing effects."

We also noted that complex feathery shading was generally less acceptable than linear forms that aligned with the architecture. Fractals seemed a useful way to explore this idea as architecture is build from elements that have an integer fractal dimension (i.e., one lines, two areas, three volumes). Architectural features with soiling patterns of noninteger fractal dimension would potentially confuse our reading of architectural form and thus be less acceptable. We measured the fractal dimension of the perimeter of our images of the pedimented window (using the box counting approach in software by Volodymyr Kindratenko available at http://www.ncsa.uiuc. edu/~kindr/phd/software.html#ref2). The fractal dimension was determined and is shown in Figure 1.9, which suggests that patterns with fractal dimensions closer to unity tend to be more acceptable.

1.9   FUTURE DISCOLORATION OF BUILDINGS

Our most valued buildings are expected to last long periods of time. The appearance of building crusts is likely to be different in the future with the changes in deposition of elemental carbon and rainfall patterns. Davidson et al. (2000) traced the evolution of blackening patterns on the "Cathedral of Learning" over the period 1930–1995. When it was built in Pittsburgh in the 1930s the surfaces rapidly soiled because of the intense coal smoke, but loss of the steel industry meant that the air and ultimately the building became cleaner.

The mechanism and dynamics of the removal of black crusts from buildings over time as pollution decreases is poorly understood. This is certainly an important area for future investigation, but lower particulate concentrations in the atmosphere will lead to lighter surfaces. Nevertheless, the removal of soot by rain occurs in ways that can create new patterns of rain streaking that disfigure buildings and promote public disquiet (Grossi and Brimblecombe, 2007).

In the future brownish or yellowish colors may be more dominant and raise aesthetic issues. Although buildings are currently perceived in terms of grays and creams (Brimblecombe and Grossi, 2005), yellow colors can derive from sulfation, oxidation of iron in light-colored stone, surface texture, and the presence of organic substances (Grossi et al., 2007). As noted earlier, the oxidation of organic rich soot can produce buildings with a warmer tone.

The patterns of discoloration also affect the appreciation of architecture so it is important to gain insight into spatial factors that may influence the acceptability of soiling. Those who manage buildings may find it necessary to consider the distribution of soiling in addition to the level of blackening when trying to gauge the strength of public reaction. They will have to anticipate future blackening patterns, which will represent a balance between accumulation and redistribution of carbon-containing materials on the surfaces. Climate change means that heavy rainfall and temperature increases may also encourage the growth of microorganisms, which can also produce discoloration (Viles, 2002). Public perceptions are not easy to integrate with considerations such as available finance, concerns over physical damage due to crust buildup, or the importance of having a clean building for specific events or celebrations. Nevertheless, this becomes an increasingly important part of the management of heritage.

1.10   CONCLUSIONS

In the coming century our buildings will confront great changes in the nature of the urban atmosphere and climate. Traditional acidic air pollutants will be less important agents of decay than a range of new pollutants. Carbon-containing particulate material is likely to adhere to and discolor urban surfaces. Additionally, climate change will mediate the appearances of the deposits on buildings. In the near future the urban air quality will be better, but still likely to be rich in organic pollutants, so buildings are likely to take on a warmer tone when these components undergo transformation. The changes under way have both aesthetic and economic implications for the cleaning and maintenance of historic buildings. Management will depend on integrating both the science and the aesthetics. Much has yet to be learned about the accumulation and removal of particulate matter from complex architectural geometries. This seems to be a fruitful area for research considering the wealth of heritage to be managed.

ACKNOWLEDGMENTS

This chapter derives from work within ENV4-CT-2000-0002 CARAMEL and CT-2003-501837-NOAH'S ARK funded by DG Research of the European Commission and has benefited greatly from discussions with Alessandra Bonazza.

REFERENCES

Andrew, C. (1992). Towards an aesthetic theory of building soiling. In R.G.M. Webster (ed.), *Stone Cleaning and the Nature, Soiling and Decay Mechanisms of Stone*, Donhead, London, U.K., pp. 63–81.

Ball, J., Laing, R., and Young, M. (2000). Stone cleaning: Comparing perception with physical and financial implications. *Journal of Architectural Conservation*, 2, 47–62.

Bonazza, A., Sabbioni, C., and Ghedini, N. (2005). Quantitative data on carbon fractions in interpretation of black crusts and soiling on European built heritage. *Atmospheric Environment*, 39, 2607–2618.

Bonazza, A., Brimblecombe, P., Grossi, C.M., and Sabbioni, C. (2007). Carbon in black layers from the Tower of London. *Environmental Science and Technology*, 41, 4199–4204.

Bowler, C. and Brimblecombe, P. (2000). Environmental pressures on the design of Manchester's John Rylands Library. *Journal of Design History*, 13, 175–191.

Brimblecombe, P. and Grossi, C.M. (2004). The rate of darkening of material surfaces. In C. Saiz-Jimenez (ed.), *Air Pollution and Cultural Heritage*. Balkema, Rotterdam, the Netherlands, pp. 193–198.

Brimblecombe, P. and Grossi, C.M. (2005). Aesthetic thresholds and blackening of stone buildings. *Science of the Total Environment*, 349, 175–189.

Davidson, C., Tang, W., Finger, S., Etyemezian, V., Striegel, F., and Sherwood, S. (2000). Soiling patterns on a tall limestone building: Changes over 60 years. *Environmental Science and Technology*, 34, 560–565.

Del Monte, M., Ausset, P., Forti, P., Lèfevre, R.A., and Tolomelli, M. (2001). Air pollution records on selenite in the urban environment. *Atmospheric Environment*, 35, 3885–3896.

Esbert, R.M., Diaz-Pache, F., Grossi, C.M., Alonso, F.J., and Ordaz, J. (2001). Airborne particulate matter around the Cathedral of Burgos (Castilla y LeoH n, Spain). *Atmospheric Environment*, 35, 441–452.

Graber, E.R. and Rudich, Y. (2005). Atmospheric HULIS: How humic-like are they? A comprehensive and critical review. *Atmospheric Chemistry and Physics*, 5, 9801–9860.

Grossi, C.M. and Brimblecombe, P. (2004a). Aesthetics and perception of soiling. In C. Saiz-Jimenez (ed.), *Air Pollution and Cultural Heritage*. Balkema, Rotterdam, the Netherlands, pp. 199–208.

Grossi, C.M. and Brimblecombe, P. (2004b). Aesthetics of simulated soiling patterns on architecture. *Environmental Science and Technology*, 38, 3971–3976.

Grossi, C.M. and Brimblecombe, P. (2005). The White Tower and the perception of blackening. Cleaning techniques in conservation practice. Special issue of *Journal of Architectural Conservation*, 11(3), 33–44.

Grossi, C.M. and Brimblecombe, P. (2007). Effects of long term changes in air pollution and climate on the decay and blackening of European stone buildings. In B. Smith and R. Prikryl (eds.), *Building Stone Decay: From Diagnosis to Conservation*. Geological Society, London, U.K., Special Publications, vol. 271, pp. 117–130.

Grossi, C.M., Brimblecombe, P., Bonazza, A., Sabbioni, C., and Zamagni, J. (2006). Sulfate and carbon compounds in black crusts from the Cathedral of Milan and the Tower of London. In R. Fort, M. Alvarez de Buergo, M. Gómez-Heras, and C. Vázquez-Calvo (eds.), *Heritage, Weathering and Conservation*. Taylor & Francis Group, London, U.K., pp. 441–446

Grossi, C.M., Brimblecombe, P., Esbert, R.M., and Alonso, J. (2007). Color changes on building limestone surfaces. *Color Research and Application*, 32, 320–331.

Hermosin, B., Gaviño, M., and Saiz-Jimenez, C. (2004). Organic compounds in black crusts from different European monuments: A comparative study. In C. Saiz-Jimenez, (ed.), *Air Pollution and Cultural Heritage*. Balkema, Rotterdam, the Netherlands, pp. 47–55.

Johansson, L.-G., Lindqvist, O., and Mangia, R.E. (1986). Corrosion of calcareous stones in humid air containing SO_2 and NO_2. Preprints of *Air and Pollution Symposium*. Rome, October, 15–17.

Sabbioni, C., Bonazza, A., Zamagni, J., Ghedini, N., Grossi, C.M., and Brimblecombe, P. (2004). The Tower of London: A case study on stone damage in an urban area. In C. Saiz-Jimenez (ed.), *Air Pollution and Cultural Heritage*. Balkema, Rotterdam, the Netherlands, pp. 57–62.

Valach, J., Bryscejn, J., Drdácký, M., Slížková, Z., and Vavřik, D. (2006). Public perception and optical characterization of degraded historic stone and mortar surfaces. In R. Fort, M. Alvarez de Buergo, M. Gómez-Heras, and C. Vázquez-Calvo (eds.), *Heritage, Weathering and Conservation*. Taylor & Francis Group, London, U.K., pp. 827–832.

Viles, H.A. (2002). Implications of future climate change for stone consolidation. In S. Siegesmund, T. Weiss, and V. Vollbrecht (eds.), *Natural Stone, Weathering Phenomena, Conservation Strategies and Case Studies*. Geological Society, London, U.K., Special Publications, vol. 205, pp. 407–418.

Viles, H.A. and Gorbushina, A.A. (2003). Soiling and microbial colonisation on urban roadside limestone: A three year study in Oxford, England. *Building and Environment*, 38, 1217–1224.

Warscheid, Th. and Braams, J. (2000). Biodeterioration of stone: A review. *International Biodeterioration and Biodegradation*, 46(4), 343–368.

2 Atmospheric Boundary Layer: Concepts and Measurements

Gilberto Fisch

CONTENTS

2.1 INTRODUCTION

The atmosphere is a gaseous layer that encompasses the Earth. It extends from the surface up to a height around 500–600 km, and it is a vital medium for the living conditions of the planet. Although its extension, the lower part of the atmosphere (called "troposphere") is the most important layer as almost all the atmospheric processes (especially the hydrologic cycle and its components like evaporation, convection, clouds, rain, etc.) occur inside this layer. Also, as the lowest part, it is where the human beings live. The troposphere can be split into two layers: the lowest part known as atmospheric boundary layer (ABL) (sometimes also referred to as planetary boundary layer—PBL) and the free atmosphere (FA), which lies between the top of the ABL and the top of troposphere (which is around 15–17 km in the tropics). However, for the air pollution issue, the ABL is the crucial layer to be understood as the pollutants are released from tall chimneys or stacks at the surface, mainly due to the anthropogenic sources. At the ABL, the pollutants are transported, diffused, advected, mixed, dry and wet deposited, as well as chemically modified. Another important difference between the FA and the ABL is the nature of the flow: inside ABL the flow is predominantly turbulent (and then provoking a well-mixed layer (ML) during daytime or at unstable conditions) while the FA is mainly laminar. The ABL plays an important role in meteorology, including studies of air pollution dispersion and aeronautical and aviation procedures regarding aircraft landing and takeoff. Also, the characteristics and behavior of the ABL by its turbulent exchange processes is very important and crucial in determining the local weather. Most of the numerical weather models have an ABL submodel in order to quantify the energy partition at the surface and the turbulent exchange of momentum, heat/water, and matter between the surface and the FA. The ABL connects these two layers.

Traditionally, micrometeorology is one of the branches of meteorology, and it deals with the atmospheric phenomenon that occurs at the ABL, mainly at the surface. Over the last decades, the subject of boundary layer meteorology has appeared at the scientific community and used with or in connection with micrometeorology. It deals with the processes that occur above the surface layer, connecting the surface and the FA. The clouds are a good example how to link the surface and FA by different processes (evaporation, convection, condensation, precipitation, etc.). All these processes involve transport of energy. Garstang and Fitzjarrald (1999) described very well these interactions, especially considering the sea–atmosphere interface.

The goal of this chapter is to introduce this important layer (ABL), the actual devices available for the measurements with their advantages and disadvantages, presenting some results for the field measurements in Amazonia.

2.2   DEFINITION OF THE ATMOSPHERIC BOUNDARY LAYER

The ABL is the lower part of the troposphere and from the meteorological point of view it plays an important role connecting the surface with the FA (above 1–2 km). The FA is governed by atmospheric processes at synoptic or large scale like the horizontal pressure gradients, cold fronts, atmospheric disturbances, etc. At this layer, because the time and space scale of the phenomena, the Coriolis effect (due to the movement of the Earth) must be considered. By the other side, the ABL is mainly determined by local processes such as topography, surface roughness variation, vegetation contrast, and soil moisture.

Following the definition by Garrat (1992), the ABL is the part of the atmosphere that is directly influenced by the surface processes (like friction and the diurnal cycle of heating and cooling) and responds to this external forces of about a timescale less than a day. Stull (1988) gave a similar definition but emphasized the timescale being less than 1 h. Associated with the external forces, there is significant turbulent fluxes of momentum, heat, and matter carried by turbulent motions (made by eddies) on a spatial scale of 1–2 km (largest eddies). The characteristics and time evolution of the ABL are governed by the turbulence and much of this turbulence is generated from the forces from the ground (thermals, frictional drag–shear, obstacle).

Consequently, the ABL can be split into two layers: an atmospheric surface layer (ASL) extending from the surface up to a height of 120–150 m and an outer layer from the top of the surface layer up to the FA. The ASL is the region at the bottom of the ABL where turbulent fluxes are almost constant (varies less than 10% of their magnitude) and turbulence is continuously being generated and/or dissipated. Besides that, it is also part of the diurnal cycle. The outer layer, which has a pronounced diurnal cycle, can be called convective boundary layer (CBL) or stable boundary layer (SBL) depending on the predominant atmospheric stability. The CBL is produced by the strong heating of the surface, which produces thermal instability in the form of thermals and plumes. Because of that, it occurs predominantly at daytime. The top is very well defined, and it can be easily determined by measurements (remote sensing or soundings) and this will be discussed in Section 2.3. Inside the CBL, the pollutants are trapped and vertically mixed by convection. Consequently, this layer is dirtier than the FA in episodes of air pollution. The maximum extension of the CBL is at late afternoon, ranging from 0.8 km up to 2–3 km. The CBL presents a time evolution of its depth, which is dependent on the surface sensible heat flux. This is a key parameter in the development of the CBL: for wetter surfaces, the partition of energy is more pronounced for latent heat fluxes, so the height of the CBL is lower (e.g., oceanic and wet soil conditions). On the contrary, for dry surfaces (continental areas and deserts), the partition energy occurs predominantly at sensible heat fluxes, so the depth of the CBL is very high. Sometimes, the CBL is called ML because the properties (mainly temperature and humidity) are vertically well mixed so the profiles are constant with height. At the top of the CBL, there is a small layer (called entrainment zone—ZE) that connects the CBL and FA through the convective penetration (Stull, 1988). This zone is not very well understood so far due to the difficulties in obtaining data (mainly by aircrafts). However, the large-eddy simulation (LES) technique appears to be a solution in understanding this important zone better (Wyngaard, 1998).

The SBL occurs mostly (but not exclusively) at night due to the nocturnal cooling of the surface. Consequently, it is sometimes also referred to as nocturnal boundary layer. The top of the SBL is not well defined as the CBL and the depth of the SBL can be computed in many different ways: at a height where there is no turbulence or at a height where the surface nocturnal cooling does not affect anymore. The extension of the SBL is much lower than that of the CBL, ranging from 0.1 km up to 0.3 km. This time evolution depends strongly on the winds. For a windy night, the SBL will extend deeper due to the generated mechanical turbulence. For a calm wind with low turbulence, this depth is very shallow. During nighttime, although there is a medium between the top of the SBL and the FA, this layer is inactive and collapsing. The better characterization for it is like a residual layer (RL). The structure is similar to the well-mixed CBL, but there is no active turbulence. Figure 2.1 presents a schematic design of the layers.

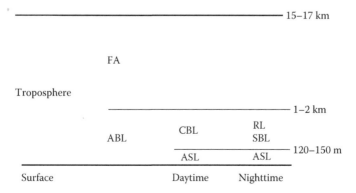

FIGURE 2.1   A schematic description of the atmosphere.

2.3 MEASUREMENTS

The ABL can be studied by three different approaches: by measurements, by numerical modeling, and by simulations using wind tunnel. However, only the direct measurements of the ABL will be discussed in detail. The others will be discussed only briefly.

The direct measurements of the ABL can be classified as surface *in situ* measurements made by masts or towers, balloon-borne measurements made by rawinsoundings, and/or tethered balloons. For the surface layer, usually masts/small towers with several levels of temperature, air humidity, and winds (speed and direction) are installed in order to characterize its structure. The required vertical spacing is dependent on the height and the goals of the tower. Usually, at the surface layer, the sensors should be installed in a logarithmic scale (to obey the Monin–Obhukov Universal Theory) and linearly spaced above. They offer convenient platforms for observing the mean and turbulent properties providing a useful dataset for characterization of the ABL. A detailed analysis can also be made and used for testing models/parameterization used in numerical models. There are several tall towers in the world used for micrometeorology research such as the BAO, the Cabauw, and the MOL towers (see Table 2.1). The Lindenberg Meteorological Observatory has several other equipments for boundary layer measurements (e.g., SODAR, LIDAR, Windprofilers) located together, and a good review of their research and principal results has been made by Beyrich and Engelbart (2008). In Brazil, there are a few tall towers, and they are mostly used to conduct research in the field of wind engineering or for assessing wind energy. For atmospheric research, the Alcantara Anemometric Tower (Figure 2.2) with six levels of direction and windspeed measurements (installed at 6, 10, 18, 26, 43, and 70 m) is being used to describe the main structure of the surface layer at the Alcantara Space Center (Fisch, 1999; Roballo and Fisch, 2008). For weather forecast and air quality monitoring, small masts measuring winds at 10 m and temperature and humidity at a lower level (2 m) are enough. Also, it is very convenient to have another measurement of temperature (10 m) in order to derive the atmospheric stability. An interesting issue to be studied and avoided when installing sensors at the towers is their structure influence and exposure to the prevailing wind. In order to avoid the influence of the wakes due to the structure, it is suggested that the sensors should be mounted at the end of booms with a length of 1.5–2.0 times the largest lateral dimension. Also, the sensors should be orientated to the prevailing

TABLE 2.1

Examples of Anemometric Towers Used for Atmospheric Research

Description	Height/Year	Reference
Boulder Atmospheric Observatory—BAO/NOAA (www.esrl.noaa.gov/psd)	300 m/1977	Kaimal and Gaynor (1983)
Ground-based observations at Cabauw—CESAR/KNMI (www.knmi.nl/kodac/ground_based_observations_climate/cabauw.html)	213 m/1972	Van Ulden and Wieringa (1996)
Lindenberg Meteorological Observatory—MOL/DWD (www.dwd.de)	99 m/1986	Neisser et al. (2002)

(a)   (b)

FIGURE 2.2   The Alcântara Anemometric Tower.

wind, allowing a range of directions (around 30° for each side). In case there is no prevailing wind dominant, another opposite boom should be installed. This is the case of the Cabauw Tower and the Kennedy Space Center Tower (which has an approximate height of 150 m), which have two booms (opposite) at each level.

The sensors installed at the towers should measure the mean and fluctuating components. For the mean values that characterize the profiles (temperature, humidity, and winds), the absolute accuracy is the most important issue to be considered (Kaimal, 1986). The time constant is in the range of 10–60 s, characterizing a slow response type of sensor. The most used sensors for temperature and humidity (using psychrometric technique) are platinum resistance thermometers, which has a time constant of around 60 s. It is essential to protect the sensors against solar radiation (radiation shields) with an aspiration rate of 5 m/s. For the winds, the cup anemometer with a wind vane is usually used for wind measurements as they are separated apart (each sensor does not influence the other), they are very simple and rugged. However, they need periodic calibrations (at least yearly) that are conducted at wind tunnels. For wind engineering purposes (especially associated with wind energy issues), this is the sensor accepted by the users. Another low time response sensor used for horizontal winds is the propeller aerovane (Figure 2.3), which has the windspeed and vane at the same body. The aerodynamic of the body rotates the sensor toward the prevailing wind. So the propellers are always pointing to the mean wind direction. If the wind direction rotates too fast, a cosine response function should be applied to the raw data. For weak wind conditions, the threshold windspeed value has to be considered. For both sensors (cup and propeller), it is around 0.5 m/s. Besides the profile, radiation equipments (incident and reflected short and long wave components measured by radiometers) are installed at the top of the mast/towers. All these equipments are measured at the sample rate of less than 0.5 Hz and 1–10 min averages are saved in

FIGURE 2.3 A detailed view of a propeller at Alcântara Anemometric Tower.

a data acquisition system. For the engineering purposes, 10 min averages are recommended, especially for the winds. Within this timescale, the average windspeed, the mean wind vector (made with the mean horizontal [zonal and meridional] windspeed components), gust factor, and turbulence intensity can be computed.

For the fluctuations measurements, the time response of the sensor is the most important issue to be considered. In order to measure the turbulent fluxes using the eddy covariance technique (see Arya, 2001 or Garrat, 1992), a sample rate in the range 10–20 Hz should be used. Usually, a tridimensional sonic anemometer (with or without a direct measurement of the air temperature) can be installed at 10 m (Figure 2.4) and the fluxes (momentum and sensible heat fluxes) as well as the spectrum can be computed. Figure 2.4 presented a three-dimensional sonic anemometer installed at the Alcântara Anemometric Tower, at the same level (10 m) as the

FIGURE 2.4 Details of a three-dimensional sonic anemometer at Alcântara Anemometric Tower.

propeller. Both the sensors (fast/sonic and slow/propeller) give the complete description of the wind field. The wind spectrum is particularly important for wind engineering and/or turbulence analysis in order to characterize the size of the eddies. Figure 2.5 shows an example of horizontal and vertical spectrum, and the inertial subrange and the −5/3 slope can be clearly seen, where both the energy production and the dissipation are negligible. The air temperature is usually measured with a very thin thermocouple, which is very vulnerable to strong winds and rain.

For the measurements of the CBL and/or SBL, rawinsoundings or tethered balloons are usually used. They measure horizontal winds (speed and direction), air temperature, relative humidity, and atmospheric pressure (which will be used

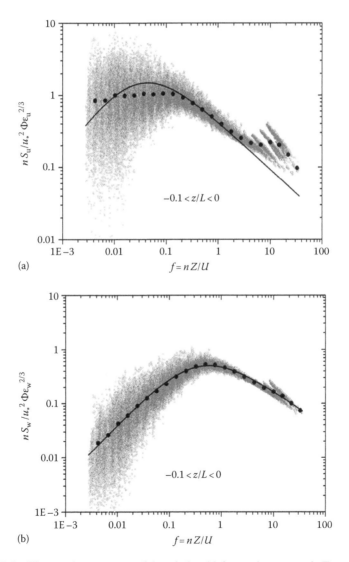

FIGURE 2.5   The u and w spectrum of the wind at Alcântara Anemometric Tower.

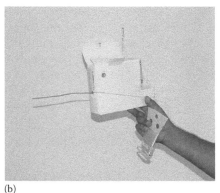

(a)   (b)

FIGURE 2.6   A Radiosonde (RS80-15G) and a release of a rawinsounding during daytime.

to compute the height by the hypsometric equation). The rawinsoundings (or radio-soundings) is a 350 g balloon filled with helium and carries a small electronic packet including the sensors for temperature, humidity, and pressure, measured at a sample rate of 0.5 Hz (Figure 2.6). The balloon should be lifted with an ascension rate around 300 m/min, which gives a 10 m vertical space resolution. The winds are determined using the Global Positioning System (GPS), and they have a resolution around 0.5 m/s for the windspeed and less than 1° for the direction of the wind. Due to this very high ascension lift, this equipment is usually suitable to measure the structure of the CBL. An important parameter for the air quality models is the determination of the ML height. Usually it determines the top of the CBL and the volume for the pollutants dispersion. Seibert et al. (2000) did a detailed analysis of the instruments (both direct and remote sensing) and methods (simple equations and complex models) used for its determination. Following their conclusions and recommendations, the profile measurements are the preferred option, if suitable data is available. Otherwise, simple methods like parcel and/or Richardson number can be used with care and tested against local data before they become operational.

During a field campaign, several profiles of rawinsoundings are made in order to have the time evolution of the CBL: an initial sounding should be made at 2–3 h after sunrise in order to measure the shallow CBL and a sounding at 1–2 h before sunset should be made in order to characterize the maximum development of the CBL during that particular day. The costs of the soundings are very high (each sounding costs around $200) and an intelligent and strategic plan must be made in order to optimize costs and measurements. At least two other soundings should be made at local noon (or the time of maximum turbulence) and at night (to capture the RL). Figure 2.7 shows an example of the profile of the CBL for Amazonia. The ML formed and the thermal inversion located at around 2300 m is very clear. As the processes mixing the properties

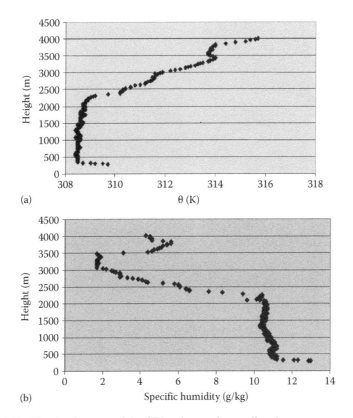

FIGURE 2.7   The development of the CBL using rawinsounding dataset.

(temperature and humidity) are the same and due to the same eddies, the height of the CBL derived from both profiles should match. For the structure of the SBL, the tethered balloon (Figure 2.8) is the suitable device as it measures the same climatic variables (horizontal winds, temperature, humidity, and pressure) but its ascension rate is very low (around 1 m/s). During the lift, the detailed structure of the SBL can be measured. It should be emphasized that the rawinsoundings is not a suitable device to carefully measure the structure of the SBL. Figure 2.9 shows the profile of potential temperature made at a tropical forest during the LBA TRMM experiment (see Fisch et al., 2004) for soundings at 6:00, 7:00, and 8:00 LT (local time). The sunrise is at 6:30 LT. The profiles at 6:00 and 7:00 LT are almost the same; the difference in the thermal structure can be seen only very close to the surface. For the profile at 8:00 LT (about 1:30 h after the sunrise), the lower structure (up to 220 m in depth) has been heated by the sensible heat fluxes from the surface. The erosion of SBL is an important issue for air quality problems because the pollutants released during the night are trapped by the SBL and its dispersion will depend on how fast this SBL will be eroded at early morning. This is an issue that is being investigated by the scientists.

During the last two decades, different techniques to measure both CBL and SBL have been developed, especially using the remote-sensing technique. The new sensors, which are both based on transmitting a sound (SODAR) and light (LIDAR) wave

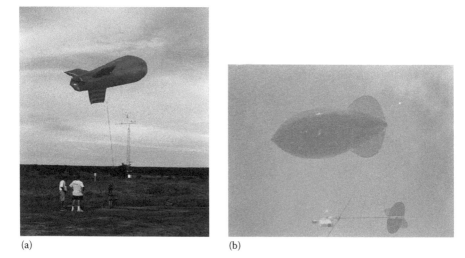

(a)   (b)

FIGURE 2.8   The tethered balloon and details of the probe.

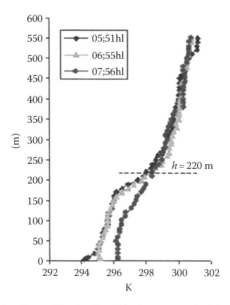

FIGURE 2.9   Example of a profile of tethered balloon at tropical forest during experiment LBA/TRMM 1999.

detecting its backscattering and reflections, are becoming commercially available for routine measurements (not only for scientific research). Their advantages related to the rawinsoundings are due to the availability of continuous measurements, but their disadvantages are the costs (still expensive) and the quality data software. There are few equipments being used nowadays, and it is a problem to choose the installations properly (for instance, SODAR is not efficient at noisy environments like edges of highways or urban areas). Figure 2.10 shows 36 h wind measurements, and it is

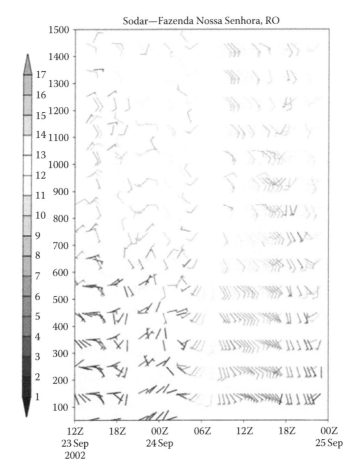

FIGURE 2.10 Detailed structure of the atmospheric boundary layer with a SODAR at the LBA RACCI 2002 experiment.

possible to see the detailed structure up to 1000–1500 m. Also, the occurrence of a nocturnal jet reaching the surface during that particular night can also be detected.

The air movement is a fully turbulent process. Although the field measurement can be considered as the true or the real value, it is hard to split and separate the simultaneous processes. Consequently, there are two alternatives to be used in order to isolate cause–effect for sensitivity analysis: the numerical simulations and the wind tunnel measurements. The numerical simulations are conducted based on the instantaneous Navier–Stokes (N–S) equation, which contains all details of the turbulence in the flow. In principle, if the N–S is resolved numerically for the whole range of scales, a direct numerical solution (DNS) is made. However, large computers need to be used, which are not often available. An alternative is to use the LES, which attempts to simulate only the scales of motion in a certain range of scales between the smallest (responsible for the heat/momentum dissipation) and the larger eddies (responsible for the transport). LES basically involves the resolved scales and subgrid scales (which are parameterized and estimated by models), especially as there are few (or none) traditional data of mean/fluctuations. This technique has been used successfully in the last two to three decades, and has given insights into the turbulent processes by estimating the pressure field (very difficult to measure) and/or isolating physical processes for developing new (or even improving) parameterization schemes. It is also used to improve the knowledge of the entrainment zone, which data is very hard to get. Nowadays, the use of LES has become the best source of data for ABL modeling. A good review of the use of this technique can be found in Moeng (1998).

The wind tunnel measurements are used to study the flow inside a chamber. The flow inside it is originally laminar and the turbulent pattern is achieved by the use of spires and/or a rough carpet. One of the weaknesses of the use of wind tunnels is the low Reynolds number achieved (usually 10^4 or 10^5 while at the atmosphere it is between 10^6 and 10^7) as well as the neutral conditions. For situations where the windspeed is very high (say higher than 8.0–9.0 m/s), the atmosphere can be considered as neutral (Loredo-Souza et al., 2004). This is the case of Alcantara Space Center, where the winds are very strong (usually higher than 6–7 m/s) due to the coupling between the trade winds and the sea breeze. Figure 2.11 shows a small wind tunnel at CTA/ITA (http://www.aer.ita.br/Laboratorios.html) used for studies about the atmospheric turbulence (Figure 2.11a) and the spires (Figure 2.11b) to create and develop the turbulence. Figure 2.12

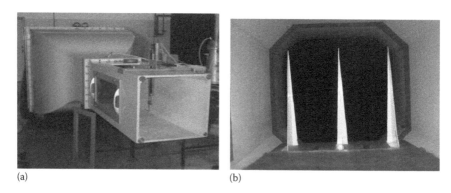

(a)   (b)

FIGURE 2.11   The ITA/CTA wind tunnel and spires.

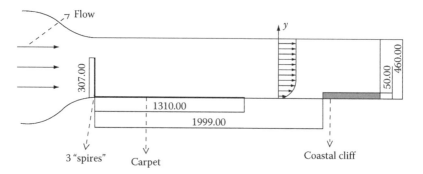

FIGURE 2.12   The schematic design of the coastal cliff at the Alcântara Space Center.

shows the schematic design of a turbulence study representing a coastal cliff at the Alcantara Space Center. The tests conducted at this facility using hot-wire and particle image velocimetry techniques have confirmed the existence of a recirculation zone with a high value of vorticity (turbulence) (Figure 2.13). This wind tunnel is also used for wind calibrations.

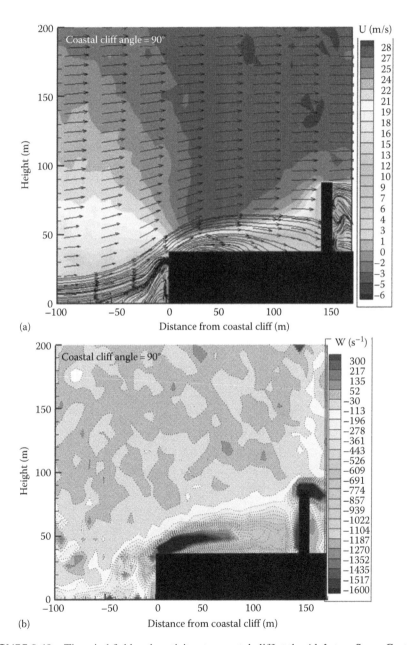

FIGURE 2.13   The wind field and vorticity at a coastal cliff at the Alcântara Space Center.

2.4 CASE STUDY: AMAZONIA

In Sections 2.2 and 2.3, the principal concepts of the ABL have been described. Understanding the behavior of the ABL is thus a critical prerequisite to know how changes in the land surface will translate into changes in the dynamics and thermodynamics of the large-scale circulation and vice versa. The Amazon region is suffering a high rate of deforestation and tropical forest is being replaced by agricultural landscape, mainly for cattle activities. Field measurements are being conducted under the LBA Program (http://lba.cptec.inpe.br/lba/site/) over tropical and pasture areas in Amazonia in order to identify this land use, land change, and its effect on the ABL structure. These two sites (Figure 2.14) are close together (the distance is less than 80 km) at the southwest of Amazonia (Rondônia State) and the observations have been made for dry and wet conditions and the main results are presented in Table 2.2 for the unstable/daytime conditions (CBL) using rawinsoundings and in Table 2.3 for stable/nighttime conditions (SBL) by tethered balloons. The dry season data has been made during the ABRACOS/RBLE experiment (in 1993) and the wet

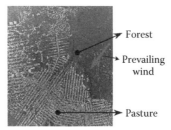

FIGURE 2.14 The location of the forest and pasture experimental site at southwest Amazonia.

TABLE 2.2

The Time Evolution (Local Time) of the Heights (m) of the CBL over the Pasture and Forest for Dry and Wet Season

	Pasture		Forest	
	Dry	Wet	Dry	Wet
08:00	62	94	75	124
11:00	517	475	267	491
14:00	1471	775	902	813
17:00	1641	927	1094	1002

Source: Adapted from Fisch, G. et al., *Theor. Appl. Climatol.*, 78, 47, 2004.

TABLE 2.3

The Time Evolution (Local Time) of the Heights (m) of the SBL over the Pasture and Forest for Dry and Wet Season

	Pasture		Forest	
	Dry	**Wet**	**Dry**	**Wet**
18:00	120	—	180	150
19:00	110	210	240	160
21:00	260	210	300	250
24:00	230	250	330	260
05:00	320	240	420	280

Source: Adapted from Fisch G. et al., *Theor. Appl. Climatol.*, 78, 47, 2004.

season during the LBA/TRMM experiment (in 1999). Fisch et al. (2004) give more details of the sites, instruments used, and the results obtained.

The CBL development shows different patterns for different surface soil moisture conditions: for forests the CBL grows up to approximately 1000 m, independent of the season (dry or wet). In contrast, the CBL at pasture shows strong seasonality with a height of 1650 m during the dry season and around 1000 m in the wet season. The soil moisture conditions determine the partition of surface energy and hence the sensible heat fluxes and consequently the height of CBL. For both sites, the highest growth rates occur between 11:00 and 14:00 LT, when the atmosphere has strong buoyancy and the sensible heat flux is very high. It should be noticed that the wind-speed is reasonably low so the main force for convection is the buoyancy (this situation can be considered as a pure thermal convection). However, there is a significant change in the growth rate from the dry to the wet season, especially in the pasture: during the dry season this rate is 320 m/h but diminishes to 100 m/h for the wet season (a factor of 3). For the other times of the day both sites present equivalent growth rates (around 100 m/h in the early morning and about 60 m/h in the late afternoon). Considering only the wet season, the results indicate that the CBL structure and dynamics are triggered by the same mechanisms (e.g., large-scale circulations) at both sites (tropical forest and pasture).

For the SBL, the heights are approximately the same for the wet season, although that for the forest is slightly higher. For the dry conditions, the SBL over the forest is always greater than the pasture and both heights (forest and pasture) are higher than during the wet conditions. The mechanical turbulence due to the wind shear is stronger at the forest, because of its higher aerodynamic roughness. The biodiversity of the trees produces very irregular heights of the trees creating roughness. The growth of the SBL is dependent on the time as well as the windspeed and nocturnal radiation budget. This produces an almost stationary curve with heights around 320–420 m before sunrise. It should also be emphasized that this height is 1/3 of the CBL at forest and 1/5 for the pasture. The layer above is the RL.

2.5 CONCLUDING REMARKS

This chapter reviewed some of the concepts involved in the ABL as well as the measurements and equipments used. For readers interested in this subject, the books by Stull (1988), Garrat (1992), and Arya (2001) present the main concepts of the ABL and are very well recommended. For the issue of measurements, the books by Kaimal and Finnigan (1994) and Lee et al. (2004) cover the most important topics, including the most actual software and procedures for the raw data collected by the eddy covariance technique. The recently published book by Foken (2008) covers both the concepts and the actual equipments and measurements. It is an updated and state-of-the-art book about micrometeorology and boundary layer meteorology. Finally, some history, philosophy, and sociology of the boundary layer modeling can be found in Wyngaard (1998). Within the scope of this book, the winds play an important role. So the books from Plate (1982) and Arya (1999) that describe the subject of wind engineering and air pollution meteorology from the perspective of the engineering community are also recommended.

REFERENCES

Arya, S.P. 1999. *Air Pollution Meteorology and Dispersion*. New York: Oxford University Press.

Arya, S.P. 2001. *Introduction to the Micrometeorology*, 2nd edn. San Diego, CA: Academic Press.

Beyrich, F. and Engelbart, D.A.M. 2008. Ten years of operational boundary layer measurements at the Richard Assmann Observatory Lindenberg: The role of remote sensing. *IOP Conference Series: Earth and Environmental Sciences*, vol. 1, 012026 (doi:10.1088/1755–1307/1/1/012026).

Fisch, G. 1999. Características do Perfil Vertical do Vento no Centro de Lançamento de Foguetes de Alcântara (CLA). *Revista Brasileira de Meteorologia* **14**:11–22.

Fisch, G., Tota, J., Machado, L.A.T. et al. 2004. The convective boundary layer over pasture and forest in Amazonia. *Theoretical and Applied Climatology* **78**:47–59.

Foken, T. 2008. *Micrometeorology*. Berlin: Springer-Verlag.

Garrat, J.R. 1992. *The Atmospheric Boundary Layer*. Cambridge, MA: Cambridge University Press.

Garstang, M. and David, F. 1999. *Observations of Surface to Atmosphere Interactions in the Tropics*. New York: Oxford University Press.

Kaimal, J. 1986. Flux and profile measurements from towers in the boundary layer. In *Probing the Atmospheric Boundary Layer*, D.H. Lenschow (ed.), pp. 19–28. Boston, MA: American Meteorological Society.

Kaimal, J.C. and Gaynor, J.E. 1983. The Boulder Atmospheric Observatory. *Journal Applied Meteorology* **22**:863–880.

Kaimal, J.C. and Finnigan, J.J. 1994. *Atmospheric Boundary Layer Flows: Their Structure and Measurements*. New York: Oxford University Press.

Lee, X., Massman, W., and Beverly, L. 2004. *Handbook of Micrometeorology: A Guide for Surface Flux Measurement and Analysis*. Dordrecht, the Netherlands: Kluwer Academic Publishers.

Loredo-Souza, A.M., Schettini, E.B.C., and Paluch, M.J. 2004. *Simulação da Camada Limite Atmosférica em Túnel de Vento*. Paper presented at annual meeting of the Associação Brasileira de Engenharia e Ciências Mecânicas—ABCM, Rio de Janeiro.

Moeng, C.H. 1998. Large eddy simulation of atmospheric boundary layer. In *Clear and Cloudy Boundary Layers*, A.A.M. Holtslag and P.G. Duynkerke (eds.), pp. 67–83. Amsterdam, the Netherlands: Royal Netherlands Academy of Arts and Sciences.

Neisser, J., Adam, W., Beyrich, F., Leiterer, U., and Steinhagen, H. 2002. Atmospheric boundary layer monitoring at the Meteorological Observatory Lindenberg as a part of "Lindenberg Column" Facilities and selected results. *Meteorologische Zeitschrift* 11:241–253

Plate, E.J. 1982. Wind tunnel modeling of wind effects in engineering. In *Engineering Meteorology: Fundamentals of Meteorology and their Application to Problems in Environmental and Civil Engineering*, E.J. Plate (ed.), pp. 573–640. Amsterdam, the Netherlands: Elsevier Scientific Publishing Company.

Roballo, S.T. and Fisch, G. 2008. Escoamento Atmosférico no Centro de Lançamento de Alcântara (CLA): Parte I—Aspectos Observacionais. *Revista Brasileira de Meteorologia* **23**:510–519.

Seibert, P., Beyrich, F., Gryning, S.E., Jofre, S., Rasmussen, A., and Tercier, P. 2000. Review and intercomparison of operational methods for the determination of the mixing height. *Atmospheric Environment* **34**:1001–1027.

Stull, R.B. 1988. *An Introduction to the Boundary Layer Meteorology*. Amsterdam, the Netherlands: Kluwer Academic Publishers.

Van Ulden, A.P. and Wieringa, J. 1996. Atmospheric boundary layer research at Cabauw. *Boundary Layer Meteorology* **78**:39–69.

Wyngaard, J. 1998. Boundary layer modelling: history, philosophy, and sociology. In *Clear and Cloudy Boundary Layers*, A.A.M. Holtslag and P.G. Duynkerke (eds.), pp. 325–332. Amsterdam, the Netherlands: Royal Netherlands Academy of Arts and Sciences.

3 Turbulence and Dispersion of Contaminants in the Planetary Boundary Layer

Gervásio Annes Degrazia, Antonio Gledson Oliveira Goulart, and Debora Regina Roberti

CONTENTS

3.1 INTRODUCTION

We consider here turbulent dispersion in geophysical flows, such as the planetary boundary layer (PBL), where Reynolds numbers are very large ($\approx 10^7$, Wyngaard, 1982), so that all or some of the possible symmetries permitted by the equations (and the boundary conditions) are restored in a statistical sense and turbulence is known as fully developed turbulence (FDT) (Frisch, 1995). Furthermore, the FDT in the PBL shows scale-invariance (small eddies are as space-filling as large ones) and self-similarity (the general aspect of turbulent signal is independent of where the window is positioned) at inertial-range scales. On the other hand, the FDT is dissipative, the mean energy dissipation per unit mass $\varepsilon(\nu)$ tends to a finite positive limit in the infinite Reynolds number limit ($\nu \to 0$, where ν is the kinematic viscosity).

In order to study turbulent dispersion, it is important to define a fluid particle. By the notion fluid particle, we mean a very small control volume of characteristic dimension much larger than the molecular scale but much smaller than the Kolmogorov microscale (Saffman, 1960; Hunt, 1982). The continuum forming such a fluid particle remains intact at the least during a time interval sufficiently large compared with the interval to be considered during the transport process. Any exchange with its direct surroundings is purely molecular in nature. The dimension of the fluid particle implies that it can be regarded as part of the fluid continuum, and its centroid responds to all scales of turbulent motion (Venkatram, 1988). The turbulent dispersion is distinct from the molecular diffusion process according to the kinetic theory of gases at least for the following two reasons: first, because of the intensive interactions between the fluid particles, there may occur a continuous exchange of a transferable property; second, as will be shown, there is a correlation in time between properties of a fluid particle at subsequent instants. Because of this memory behavior the turbulence diffusion process may not be considered as a Markov process (a stochastic process that has zero memory of the past and a future that is a function of the present and some statistical rule for the transition).

The aim of this chapter is to present and discuss some general characteristics of a FDT. The present analysis is based on plausible hypothesis, which were experimentally observed, that was mathematically represented by Kolmogorov. Phenomenological characteristics (associated to a FDT), known as self-similarity, scale-invariance within the inertial subrange, localness of interaction, and the turbulent energy spectrum, will be assumed and employed for the derivation of the different turbulent parameters. Based on these characteristics associated to a FDT and on the Taylor statistical diffusion theory, turbulent eddy diffusivities for a convective boundary layer (CBL) will be derived.

Furthermore, in this chapter we study two phenomena associated to the turbulence and that frequently play a fundamental role in describing contaminants dispersion in a PBL. One of these phenomena concerns the turbulence decaying in a CBL. This dynamical process, which occurs in the characteristic sunset transition time (1 h), is responsible for sustaining turbulence in the residual layer (RL). Some contaminant sources (stacks) are localized in this RL and consequently a description of the turbulent diffusion in this environment of decaying turbulence is of fundamental importance to parameterize turbulent terms that appear in analytical and numerical mathematical models of air pollution.

The second phenomenon occurs in situations characterized by low wind speed (LWS). The study of LWS conditions is of interest, partly because the simulation of airborne pollutant dispersion in these conditions is rather difficult. In fact, most of existing regulatory dispersion models become unreliable as the wind speed U approaches zero, so that their application is generally limited to $U > 2.0$ m/s. In these conditions, the contaminant plume is unlikely to have any definable travel and dispersion is governed by meandering (low-frequency horizontal wind oscillations). As a consequence, the more the wind speed decreases the more the standard deviation of the wind direction increases, which makes it more difficult to define a mean plume direction. Even when the stability (during nighttime) reduces the vertical dispersion and the instantaneous plume may be thin, meandering disperses the plume over a rather wide angular sector. As a consequence, horizontal diffusion is enhanced because of the meander. Since the meandering phenomenon occurs frequently, their description is of fundamental importance in air dispersion modeling in the atmospheric boundary layer.

Based on fact that the Navier–Stokes (N–S) equations describe the multitude of phenomena associated to the turbulence processes, the investigation of both cases (decaying turbulence in the CBL and the low-frequency horizontal wind oscillations) will be accomplished employing these conservation equations. In the decaying turbulence case, the N–S equations will be used to derive eddy diffusivities. In the LWS case, the N–S equations will be utilized to explain the low-frequency horizontal wind oscillations.

3.1.1 Taylor's Model

By considering Taylor's classical paper (Taylor, 1921) on diffusion by continuous movements, we assume the motion of fluid particles (massless) in a turbulent flow field by velocity fluctuations.

We take the simplest case of dispersion in one direction only, where X_i corresponds to an arbitrary direction associated to the i-component of the velocity of the fluid particle ($i = u, v, w$). If the fluid particle leaves the origin at $t = 0$, its position X_i at time t is given by

$$X_i(t) = \int_0^t v_i(t')dt' \tag{3.1}$$

An eddy diffusivity can be obtained by multiplying Equation 3.1 by $v_i(t)$

$$X_i(t)v_i(t) = X_i(t)\frac{dX_i}{dt} = \frac{d}{dt}\left(\frac{1}{2}X_i^2\right) = \int_0^t v_i(t)v_i(t')dt' \tag{3.2}$$

and taking an ensemble-mean over many realizations (i.e., consider a large number of fluid particles that are assumed to start in succession from a fixed time $t = 0$), we obtain

$$\frac{d}{dt}\left(\frac{1}{2}\overline{X_i^2}\right) = \int_0^t \overline{v_i(t)v_i(t')}dt' \tag{3.3}$$

In the above equation, both sides have dimensions of an eddy diffusivity ($m^2 s^{-1}$).

Taylor's theory applies to dispersion in a field of homogeneous and stationary turbulence, that is, turbulence whose statistical properties have quantitatively the same structure in all parts of the flow field and the statistical properties of variables do not change with time, so that the correlation function R_{L_i} in the integrand of Equation 3.3 is an even function of the time difference $\tau = t - t'$. For an arbitrary turbulent velocity component the form of the function $R_{L_i}(\tau)$ is given by

$$R_{L_i}(\tau) = \overline{v_i(t')v_i(t'+\tau)} = \overline{v_i^2}\rho_{L_i}(\tau) \tag{3.4}$$

Equation 3.4 defines the correlation between the particle velocity at one time $v_i(t')$ and at some later time $v_i(t'+\tau)$. The dimensionless form of the function $\rho_{L_i}(\tau)$ is called a correlation coefficient and satisfies $\rho_{L_i}(0) = 1$. The subscript L refers to the fact that these are Lagrangian correlations, and measurements are made by following a fluid particle as it is carried by the turbulence.

The substitution of Equation 3.4 into Equation 3.3 yields

$$\frac{d}{dt}\left(\frac{1}{2}\overline{X_i^2}\right) = \int_0^t R_{L_i}(\tau)d\tau = \overline{v_i^2}\int_0^t \rho_{L_i}(\tau)d\tau \tag{3.5}$$

This equation can be integrated to yield

$$\overline{X_i^2} = 2\overline{v_i^2}\int_0^t\left(\int_0^{t'}\rho_{L_i}(\tau)d\tau\right)dt' \tag{3.6}$$

Equation 3.6 may be written somewhat differently by carrying out an integration by parts

$$\int_0^t dt'\int_0^{t'}\rho_{L_i}(\tau)d\tau = \left|t'\int_0^{t'}d\tau\,\rho_{L_i}(\tau)\right| - \int_0^t t'\rho_{L_i}(t')dt' = t\int_0^t d\tau\,\rho_{L_i}(\tau) - \int_0^t \tau\,\rho_{L_i}(\tau)d\tau$$

Then Equation 3.6 reads as

$$\overline{X_i^2} = 2\overline{v_i^2}\int_0^t (t-\tau)\rho_{L_i}(\tau)d\tau \tag{3.7}$$

The expressions 3.5 and 3.6 characterize turbulent dispersion in terms of the particle's ability to remember its velocity between 0 and t. Of fundamental interest is the behavior of these equations for large values of t. If we consider very long periods of time, such that $t \gg t^*$, where t^* is the time for which $\rho_{L_i}(t^*) \approx 0$, the relation in Equation 3.7 gives

$$\overline{X_i^2} = 2\overline{v_i^2}\left[t\int_0^{t^*}\rho_{L_i}(\tau)d\tau - \int_0^{t^*}\rho_{L_i}(\tau)\tau\,d\tau\right]$$

For $t \gg t^*$, the second term on the right-hand side will become very small with respect to the first term and may be neglected so that

$$\overline{X_i^2} = 2\overline{v_i^2}\, t \int_0^{t^*} \rho_{L_i}(\tau)d\tau \tag{3.8}$$

For the constant value of the above integral we can write

$$T_{L_i} = \int_0^\infty \rho_{L_i}(\tau)d\tau \tag{3.9}$$

where the Lagrangian integral timescale T_{L_i} is usually considered a measure of the longest time during which, on the average, a fluid particle persists in a motion in a given direction (Hinze, 1975). With the definition of T_{L_i}, for long times, $t \gg T_{L_i}$, so that $\rho_{L_i}(\tau) \approx 0$ and Equation 3.8 can be written as

$$\overline{X_i^2} = 2\overline{v_i^2}\, t\, T_{L_i} \tag{3.10}$$

In this time limit, $\sqrt{\overline{X_i^2}}$ grows parabolically with t, which is a diffusive type of behavior.

For $\tau \gg T_{L_i}$, the eddy diffusivity in Equation 3.5 can be approximated by

$$\frac{d}{dt}\left(\frac{1}{2}\overline{X_i^2}\right) = \sigma_i^2 \int_0^\infty \rho_{L_i}(\tau)d\tau = \sigma_i^2 T_{L_i} \tag{3.11}$$

where $\sigma_i^2 \equiv \overline{v_i^2}$ is the velocity fluctuation variance. The relation in Equation 3.11 may also be written as

$$\frac{d}{dt}\left(\frac{1}{2}\overline{X_i^2}\right) = \sigma_i\, l_{L_i} \tag{3.12}$$

with

$$l_{L_i} = \sigma_i\, T_{L_i} \tag{3.13}$$

where the Lagrangian length scale (l_{L_i}) may be interpreted as a space scale in which the particle moves substantially in only one direction. Equations 3.5 and 3.11 define eddy diffusivities. The eddy diffusivity Equation 3.5 depends upon the travel time t from the source. Thus, the eddy diffusivity for the fluid particles emanating from a continuous point source differ from that Equation 3.11 for temperature and water vapor diffusing in the same flow, since the latter has an effectively infinite area source at the surface. In fact Equation 3.5 for large travel times is identical to Equation 3.11 and in this case the eddy diffusivity become independent of the travel time (or distance) from the source and just a function of the turbulence (e.g., large eddy length and velocity scales). The eddy diffusivity Equation 3.5 can accurately represent the near source diffusion in weak winds. For this, eddy diffusivities should be considered as functions of not only turbulence but also distance from the source (Arya, 1995).

For short times ($t << T_{L_i}$), $\rho_{L_i}(\tau) \approx 1$, and

$$\overline{X_i^2} = \overline{v_i^2} t^2 \tag{3.14}$$

the fluid particle plume growth is linear with time.

Kampé de Fériet expressed Taylor's equation in the form of Equation 3.7 which can be recast into

$$\overline{X_i^2} = 2\overline{v_i^2} T_{L_i}^2 \int_0^{X/UT_{L_i}} \left(\frac{X}{UT_{L_i}} - \xi\right) \rho_{L_i}(\xi) d\xi \tag{3.15}$$

where
  U is the horizontal mean wind velocity
  X is the particle mean displacement $X = Ut$
  ξ is the rescaled time τ/T_{L_i}.

From Equation 3.15 we obtain the nondimensionalized formula

$$\frac{\overline{X_i^2}}{2\overline{v_i^2} T_{L_i}^2} = \int_0^{t/T_{L_i}} \left(\frac{t}{T_{L_i}} - \xi\right) \rho_{L_i}(\xi) d\xi \tag{3.16}$$

Equation 3.16 yields the nondimensionalized mean square generalized displacement $\overline{X_i^2}$ as a universal function of the normalized time t/T_{L_i} (Degrazia et al., 1991)

$$\frac{\overline{X_i^2}}{2\overline{v_i^2} T_{L_i}^2} = g_i\left(\frac{t}{T_{L_i}}\right) \tag{3.17}$$

It is worth recalling at this point that Equation 3.17 yields the form of the universal function for the nondimensionalized dispersion as proposed earlier by Draxler (1976). Notice also that the left-hand side of Equation 3.17 is normalized in terms of the velocity fluctuation variance and its Lagrangian integral timescale. The right-hand side of Equation 3.17 yields the form of the universal function $g_i(t/T_{L_i})$. In this latter form, Taylor's model suggests that the relevant parameters for the determination of $\overline{X_i^2}$ are the Lagrangian quantities $\overline{v_i^2}$ and T_{L_i}.

3.1.1.1 Some Considerations about Taylor's Model

In the analysis of the diffusion of tracers in the PBL, Equation 3.7 can be explored in two complementary ways. In a first approach, one aims at predicting values and/or forms for the mean square generalized displacement $\overline{X_i^2}(t)$ by assuming (physically) reasonable expressions for the Lagrangian velocity variances (σ_i^2) and for the velocity autocorrelation coefficients $\rho_{L_i}(\tau)$. A second important utilization of Equation 3.7 concerns the derivation, from the actually measured data, of the turbulent velocity field. This kind of application has already been partly initiated by Frenkiel (1953) early in the fifties. More recently Taylor's diffusion theorem has been used in the limits of small and large diffusion times, Equations 3.14 and 3.10 respectively, to derive by employing Equation 3.17 an analytical formula for the lateral dispersion parameter $\sigma_y = \sqrt{X_v^2}$ adjusted for a specific experiment (in that case, the Hanford-67 experiment, Degrazia, 1991). This dispersion parameter admits a much better representation when renormalized (rescaled) by the Lagrangian velocity and Lagrangian timescales. The closer packing of the experimental data obtained by this normalization allows an analytical fitting of the results. This in turn helps to determine both the values of relevant parameters like $\sigma_v = \sqrt{\sigma_v^2}$ and the timescale T_{Lv}, besides allowing for a derivation of the form of the lateral velocity autocorrelation coefficient ρ_{Lv}. We finally stress that this derivation was applied to the Hanford-67 data series but should be equally valid for other experiments as long as the underlying hypotheses of Taylor's theorem are applicable. This refers mainly to the homogeneity and stationarity character of the turbulence. Whenever these features are warranted, the same procedure should lead to the local values of the Lagrangian quantities (σ_i^2 and T_{L_i}) and to a functional form of the velocity autocorrelation coefficient. The appropriated adjusting constants should be obtained by fitting an analytical curve to the experimental data.

3.2   THE WIENER–KHINCHIN THEOREM: SELECTING ENERGY-CONTAINING EDDIES

It is said that turbulent flow consists of a superposition of eddies. All these various-sized eddies of which a turbulent motion is composed have a certain kinetic energy, quantified by their vorticity or by the magnitude of the velocity fluctuation of the corresponding frequency. These eddies interact continuously with the turbulent forcing mechanism, from which they extract their energy, and also with each other. An interesting question is how the kinetic energy of turbulence is distributed according to various scales (frequencies) of eddy motion. From a practical point of view, it is still fundamental to identify the scales (frequencies) associated to the large energy-containing eddies. These contain most of the kinetic energy and are responsible for the turbulent transport in the PBL. In this context, it seems appropriate to introduce the concept of Lagrangian energy spectrum.

Following Sorbjan (1989), the spectrum measures the distribution of the variance of a variable over frequencies or wavelengths. If the variable is a fluid particle turbulent velocity component, the spectrum also describes the distribution of kinetic energy over frequencies or wavelengths.

Using the Fourier transform, we can define (Panofsky and Dutton, 1984)

$$\Phi_{L_i}(\omega) = \frac{1}{\pi} \int_{-\infty}^{+\infty} R_{L_i}(\tau) e^{i\omega\tau} d\tau \qquad (3.18)$$

where the inverse transform will be

$$R_{L_i}(\tau) = \frac{1}{2} \int_{-\infty}^{+\infty} \Phi_{L_i}(\omega) e^{-i\omega\tau} d\omega \qquad (3.19)$$

or

$$R_{L_i}(0) = \frac{1}{2} \int_{-\infty}^{+\infty} \Phi_{L_i}(\omega) d\omega \qquad (3.20)$$

so that $\Phi_{L_i}(\omega)$ shows how the turbulent kinetic energy (TKE) is distributed with respect to frequency.

$\omega = 2\pi/T = 2\pi n$, where T is the period of a sinusoidal oscillation and n is the frequency in cycles/time or Hertz.

Equations 3.18 and 3.19 define the Wiener–Khinchin theorem (Gardiner, 1983) and establish a fundamental result that relates the Fourier transform of the autocorrelation function to the spectrum. It means that one may directly measure the autocorrelation function of either a signal or the spectrum, and convert back and forth, which by means of the fast Fourier transform and computer is relatively straightforward.

We can simplify the expressions (3.18) and (3.19) by noting that because of stationarity of the turbulent velocity field, $R_{L_i}(\tau) = R_{L_i}(-\tau)$; that is, $R_{L_i}(\tau)$ is an even function. From this property and from Equation 3.18 we obtain

$$\Phi_{L_i}(\omega) = \frac{1}{\pi} \int_{-\infty}^{+\infty} R_{L_i}(\tau)(\cos \omega\tau + \sin \omega\tau) d\tau = \frac{2}{\pi} \int_{0}^{\infty} R_{L_i}(\tau) \cos \omega\tau \, d\tau \qquad (3.21)$$

since $\sin \omega\tau$ is an odd function.

This shows that $\Phi_{L_i}(\omega) = \Phi_{L_i}(-\omega)$, which allows Equations 3.19 and 3.20 to be written as

$$R_{L_i}(\tau) = \int_{0}^{\infty} \Phi_{L_i}(\omega) \cos \omega\tau \, d\omega \qquad (3.22)$$

and

$$R_{L_i}(0) = \int_{0}^{\infty} \Phi_{L_i}(\omega) d\omega \qquad (3.23)$$

The transform of $R_{L_i}(\tau)$, $\Phi_{L_i}(\omega)$ is called the energy spectral density (ESD) function in analogy with the spectra of light studied in physics. The product $\Phi_{L_i}(\omega) d\omega$ is the contribution to variance made by fluctuations in the interval of width $d\omega$ centered at ω.

Now, changing from frequency ω expressed in radians per second to frequency $n = \omega/2\pi$ in cycles per second, a new spectral density $S_{L_i}(n) = 2\pi\,\Phi_{L_i}(2\pi n)$ can be introduced, so that

$$R_{L_i}(\tau) = \int_0^\infty \Phi_{L_i}(2\pi n)\cos(2\pi n\tau)\,2\pi\,dn = \int_0^\infty S_{L_i}(n)\,\cos(2\pi n\tau)\,dn \qquad (3.24)$$

For $\tau = 0$, Equation 3.24 becomes

$$\sigma_i^2 = R_{L_i}(0) = 2\pi \int_0^\infty \Phi_{L_i}(2\pi n)\,dn = \int_0^\infty S_{L_i}(n)\,dn \qquad (3.25)$$

This confirms that (twice) the kinetic energy per unit of mass is obtained if the spectrum is integrated over all frequencies.

On the other hand, in terms of the frequency n, Equation 3.21 can be written as

$$2\pi\Phi_{L_i}(2\pi n) = 4\int_0^\infty R_{L_i}(\tau)\cos 2\pi n\tau\,d\tau$$

and yields

$$S_{L_i}(n) = 4\int_0^\infty R_{L_i}(\tau)\cos 2\pi n\tau\,d\tau \qquad (3.26)$$

Equation 3.25 above can be expressed in a slightly changed form:

$$\sigma_i^2 = \int_0^\infty S_{L_i}(n)\,dn = \int_0^\infty n S_{L_i}(n)\,d(\ln n) \qquad (3.27)$$

This shows that the area under the Lagrangian one-dimensional (1-D) energy spectrum $S_{L_i}(n)$ plotted against n is the same as the area under the curve $n S_{L_i}(n)$ plotted against $(\ln n)$.

The use of $n S_{L_i}(n)$ has the advantage that this quantity has unit independent of the unit of frequency selected. The quantity $n S_{L_i}(n)$ has the dimension of the variance. If the variance σ_i^2 is finite then from Equation 3.27 it follows that $n S_{L_i}(n) \rightarrow 0$ for $n \rightarrow \infty$. From this and the obvious fact that $n S_{L_i}(n) \rightarrow 0$ when $n \rightarrow 0$, it can be concluded that $n S_{L_i}(n)$ vanishes at both the small and large frequencies and must have its maximum somewhere in between.

Reciprocals of the frequency at maxima obtained from the plot of the function $n S_{L_i}(n)$ are interpreted as the principal timescales of the turbulent flow. Then $1/n_{max}$ is a useful integral timescale typically six times larger than the Lagrangian integral timescale (Hanna, 1981).

Considering now $n = 0$ in Equation 3.26, we obtain

$$S_{L_i}(0) = 4\int_0^\infty R_{L_i}(\tau)\,d\tau = 4\overline{v_i^2}\,T_{L_i} \tag{3.28}$$

The eddy diffusivity $\overline{v_i^2}\,T_{L_i}$ was encountered before in the asymptotic diffusion Equations 3.10 and 3.11. As a consequence, we conclude that the eddy diffusivity independent of the travel time is just a function of the turbulence and depends on the large eddies.

This consideration raises the question how different frequency components contribute to the turbulent diffusion of fluid particles. In order to answer this question we return to the general equation for $\overline{X_i^2}$ (Equation 3.7). Substituting $\rho_{L_i}(\tau)$ in Equation 3.7 by its Fourier transform, Equation 3.24 yields

$$\overline{X_i^2} = 2\overline{v_i^2}\int_0^t (t-\tau)\left[\int_0^\infty F_{L_i}(n)\cos 2\pi n\tau\,dn\right]d\tau$$

$$\overline{X_i^2} = 2\overline{v_i^2}\int_0^\infty\left[\int_0^t (t-\tau)\cos 2\pi n\tau\,d\tau\right]F_{L_i}(n)\,dn$$

$$\overline{X_i^2} = \overline{v_i^2}\int_0^\infty F_{L_i}(n)\left[\frac{1-\cos 2\pi nt}{2(n\pi)^2}\right]dn$$

$$\overline{X_i^2} = \overline{v_i^2}\,t^2\int_0^\infty F_{L_i}(n)\,\frac{\sin^2(n\pi t)}{(n\pi t)^2}\,dn \tag{3.29}$$

where $F_{L_i}(n) = S_{L_i}(n)/\overline{v_i^2}$ is the value of the Lagrangian spectrum of energy normalized by the velocity variance.

The eddy diffusivity (Equation 3.5) is related to the generalized dispersion parameter (Equation 3.29) by the following derivation (Batchelor, 1949),

$$\frac{d}{dt}\left(\frac{1}{2}\overline{X_i^2}\right) = \frac{1}{2}\frac{\overline{v_i^2}}{\pi^2}\frac{d}{dt}\left[\int_0^\infty F_{L_i}(n)\,\frac{\sin^2(n\pi t)}{n^2}\,dn\right] = \frac{\pi\overline{v_i^2}}{\pi^2}\int_0^\infty \frac{F_{L_i}(n)\sin(n\pi t)\cos(n\pi t)}{n}\,dn$$

$$\frac{d}{dt}\left(\frac{1}{2}\overline{X_i^2}\right) = \frac{\overline{v_i^2}}{2\pi}\int_0^\infty F_{L_i}(n)\,\frac{\sin(2\pi nt)}{n}\,dn \tag{3.30}$$

On the other hand, for large diffusion travel times ($t \gg T_{L_i}$), the following analysis can be developed. Considering that $F_{L_i}(n)$ and $\sin(2\pi nt)/n$ are even functions of n, Equation 3.30 can be written as

$$\frac{d}{dt}\left(\frac{1}{2}\overline{X_i^2}\right) = \frac{\overline{v_i^2}}{4}\int_{-\infty}^{\infty}\frac{F_{L_i}(n)\sin(2\pi nt)}{\pi n}\,dn$$

Defining $g = 2\pi n$, the above equation can be rewritten in the form:

$$\frac{d}{dt}\left(\frac{1}{2}\overline{X_i^2}\right) = \frac{\overline{v_i^2}}{4}\int_{-\infty}^{\infty}\frac{F_{L_i}\left(\dfrac{g}{2\pi}\right)\sin(gt)}{\pi g}\,dg$$

For $t \to \infty$ yields

$$\frac{d}{dt}\left(\frac{1}{2}\overline{X_i^2}\right) = \frac{\overline{v_i^2}}{4}\int_{-\infty}^{\infty}F_{L_i}\left(\frac{g}{2\pi}\right)\lim_{t\to\infty}\frac{\sin(gt)}{\pi g}\,dg$$

where $\displaystyle\lim_{t\to\infty}\frac{\sin(gt)}{\pi g}$ is a well-known representation of a Dirac delta function. Finally,

$$\frac{d}{dt}\left(\frac{1}{2}\overline{X_i^2}\right) = \frac{\overline{v_i^2}}{4}\int_{-\infty}^{\infty}F_{L_i}\left(\frac{g}{2\pi}\right)\delta(g)\,dg = \frac{\overline{v_i^2}F_{L_i}(0)}{4} \tag{3.31}$$

where $\delta(g)$ is a Dirac delta function.

Integration of Equation 3.31 yields the classical result obtained from Taylor's statistical diffusion theory for $t \to \infty$, that is, Equation 3.10. The above analysis shows that, as time proceeds, the filter function begins to remove the energy associated to high frequencies in the turbulent spectrum. This behavior becomes evident by the presence of Dirac's delta function, an extremely restrictive operator, which selects only the very low-frequency components of the turbulent spectrum.

The formula (3.31) represents a parameterization for the eddy diffusivities in terms of the spectrum at the origin. In this case, we write $K_\alpha = (\overline{v_i^2}\,F_{L_i}(0)/4)$, with $\alpha = x, y, z$ in terms of the energy-containing eddy characteristics. Considering the asymptotic Taylor's model for large travel times:

$$K_\alpha = \frac{\overline{v_i^2}\,F_{L_i}(0)}{4} = \overline{v_i^2}\,T_{L_i} \tag{3.32}$$

yields

$$T_{L_i} = \frac{F_{L_i}(0)}{4} \tag{3.33}$$

and

$$L_i = \frac{\sqrt{v_i^2}\, F_{L_i}(0)}{4} \tag{3.34}$$

where T_{L_i} and L_i are time and length characteristic scales for a fully developed turbulent.

3.3   RELATIONS BETWEEN LAGRANGIAN AND EULERIAN STATISTICS

The spectrum and autocorrelation function in the above expressions are Lagrangian. In practice, only Eulerian statistics parameters are measured. An Eulerian measurement is one made by an instrument whose position is fixed in one way or another, for example, an anemometer on a tower or a pitot tube on an airplane. In turbulent dispersion problems we must describe Lagrangian diffusion using Eulerian measurements, so that one of the fundamental questions in turbulence and diffusion is the relation between the Lagrangian and Eulerian frames of reference for measuring turbulence (Hanna, 1981).

In stationary and homogeneous turbulence, the Eulerian and Lagrangian velocity variance σ_i^2 are equal (Corrsin, 1963). This assumption is commonly made and is based upon the fact that TKE is the same for both approaches (Hanna, 1982). In contrast, Lagrangian and Eulerian spectra or correlation functions differ systematically from each other. In the atmosphere, it is almost impossible to obtain ideal Lagrangian series because most tracers follow the air quite imperfectly. The most complete experiments were reported by Angell et al. (1971) and Hanna (1981). In both cases, Eulerian data were obtained from towers and Lagrangian observations made by following tetrahedral constant-level balloons by radar. Generally, velocities of particles following the turbulent flow (Lagrangian) are more slowly varying than those measured by a fixed instrument (Eulerian), so that Eulerian time series fluctuate more frequently with time than Lagrangian series. As a consequence, Lagrangian spectra are concentrated at lower frequencies than Eulerian spectra.

Gifford (1955) and Hay and Pasquill (1959) made the very useful assumption that the Lagrangian and Eulerian autocorrelation functions were similar in shape but were displaced by a scale factor β_i. The same assumption is valid for the energy spectra. Mathematically, this assumption can be stated as

$$\rho_{L_i}(\beta_i \tau) = \rho_i(\tau) \tag{3.35}$$

where $\rho_i(\tau)$ is the Eulerian autocorrelation coefficient and β_i is the scale factor for the i-component of the velocity, defined formally as the ratio of the Lagrangian and Eulerian timescales, that is,

$$\beta_i = \frac{T_{L_i}}{T_i} \tag{3.36}$$

with T_i the Eulerian integral timescale.

The relation between Lagrangian and Eulerian spectra can be found by considering the following expression (Equation 3.26):

$$F_i(n) = 4 \int_0^\infty \rho_i(\tau) \cos 2\pi n\tau \, d\tau \qquad (3.37)$$

where $F_i(n)$ is the Eulerian spectrum normalized by σ_i^2.

The substitution of Equation 3.35 in Equation 3.37 yields

$$F_{L_i}(n) = 4\beta_i \int_0^\infty \rho_{L_i}(\beta_i\tau) \cos 2\pi n\beta_i\tau \, d\tau$$

from Equations 3.35 and 3.37 we obtain

$$n F_{L_i}(n) = \beta_i n F_i(\beta_i n) \qquad (3.38)$$

By using Equation 3.38, the generalized dispersion parameter and the eddy diffusivity, respectively Equations 3.29 and 3.30, can be related to the scale factor β_i by the following relationships:

$$\overline{X_i^2} = \sigma_i^2 t^2 \int_0^\infty \beta_i F(\beta_i n)_i \frac{\sin^2(n\pi t)}{(n\pi t)^2} \, dn \qquad (3.39)$$

and

$$\frac{d}{dt}\left(\frac{1}{2}\overline{X_i^2}\right) = \frac{\sigma_i^2}{2\pi} \int_0^\infty \beta_i F_i(\beta_i n) \frac{\sin(2\pi n t)}{n} \, dn \qquad (3.40)$$

Equations 3.39 and 3.40 can be transformed to (Batchelor, 1949; Pasquill and Smith, 1983; Degrazia and Moraes, 1992)

$$\overline{X_i^2} = \frac{\sigma_i^2 \beta_i}{\pi^2} \int_0^\infty F_i(n) \frac{\sin^2(n\pi t/\beta_i)}{n^2} \, dn \qquad (3.41)$$

and

$$K_\alpha = \frac{d}{dt}\left(\frac{1}{2}\overline{X_i^2}\right) = \frac{\sigma_i^2 \beta_i}{2\pi} \int_0^\infty F_i(n) \frac{\sin(2\pi n t/\beta_i)}{n} \, dn \qquad (3.42)$$

Equation 3.42 has an asymptotic behavior when travel time becomes large ($\lim \tau \to \infty$) that has the effect of selecting $F_i(n)$ at the origin of the frequency space (Degrazia and Moraes, 1992). As a consequence, the rate of dispersion becomes independent of travel time from the source and can be expressed as a function of the local properties of turbulence as follows:

$$\frac{\mathrm{d}}{\mathrm{d}t}\left(\frac{1}{2}\overline{X_i^2}\right) = \frac{\sigma_i^2 \beta_i F_i(0)}{4} \tag{3.43}$$

Therefore, from Equations 3.11 and 3.43, a Lagrangian decorrelation timescale for homogeneous or nonhomogeneous turbulence can be expressed as

$$T_{L_i} = \frac{\beta_i F_i(0)}{4} \tag{3.44}$$

Furthermore, from Equations 3.13 and 3.44, the Lagrangian length scale can be expressed as

$$l_{L_i} = \frac{\sigma_i \beta_i F_i(0)}{4} \tag{3.45}$$

Hypothetically, we will assume here that the usual mixing length scale is given by the Lagrangian length scale (Equation 3.45). This hypothesis has been assumed in several other works (Tennekes and Lumley, 1972).

It is important to point out the benefits of using the parameterization given by Equations 3.43 through 3.45. Taylor's theory is valid only for homogeneous turbulence, whereas Equations 3.43 through 3.45 are more general and can be also applied in nonhomogeneous turbulence.

3.4   A HEURISTIC FORMULATION FOR THE SCALE FACTOR β_i

In the inertial subrange, the dynamic of turbulence is dominated by the inertia terms in the NS equation; that is, all but the viscous and forcing terms. In this range, where eddies are small compared with the energy-containing eddies, energy neither enters the system nor is dissipated. It is merely transmitted at rate ε from large-scale toward small-scale motion. It is natural to assume that such small-scale turbulence, far from solid bodies, is homogeneous and isotropic (Monin and Yaglom, 1975). Isotropy implies that the turbulent velocity field is independent of rotation and reflection about the spatial axes. Even though isotropy assumption does not apply to the large eddies in the energy-containing range, we can assume that the small-scale turbulence structure in the inertial subrange is effectively isotropic. This local isotropy is fundamental for the derivation of small-scale turbulence quantities. It is found that several important results concerning the local properties can be obtained from similarity arguments (Kolmogorov, 1941). Kolmogorov, who first conceived the existence of an inertial subrange separating the energy-containing and dissipation ranges (statistical independence of the small and large scales of turbulence), derived from dimensional arguments that the Eulerian three-dimensional (3-D) ESD function and the Eulerian velocity structure function are, respectively, the following (Monin and Yaglom, 1975):

$$E(k) = \alpha_t \varepsilon^{2/3} k^{-5/3} \tag{3.46}$$

and

$$D_i(\tau) = \overline{[v_i(t+\tau) - v_i(t)]^2} = \alpha_i c_S (\varepsilon U)^{2/3} \tau^{2/3} \tag{3.47}$$

where
ε is the mean dissipation of energy per unit time per unit mass of fluid (for a FDT ε is the energy flux)
$k = 2\pi n/U$ is the wave number
α_t, α_S, and c_S are numerical constants

The Eulerian velocity structure function must be distinguished from the Lagrangian velocity structure function that is described in terms of the variation in velocity of a fluid particle as it moves about in the turbulent flow. This variation can evidently depend only on ε, which determines the local structure of the turbulence, and of course on τ itself. Forming the only combination of ε and τ that has the correct dimensions, we obtain for the Lagrangian velocity structure function the following relation:

$$D_{L_i}(\tau) = \overline{[v_i(t+\tau) - v_i(t)]^2} = C_{0_i} \varepsilon \tau \tag{3.48}$$

where C_{0_i} is the Kolmogorov constant. By the isotropy condition in the inertial sub-range the 1-D spectra must have the same power law dependence on ε and k of Equation 3.46, that is,

$$E_i(k) = \alpha_i \alpha_u \varepsilon^{2/3} k^{-5/3} \tag{3.49}$$

where α_u is determined experimentally to be about 0.5 ± 0.05 for the u-spectrum and $\alpha_i = 1, 4/3, 4/3$, for u, v, and w components, respectively (Monin and Yaglom, 1975; Champagne et al., 1977; Sorbjan, 1989; Kaimal and Finnigan, 1994). The relation in Equation 3.49 can be written as

$$\Phi_i(\omega) = \frac{1}{U} E_i\left(\frac{\omega}{U}\right) = \alpha_i \alpha_u (\varepsilon U)^{2/3} \omega^{-5/3} \tag{3.50}$$

and finally

$$S_i(n) = 2\pi\Phi_i(2\pi n) = \frac{\alpha_i \alpha_u}{(2\pi)^{2/3}} (\varepsilon U)^{2/3} n^{-5/3} \tag{3.51}$$

In the inertial subrange, the Lagrangian energy spectrum $\Phi_{L_i}(\omega)$ can depend only on ε and ω. Consequently, dimensional consideration require that (Tennekes, 1981)

$$\Phi_{L_i}(\omega) = B_{0_i} \varepsilon \omega^{-2} \tag{3.52}$$

where B_{0_i} is another nondimensional constant. In terms of the frequency n, Equation 3.52 can be written as

$$2\pi\Phi_{L_i}(2\pi n) = \frac{B_{0_i}}{2\pi}\varepsilon n^{-2} \tag{3.53}$$

and yields

$$S_{L_i}(n) = B_i\varepsilon n^{-2} \tag{3.54}$$

where $B_i = B_{0_i}/2\pi$.

The use of the equations discussed above allows to obtain a theoretical expression for the fundamental scale factor β_i. Following Corrsin (1963), let us integrate Equations 3.51 and 3.54 to obtain an expression for the scale factor β_i:

$$\sigma_i^2 = \int_{n_E}^{\infty} S_i(n)\,dn = \frac{\alpha_i\,\alpha_u}{(2\pi)^{2/3}}(\varepsilon U)^{2/3}\int_{n_E}^{\infty} n^{-5/3}\,dn = \frac{3}{2}\frac{\alpha_i\,\alpha_u}{(2\pi)^{2/3}}\left(\frac{\varepsilon U}{n_E}\right)^{2/3} \tag{3.55}$$

and

$$\sigma_{L_i}^2 = \int_{n_L}^{\infty} S_{L_i}\,dn(n) = B_i\,\varepsilon\int_{n_L}^{\infty}\frac{dn}{n^2} = \frac{B_{0_i}\,\varepsilon}{2\pi}\frac{1}{n_L} \tag{3.56}$$

where n_E and n_L are, respectively, the Eulerian and Lagrangian initial frequencies of the inertial subrange. These characteristic frequencies may be regarded as the inverse of the timescales.

Equation 3.56 can also be rewritten in the form:

$$\varepsilon^{2/3} = \frac{(2\pi)^{2/3}}{B_{0_i}^{2/3}}n_L^{2/3}\,\sigma_{L_i}^{4/3} \tag{3.57}$$

Since the mean energy dissipation rate ε is equal for both Lagrangian and Eulerian frames of reference we can substitute Equation 3.57 in Equation 3.55 to obtain

$$\frac{n_E^{2/3}}{n_L^{2/3}} = \frac{3}{2}\alpha_i\,\alpha_u\frac{U^{2/3}}{B_{0_i}^{2/3}}\frac{\sigma_{L_i}^{4/3}}{\sigma_i^2} \tag{3.58}$$

At this point, we are going to assume that the Lagrangian and Eulerian variances of the turbulent velocity are equal ($\sigma_{L_i} = \sigma_i^2$). Here, this equivalence will be identified by substituting both variances with σ_i. Therefore, from Equation 3.58, the scale factor β_i can be expressed as (Wandel and Kofoed-Hansen, 1962; Angell et al., 1971; Pasquill, 1974; Hanna, 1981)

$$\beta_i = \frac{T_{L_i}}{T_i} = \frac{n_E}{n_L} = \gamma\frac{U}{\sigma_i} \tag{3.59}$$

where

$$\gamma = \left(\frac{3}{2}\right)^{3/2} \frac{(\alpha_i \alpha_u)^{3/2}}{B_{0_i}} \tag{3.60}$$

is a coefficient.

An estimative for the numerical coefficient γ can be obtained from α_i, α_u, and B_{0_i} constants. For the velocity components v and w, $\alpha_v = \alpha_w = 4/3$ (isotropy condition).

On the other hand, the following general relationship relating, in the Lagrangian framework, the structure function, the autocorrelation function, and the energy spectrum can be written by

$$D_{L_i}(\tau) = 2\left[R_{L_i}(0) - R_{L_i}(\tau)\right] = 2\int_0^\infty [1 - \cos(\omega\tau)]\,\Phi_{L_i}(\omega)\,d\omega \tag{3.61}$$

Changing from angular frequency ω to frequency n (by setting $\omega = 2\pi n$ and $S_{L_i}(n) = 2\pi\Phi_{L_i}(2\pi n)$, using Equation 3.54 in Equation 3.61 and integrating over n) gives

$$D_{L_i}(\tau) = \frac{B_{0_i}}{\pi}\,\varepsilon \int_0^\infty \frac{[1 - \cos(2\pi n\tau)]}{n^2}\,dn = \pi B_{0_i}\varepsilon\tau \tag{3.62}$$

By comparing this result with the Kolmogorov definition of Lagrangian structure function in the inertial subrange, that is, Equation 3.48, the relation between C_0 and B_{0_i} is given by

$$C_0 = \pi B_{0_i} \tag{3.63}$$

Let us now derive a relation between C_0 and C_S. The starting point will be Equation 3.61 written in this case in terms of Eulerian quantities in Equation 3.51:

$$D_i(\tau) = \frac{2\alpha_i\alpha_u}{(2\pi)^{2/3}}(\varepsilon U)^{2/3}\int_0^\infty \frac{[1 - \cos(2\pi n\tau)]}{n^{2/3}}\,dn \approx 4\alpha_i\alpha_u(\varepsilon U)^{2/3}\tau^{2/3} \tag{3.64}$$

where, in this case, the following approximate result is used:

$$\int_0^\infty \frac{[1 - \cos(x)]}{x^{5/3}}\,dx \cong 2$$

Comparing Equations 3.47 and 3.64 we obtain

$$\alpha_u \cong \frac{C_S}{4} \tag{3.65}$$

Finally, inserting Equations 3.63 and 3.65 into Equation 3.60 yields the following relation between C_S and C_0 for the γ coefficient:

$$\gamma = \left(\frac{3}{8}\right)^{3/2} \frac{\left(\alpha_i C_S\right)^{3/2} \pi}{C_{0_i}} \tag{3.66}$$

A study from Anfossi et al. (2000) suggests a value for $C_S \approx 1.65$, whereas Hanna's experimental work (Hanna, 1981) leads to $C_0 \approx 4.0$. The substitution of these numerical constants in the relation (Equation 3.66) sets a value for $\gamma \approx 0.58$. This result agrees with the value of $\gamma \approx 0.55 \pm 0.14$, which was estimated as a mean value obtained from a large number of theoretical and experimental works found in the literature by Degrazia and Anfossi (1998).

3.5 DERIVATION OF AN EDDY DIFFUSIVITY FOR INHOMOGENEOUS TURBULENCE IN A CONVECTIVE BOUNDARY LAYER

The aim of this section is to report a new formulation for eddy diffusivities as functions of distance (travel times) from the source in inhomogeneous turbulence. It is based on turbulent velocity spectra and the statistical diffusion theory. These eddy diffusivities, derived for convective and moderately unstable conditions, contain the characteristic velocity and length scales of energy concentration containing eddies and can describe dispersion in the near and intermediate fields of an elevated continuous point source, that is, when the scale of the plume is smaller than the scale of the turbulence.

The equation for Eulerian velocity spectra under unstable conditions can be expressed as a function of convective scales as follows (Degrazia et al., 1997):

$$\frac{nS_{ic}^{E}(n)}{w_*^2} = \frac{1.06c_i f \left(\psi_\varepsilon \dfrac{z}{z_i}\right)^{2/3}}{\left[\left(f_m^*\right)_i^c\right]^{5/3} \left\{1 + 1.5 \dfrac{f}{\left[\left(f_m^*\right)_i^c\right]}\right\}^{5/3}} \tag{3.67}$$

where

$c_v = c_w = 0.36$ and $c_u = 0.27$
$f = nz/U(z)$ is the nondimensional frequency
z is the height above the ground
$(f_m^*)_i^c$ is the normalized frequency of the spectral peak regardless of stratification
z_i is the top of the unstable boundary layer height
w_* is the convective velocity scale

The nondimensional molecular dissipation rate function is defined by $\psi_\varepsilon = (\varepsilon z_i / w_*^3)$, where ε is the mean dissipation of TKE per unit time per unit mass of fluid, with the order of magnitude of ε determined only by those quantities that characterize the large energy-containing eddies.

The analytical integration of Equation 3.67 over the whole frequency domain leads to the following Eulerian turbulent velocity variance:

$$\sigma_{ic}^2 = \frac{1.06 c_i z \left(\psi_\varepsilon \dfrac{z}{z_i} \right)^{2/3} w_*^2}{U \left[\left(f_m^* \right)_i^c \right]^{5/2}} \int_0^\infty \left[1 + 1.5 \frac{nz}{U \left[\left(f_m^* \right)_i^c \right]} \right]^{-5/3} dn \qquad (3.68)$$

and

$$\sigma_{ic}^2 = \frac{1.06 c_i \left(\psi_\varepsilon \dfrac{z}{z_i} \right)^{2/3} w_*^2}{\left[\left(f_m^* \right)_i^c \right]^{2/3}} \qquad (3.69)$$

which is used to normalize the spectrum so that the normalized Eulerian spectrum can be written as follows:

$$F_{ic}^E(n) = \frac{S_{ic}^E(n)}{\sigma_{ic}^2} = \frac{z}{U \left[\left(f_m^* \right)_i^c \right]} \left\{ 1 + 1.5 \frac{(nz/U)}{\left[\left(f_m^* \right)_i^c \right]} \right\}^{-5/3} \qquad (3.70)$$

Substituting Equation 3.69 and $\beta_{ic} = 0.55 U / \sigma_{ic}$ (Degrazia and Anfossi, 1998) in Equation 3.42 yields

$$\frac{\sigma_{ic}^2 \beta_{ic}}{2\pi} = \frac{0.09 U c_i^{1/2} w_* \left(\psi_\varepsilon \dfrac{z}{z_i} \right)^{1/3}}{\left[\left(f_m^* \right)_i^c \right]^{1/3}} \qquad (3.71)$$

and

$$\frac{2\pi t}{\beta_{ic}} \equiv a = \frac{11.76 c_i^{1/2} \left(\psi_\varepsilon \dfrac{z}{z_i} \right)^{1/3}}{\left[\left(f_m^* \right)_i^c \right]^{1/3}} \frac{z_i}{U} X \qquad (3.72)$$

where a time-to-space transposition is applied to the time dependency in Equation 3.42 to yield a spatially dependent K_α, with $X = (x w_* / U z_i)$, a nondimensional distance defined by the ratio of travel time x/U and the convective timescale z_i/w_*.

Now defining $n' = bn$, where $b = 1.5z/U(f_m^*)_i^c$, Equations 3.70 through 3.72 can be introduced into Equation 3.42 to obtain

$$\frac{K_\alpha}{w_* z_i} = \frac{0.09 c_i^{1/2} \psi_\varepsilon^{1/3} (z/z_i)^{4/3}}{\left[\left(f_m^* \right)_i^c \right]^{4/3}} \int_0^\infty \frac{\sin\left(\frac{a}{b} n' \right) dn'}{(1+n')^{5/3} n'} \tag{3.73}$$

which expands to (Degrazia et al., 2001)

$$\frac{K_\alpha}{w_* z_i} = \frac{0.09 c_i^{1/2} \psi_\varepsilon^{1/3} \left(\dfrac{z}{z_i} \right)^{4/3}}{\left[\left(f_m^* \right)_i^c \right]^{4/3}} \int_0^\infty \frac{\sin\left\{ \dfrac{7.84 c_i^{1/2} \psi_\varepsilon^{1/3} \left[\left(f_m^* \right)_i^c \right]^{2/3}}{(z/z_i)^{2/3}} X n' \right\} dn'}{(1+n')^{5/3} n'} \tag{3.74}$$

In his statistical diffusion theory, Taylor (1921) pointed out that turbulent diffusion differs in the near and the far regions from a continuous point source. In the proximity of the source, fluid particles retain their memory of their initial turbulent environment. For long travel times, this memory is lost, and particles follow only the local properties of turbulence (Batchelor, 1949). This asymptotic behavior of Equation 3.42 for large diffusion travel time when the eddy diffusivity has lost its memory of initial condition is given by Equation 3.43. The substitution of β_{ic}, σ_{ic}^2 (Equation 3.69) and $F_{ic}^E (n = 0)$ (Equation 3.70) in Equation 3.43 leads to the following asymptotic eddy diffusivity

$$K_\alpha = \frac{0.14 c_i^{1/2} \psi_\varepsilon^{1/3} \left(z/z_i \right)^{1/3} w_* z}{\left[\left(f_m^* \right)_i^c \right]^{4/3}} \tag{3.75}$$

The eddy diffusivity expressed by Equation 3.74 depends on the geometry of the source distribution and is suitable for calculation of the contaminant concentration released by elevated continuous point sources. On the other hand, their asymptotic behavior expressed by Equation 3.75 is employed to calculate the concentration of scalar and vector species released by infinite area sources (Degrazia and Moraes, 1992). As a consequence, the asymptotic eddy diffusivity, Equation 3.75 can be used to describe the transfer of heat, momentum, and contaminants in the PBL. Equations 3.74 and 3.75 are expressed in terms of the quantities ψ_ε and $(f_m^*)_i^c$. These fundamental parameters, which describe the structure of turbulence, are derived from observational data measured in the PBL. It is important to note that turbulence represents a complex physical state characterized by a diversified phenomenology and hence values of ψ_ε and $(f_m^*)_i^c$ must be necessarily obtained from heuristic arguments (observations).

In the case of horizontal homogeneity, the evolution of CBL is controlled by the vertical transport of heat. As a consequence, the present analysis will focus on the vertical eddy diffusivity. This eddy diffusivity can be derived from Equation 3.74 by assuming

$$(f_m^*)_w^c = \frac{z}{(\lambda_m)_w^c} = 0.55\left(\frac{z}{z_i}\right)\left[1-\exp\left(-4\frac{z}{z_i}\right)-0.0003\exp\left(8\frac{z}{z_i}\right)\right]^{-1} \tag{3.76}$$

where $(\lambda_m)_w^c = 1.8z_i[1-\exp(-4(z/z_i))-0.0003\exp(8(z/z_i))]$ is the value of the vertical wavelength at the spectral peak, which was obtained from empirical data by Caughey and Palmer (1979). To proceed, the vertical eddy diffusivity described in terms of energy-containing eddies and a function of the downwind distance X and the height z can be obtained from Equations 3.74 and 3.76 using $c_w = 0.36$, as follows:

$$\frac{K_z}{w^*z_i} = 0.12\psi_\varepsilon^{1/3}\left[1-\exp\left(-4z/z_i\right)-0.0003\exp\left(8z/z_i\right)\right]^{4/3}$$

$$\times\int_0^\infty\frac{\sin\left\{3.17\left[1-\exp\left(-4z/z_i\right)-0.0003\exp\left(8z/z_i\right)\right]^{-2/3}\psi_\varepsilon^{1/3}\,Xn'\right\}}{(1+n')^{5/3}}\frac{dn'}{n'}$$

$$\tag{3.77}$$

The formula (3.77) allows to describe the turbulent dispersion in the CBL in terms of the energy-containing eddies and of memory effect of the turbulent field represented by the source distance X. The dissipation function ψ_ε can be evaluated from the following expression, which was obtained from empirical fitting by Højstrup (1982)

$$\psi_\varepsilon^{1/3} = \left[\left(1-\frac{z}{z_i}\right)^2\left(-\frac{z}{L}\right)^{-2/3}+0.75\right]^{1/2} \tag{3.78}$$

where L is the Monin–Obukhov length in the convective surface layer. The behavior of the vertical eddy diffusivity for two different heights, as given by Equation 3.77, is presented in Figure 3.1. From Figure 3.1, we see that the eddy diffusivity is initially zero, increases with time at first linearly and then more slowly and finally tends to a constant value, which can be obtained from Equations 3.75 and 3.76. This turbulent parameterization describes the asymptotic vertical eddy diffusivity far from the source and can be written as (Degrazia et al., 2001)

$$\frac{K_z}{w_*z_i} = 0.19\psi_\varepsilon^{1/3}\left[1-\exp\left(-4z/z_i\right)-0.0003\exp\left(8z/z_i\right)\right]^{4/3} \tag{3.79}$$

Figure 3.2 exhibits the behavior of vertical profile of K_z/w_*z_i as given by Equation 3.79. This profile represents a well-behaved eddy diffusivity with a maximum in the

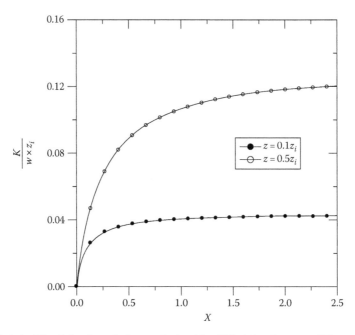

FIGURE 3.1 The behavior of the vertical eddy diffusivity for two different heights ($z/z_i = 0.1$ and $z/z_i = 0.5$), as given by Equation 3.77.

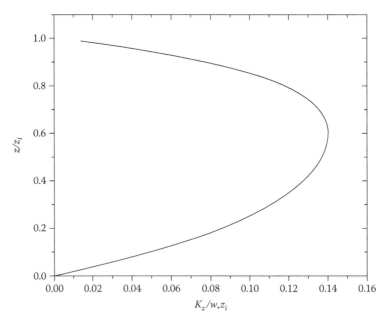

FIGURE 3.2 The behavior of the vertical profile for asymptotic eddy diffusivity (Equation 3.79).

central regions of the CBL and with small values at $z = 0$ and $z = z_i$. The formulas (3.73) and (3.77) describe an inhomogeneous turbulence and consider the memory effect contained in the autocorrelation function. These parameterizations are particularly suitable to represent turbulent transport of contaminants released from an elevated continuous point source. On the other hand, Equations 3.75 and 3.79 describe the turbulent transport of contaminants released by infinite area sources and, consequently, the parameterization Equations 3.75 and 3.79 allow to represent the vertical transfer of species as humidity, momentum, and heat in the PBL.

3.5.1 TURBULENT TRANSPORT MODELING OF CONTAMINANTS DURING THE DECAYING OF A CBL

Equations 3.30, 3.32, 3.77, and 3.79 describe the transport process associated to a stationary and FDT. This means that a turbulent field as proposed by Kolmogorov exhibits self-similar and scale-invariance characteristics. However, it is important to note that about half hour before sunset, over land, the surface heat flux (positive during the day) begins to decrease and then the thermals cease to form and along the time the turbulence tends to disappear in the CBL. The new resulting layer of air separated from the surface by the stable nocturnal boundary layer presents characteristics of the decaying convective turbulence (RL). Concerning to this layer, it is important to note that a large number of elevated stacks release contaminants around the evening transition. This sunset transition time occurs regularly on a daily basis and for this situation the derivation of eddy diffusivities in this period provides a turbulence parameterization for atmospheric diffusion models. The RL is a neutrally stratified elevated layer that is not influenced by turbulent transport of surface-related characteristics and their properties are generally observed to be initially the same as those of the recently decayed CBL.

For a homogeneous and isotropic turbulence, an approach to derive eddy diffusivities for convective decaying turbulence in the RL was proposed by Goulart et al. (2002). This model is based on the budget equation for the TKE in which buoyant contribution was disregarded and only the inertial transfer term is retained. The results of this approach were compared with a decaying vertical eddy diffusivity obtained from LES data (Nieuwstadt and Brost, 1986).

From this comparison both vertical eddy diffusivities show a good agreement for small decaying times. For larger times, as a consequence of -2 exponent obtained for the decaying vertical velocity variance calculated from LES, the LES vertical eddy diffusivity decays strongly faster.

More recently, Goulart et al. (2003) developed a theoretical model to study the TKE decaying in a CBL. This model is also based on the dynamical energy spectrum equation in which the buoyancy and inertial transfer terms are retained. Different from Goulart's papers, Degrazia et al. (2003) by employing Heisenberg's elementary decaying turbulence theory derived a vertical eddy diffusivity applied to the RL. This modeled eddy diffusivity was compared with LES data from NB, and the results showed that the model is not a good estimator of the vertical velocity variance as calculated from the LES.

Considering a nonisotropic decaying convective turbulence, the aim of this section is to derive a formulation for the eddy diffusivities in a RL.

3.5.1.1 Energy Density Spectrum Dynamical Equation

It is possible to derive a spectral form of the turbulent energy equation from the momentum conservation law, expressed through the NS equations. In the case of a homogeneous turbulent flow, the spectral form of the turbulent energy equation is (Hinze, 1975; Stull, 1988):

$$\frac{\partial E(k,t;z)}{\partial t} = M(k,t;z) + W(k,t;z) + \frac{g}{T_0} H(k,t;z) - 2\nu k^2 E(k,t;z) \qquad (3.80)$$

where

t is the time

g/T_0 is the buoyancy parameter

k is the wave number

z is the height above the ground

$E(k,t;z)$ is the 3-D energy density spectrum

$W(k,t;z)$ is the transport term composite of the energy-transfer-spectrum function that represents the contribution due to the inertial transfer of energy among different wave numbers or the time-rate-of-change per unit wave number of the energy spectrum and pressure–velocity correlation

$M(k,t;z)$ is the energy production by mechanical (shear) effects

$H(k,t;z)$ is the production or loss due to buoyancy contribution, and the last term on the right-hand side of Equation 3.80 is the energy loss due to viscous dissipation

In this chapter, we take into account the case in which the inertial energy-transfer terms are important. Under this condition, Equation 3.80 becomes

$$\frac{\partial}{\partial t} E(k,t) = W(k,t) - 2\nu k^2 E(k,t) \qquad (3.81)$$

A turbulent flow contains eddies of different size or different wavelengths. The small eddies are subjected to the stress generated by large eddies. This field increases the vorticity of small eddies and, consequently, their kinetic energy. Thus, TKE is transferred from large eddies toward smaller and smaller eddies until the Kolmogorov microscale is reached, where the energy is dissipated as heat. This process is represented by the term $W(k,t)$ of Equation 3.81. This term was parameterized according to Pao (1965) for a turbulent isotropic flow on the basis of dimensional analysis, as follows:

$$W(k,t) = -\frac{\partial}{\partial k}\left(\alpha^{-1}\varepsilon^{1/3}k^{5/3}E(k,t)\right) \qquad (3.82)$$

where

α is the Kolmogorov constant

ε is the rate of molecular dissipation of TKE

By substituting Equation 3.82 in Equation 3.81 one obtains:

$$\frac{\partial E(k,t)}{\partial t} + \alpha^{-1}\varepsilon^{1/3}k^{5/3}\frac{\partial E(k,t)}{\partial k} + \frac{5}{3}\alpha^{-1}\varepsilon^{2/3}k^{2/3}E(k,t) - 2vk^2 E(k,t) = 0 \qquad (3.83)$$

Considering the following dimensionless parameters, in which w_* is the convective velocity scale and z_i is the CBL height:

$$t_* = \frac{w_* t}{z_i}, \quad R_e = \frac{w_* z_i}{v}, \quad \psi_\varepsilon = \frac{\varepsilon z_i}{w_*^3} \qquad (3.84)$$

Equation 3.83 becomes

$$\frac{\partial E(k',t_*)}{\partial t_*} + \alpha^{-1}\psi_\varepsilon^{1/3}(k')^{5/3}\frac{\partial E(k',t_*)}{\partial k'} + \frac{5}{3}\alpha^{-1}\psi_\varepsilon^{1/3}(k')^{2/3}E(k',t_*) - \frac{2}{R_e}(k')^2 E(k',t_*) = 0 \qquad (3.85)$$

where $k' = kz_i$.

To solve Equation 3.85 the following changes of variables are considered: $t \to s$ and $k' \to m$. We have in $t = 0$ $(m,0,E(m))$:

$$\dot{s}(t) = 1 \qquad (3.86)$$

$$\dot{k}'(s) = \alpha^{-1}\psi_\varepsilon^{1/3}k'^{5/3} \qquad (3.87)$$

$$\dot{Z}(s) = -\left(\frac{5}{3}\alpha^{-1}\psi_\varepsilon^{1/3}k'^{2/3} + \frac{2}{R_e}k'^2\right)Z(s) \qquad (3.88)$$

$$s(0) = 0 \qquad (3.89)$$

$$k'(0) = m \qquad (3.90)$$

$$Z(m,0) = E(m,0) \qquad (3.91)$$

From Equation 3.86

$$\frac{ds}{dt_*} = 1 \quad \text{or} \quad s = t_* \qquad (3.92)$$

and from Equation 3.87

$$\frac{dk'}{ds} = \alpha^{-1}\psi_\varepsilon^{1/3}k'^{5/3} \qquad (3.93)$$

The solution of Equation 3.93, considering the initial conditions Equations 3.89 and 3.90, provides the relation between k' and m:

$$k' = \left(-\frac{2}{3}\alpha^{-1}\psi_\varepsilon^{1/3}s + m^{-2/3}\right)^{-3/2} \tag{3.94}$$

Substituting Equation 3.94 in Equation 3.88 and considering the initial conditions in Equations 3.89 through 3.91 yields the following equation:

$$E(m,s) = E(m,0)\left(\frac{-2/3\alpha^{-1}\psi_\varepsilon^{1/3}s + m^{-2/3}}{m^{-2/3}}\right)^{5/3} \exp\left[\left(-2/3\alpha^{-1}\psi_\varepsilon^{1/3}s + m^{-2/3}\right)^{-2} - m^{4/3}\right] \tag{3.95}$$

Substituting m given by Equation 3.94 in Equation 3.95 and considering $s = t_*$ given by Equation 3.92 results

$$E(k',t_*) = E(\xi,0)\left(\frac{k'}{\xi}\right)^{-5/3} \exp\left\{-\frac{3\alpha}{2R_e\psi_\varepsilon^{1/3}}\left((k')^{4/3} - \xi^{4/3}\right)\right\} \tag{3.96}$$

where

$$\xi = \left\{(k')^{-2/3} + \frac{2}{3}\alpha^{-1}\psi_\varepsilon^{1/3}t_*\right\}^{-3/2}$$

$\alpha = 1.5$

$E(\xi,0)$ is the initial ($t = 0$) 3-D spectrum

The TKE derived from Equation 3.96 decays as a function of time according to the power law $t_*^{-1.3}$. We note that this last exponent lies in the range usually observed for the decay of turbulent energy in the case of isotropic turbulence.

The dynamical equation describing the turbulent flow is valid just in 3-D space. Consequently, the spectrum $E(k,0)$ that represents the CBL initial condition in Equation 3.86 is the CBL turbulent 3-D spectrum.

In this work we are considering nonisotropic turbulence, and, as a consequence, we will use the formulation proposed by Kristensen et al. (1989) to determine the initial 3-D spectrum. This formulation allows determining the 3-D spectrum of a homogeneous turbulent flow from known 1-D spectra, namely,

$$E_0(k,z) = k^3 \frac{d}{dk}\frac{1}{k}\frac{dF_u(k)}{dk} + 12A_im_iB_i^{-17/6}k^4\sum_{n=0}^{3}C_n\int_{W_{1i}}^{\infty}\frac{Z_i^{3n-12}}{\left(Z_i^3 - 1\right)^5}dZ_i$$

$$-\frac{84}{9}A_im_iB_i^{-3/2}k^{4/3}\sum_{n=0}^{3}C_n\int_{1}^{W_{2i}}\frac{Z_i^{3n-12}}{\left(Z_i^3 - 1\right)^{n-5}}dZ_i \tag{3.97}$$

with

$$W_{1i} = \left(1 + \frac{1}{\sqrt{B_i s}}\right)^{1/3}, \quad W_{2i} = \left(1 + \sqrt{B_i s}\right)^{1/3}, \quad A_i = a_i\left(\frac{1}{b_i}\right)^{5/6}, \quad m_u = 2, \, m_v = m_w = -1,$$

$$C_0 = -\frac{55}{27}, \quad C_1 = \frac{70}{9}, \quad C_2 = -\frac{725}{72}, \quad C_3 = \frac{935}{216}, \quad A_i = a_i b_i^{-5/6}, \quad B_i = b_i^{-2} \quad (3.98)$$

According to Degrazia and Anfossi (1998) the initial 1-D component of spectrum can be written as

$$F_i(k,0) = \frac{a_i}{(1+b_i k)^{5/3}}, \quad i = u, v, w \quad (3.99)$$

where

$a_i = (0.98/2\pi)c_i(z/z_i)^{5/3}z_i\psi_\varepsilon^{2/3}w_*^2[(f_m^*)_i^c]^{-5/3}$

$b_i = (1.5/2\pi)(z/z_i)z_i(1/(f_m^*)_i^c)$, with $c_i = \alpha_i$ $(0.5 \pm 0.05)(2\pi\kappa)^{-2/3}$, $\alpha_i = 1, 4/3, 4/3$ for u, v, and w, respectively (Champagne et al., 1977) $w_* = (u_*)_0 (-(z_i/\kappa L))^{1/3}$, $(f_m^*)_i^c = (z/G_i z_i)$, $G_u = 1.5$, $G_v = 1.5$, and $G_w = 1.8[1 - \exp -(4z/z_i) - 0.0003 \times \exp(8z/z_i)]$

The CBL is nonisotropic only in the vertical direction, due to the heat flux in the surface. To calculate the w component of the spectrum, we consider that for a particular time instant t there is a relationship between the 1-D spectrum and the 3-D average spectrum given by the following expression:

$$F_w(k,t) = \alpha(k)\frac{\frac{1}{T}\int_0^t F_w(k,t)dt}{\frac{1}{T}\int_0^t E(k,t)dt} E(k,t) \quad (3.100)$$

where the ratio between the two integrals is a weight function that indicates that the w component takes part in the construction of the 3-D spectrum and $\alpha(k)$ is the proportionality constant.

The solution of Equation 3.100 provides the vertical component as a function of the 3-D spectrum:

$$F_w(k,t) = F_w(k,0)\exp\left[\int_0^t Q'(k,s)ds\right] \quad (3.101)$$

In this case $F_w(k,0)$ is given by Equation 3.99 where

$$Q'(k,s) = \alpha(k)Q(k,s) + \frac{1}{Q(k,s)}\frac{\partial Q(k,s)}{\partial s} \quad (3.101a)$$

and $Q(k,s)$ given by

$$Q(k,s) = \frac{E(k,s)}{\displaystyle\int_0^t E(k,s)\,\mathrm{d}s} \qquad (3.101\mathrm{b})$$

Since, to our knowledge, there are no conclusive observations of the CBL turbulence decay process, we compare our theoretical expressions with LES data (Nieuwstadt and Brost, 1986). The velocity variance of the turbulent flow is calculated from the following equation:

$$\sigma_i^2(t;z) = \int_0^\infty F_i(k,t;z)\,\mathrm{d}k \qquad (3.102)$$

Figure 3.3 shows the temporal evolution of the vertical velocity variance decay across the CBL ($0.2 < z/z_i < 0.8$), calculated from Equations 3.96, 3.99, 3.101, and 3.102 made dimensionless by w_*^2.

Figure 3.3 shows that our theoretical model (solid line) agrees very well with the results simulated by LES (crosses; Nieuwstadt and Brost, 1986).

The eddy diffusivities calculated by Batchelor (1949) can be identified with those from advection–diffusion equation when a homogeneous turbulent flow

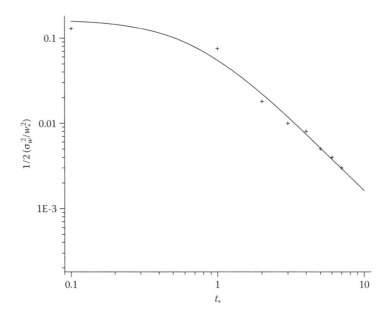

FIGURE 3.3   Temporal evolution of the vertical velocity variance decay across the CBL ($0.2 < z/z_i < 0.8$), calculated from Equations 3.96, 3.99, 3.101, and 3.102 made dimensionless by w_*^2.

is considered. For large diffusion travel times, the eddy diffusivity has the form (Hanna, 1981)

$$K_\alpha(t;z) = \frac{\pi}{3} \frac{\beta_i}{U} \frac{\sigma_i^2(t;z)}{(k_m)_i}, \quad \alpha = x, y, z \tag{3.103}$$

where

$(k_m)_i$ is the wave number of the spectral peak (associated to the energy-containing eddies)

U is the mean wind speed

β_i is defined as the ratio of the Lagrangian to the Eulerian integral timescales

Considering $\beta_i = 0.55(U/\sigma_i)$ (Hanna, 1981; Degrazia et al., 1998) in Equation 3.103 yields the following decaying eddy diffusivity

$$K_\alpha(t;z) = \frac{0.55\pi}{3} \frac{\sigma_i(t;z)}{(k_m)_i} \tag{3.104}$$

where, for our model, $\sigma_i(t;z)$ is obtained from Equations 3.96, 3.97, 3.99, 3.101, and 3.102.

Nieuwstadt and Brost (1986) report that the vertical velocity spectrum (depicted in Figure 14 of their article), computed for several dimensionless times, all have a maximum value for $(k_m)_w z_i \approx 4$. This allows calculating the vertical eddy diffusivity averaged across the boundary layer for different time t_*. In Figure 3.4, we show this decaying vertical eddy diffusivity. The crosses were calculated from Equation 3.104 using LES data for σ_w. Solid line was also calculated from Equation 3.104 using σ_w values derived from Equations 3.96, 3.97, 3.99, 3.101, and 3.102. We can see that the agreement is very good for the whole decaying time t_*. The following

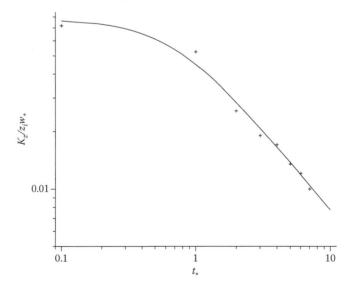

FIGURE 3.4 Decaying vertical eddy diffusivity. The crosses were calculated from Equation 3.104 using LES data for σ_w. Solid line was also calculated from Equation 3.104 using values for σ_w determined from Equations 3.96, 3.97, 3.99, 3.101, 3.102, and 3.104.

relationship represents a good fit to the decaying vertical eddy diffusivity calculated from Equations 3.96, 3.97, 3.99, 3.101, 3.102, and 3.104:

$$\frac{K_z}{z_i w_*} = \frac{0.079}{\sqrt{1 + 2t_*^{1.7}}} \qquad (3.105)$$

The developed model derived from Equation 3.80, which describes the turbulence decaying in a CBL, shows that the simple algebraic relation (Equation 3.105) can be employed in dispersion operational models to simulate the contaminants concentration field released from continuous point sources (stacks) situated in a RL.

3.6 ANALYSIS OF THE LOW-FREQUENCY HORIZONTAL WIND OSCILLATIONS EMPLOYING THE NAVIER–STOKES EQUATIONS

Generally in stable conditions, during situations of LWS ($\bar{u} \leq 1 - 2$ m/s), low-frequency horizontal wind oscillations (meandering) are observed in a PBL. The study of LWS conditions is of interest, partly because the simulation of airborne pollutant dispersion in these conditions is rather difficult. In fact, most of the existing regulatory dispersion models become unreliable as $\bar{u}$ approaches zero, so that their application is generally limited to $\bar{u} > 2$ mps. The meandering movements are clearly distinct from those associated to a FDT, which are responsible for the contaminants diffusion in a PBL. Even when the stability reduces the vertical dispersion and the instantaneous plume may be thin, meandering disperses the plume over a rather wide angular sector. As a consequence, any operational dispersion model to be reliable must take into account the transport effect provocated by the meandering.

In this study, the NS equations are employed to describe the horizontal mean wind velocity and to investigate the origin of low-frequency oscillations (wind meandering). An analytical method to solve the simplified NS equations is presented and this solution shows that the observed wind field oscillatory character is associated to the mathematical behavior of this equation system. Indeed, the low-frequency horizontal wind oscillations emerge as a phenomenon related to the structure of the NS equations, that is, when the equilibrium between Coriolis force and the pressure gradient is present in this equation system.

3.6.1 ANALYTICAL SOLUTION OF THE SIMPLIFIED NAVIER–STOKES EQUATIONS

To investigate the physical process responsible for wind meandering phenomenon and to obtain an expression for the autocorrelation function we consider the form of the NS equations in two dimensions:

$$\frac{\partial \bar{u}}{\partial t} + \bar{u}\frac{\partial \bar{u}}{\partial x} + \bar{v}\frac{\partial \bar{u}}{\partial y} = f_c \bar{v} - \frac{1}{\bar{\rho}}\frac{\partial \bar{p}}{\partial x} - \frac{\partial \overline{(uu)}}{\partial x} - \frac{\partial \overline{(uv)}}{\partial y} \qquad (3.106)$$

$$\frac{\partial \bar{v}}{\partial t} + \bar{u}\frac{\partial \bar{v}}{\partial x} + \bar{v}\frac{\partial \bar{v}}{\partial y} = f_c \bar{u} - \frac{1}{\bar{\rho}}\frac{\partial \bar{p}}{\partial y} - \frac{\partial \overline{(vu)}}{\partial x} - \frac{\partial \overline{(vv)}}{\partial y} \qquad (3.107)$$

where $\bar{u}$ and $\bar{v}$ are respectively the longitudinal and lateral components of the mean wind velocity, $\bar{\rho}$ is the air mean density, u and v are the wind velocity turbulent fluctuations, $\bar{p}$ is the mean pressure, and f_c is the Coriolis parameter.

Since the system of Equations 3.106 and 3.107 cannot be analytically solved as they are, in any particular problem it is necessary to make some appropriate simplifying assumptions allowing for finding a solution. In the present case of LWS meandering conditions, we assume that all the horizontal gradients of the two wind velocity components and of pressure can be taken as constant and that the horizontal gradients of the Reynolds stresses can be disregarded. This derives from the consideration that turbulence levels are very low in LWS and, consequently, their horizontal gradients are vanishing. If we define

$$a_1 = \frac{\partial \bar{u}}{\partial x}, \quad b_1 = +f_c - \frac{\partial \bar{u}}{\partial y}, \quad c_1 = -\frac{1}{\bar{\rho}}\frac{\partial \bar{p}}{\partial x}$$

$$a_2 = \frac{\partial \bar{v}}{\partial y}, \quad b_2 = -f_c - \frac{\partial \bar{v}}{\partial x}, \quad c_2 = -\frac{1}{\bar{\rho}}\frac{\partial \bar{p}}{\partial y} \qquad (3.108)$$

Equations 3.106 and 3.107 can be written as

$$\frac{\partial \bar{u}(t)}{\partial t} = -a_1\bar{u}(t) + b_1\bar{v}(t) + c_1 \qquad (3.109a)$$

$$\frac{\partial \bar{v}(t)}{\partial t} = -a_2\bar{v}(t) + b_2\bar{u}(t) + c_2 \qquad (3.109b)$$

Combining Equations 3.109a and b into one equation by taking the time derivative of Equation 3.109a and then substituting in Equation 3.109b, we obtain:

$$\overline{u''} + (a_1 + a_2)\overline{u'} + (a_1a_2 - b_1b_2)\bar{u} = a_2c_1 + b_1c_2 \qquad (3.110)$$

that can be written as

$$\overline{u''} + B\overline{u'} + C\bar{u} = D \qquad (3.111)$$

where

$$B = a_1 + a_2, \quad C = (a_1a_2 - b_1b_2), \quad D = a_2c_1 + b_1c_2 \qquad (3.112)$$

Equation 3.111 has a known analytical solution. There are three cases in the solution of Equation 3.111 according to the values of the roots m_1 and m_2 from the auxiliary equation $m^2 + Bm + C = 0$. The roots may be written as

$$m = \frac{-B \pm \sqrt{B^2 - 4C}}{2} \qquad (3.113a)$$

We only consider the case

$$B^2 - 4C < 0 \left[\text{i.e.}, (a_1 - a_2)^2 < -4b_1b_2 \right] \quad \text{or} \quad \left(\frac{\partial \overline{u}}{\partial x} - \frac{\partial \overline{v}}{\partial y} \right)^2 < 4 \left(f_c - \frac{\partial \overline{u}}{\partial y} \right) \left(f_c + \frac{\partial \overline{v}}{\partial x} \right)$$

(3.113b)

which yields oscillatory behavior. Setting

$$m_1 = p + qi \quad \text{and} \quad m_2 = p - qi \tag{3.114}$$

where

$$p = \frac{B}{2}, \quad q = \frac{\sqrt{B^2 - 4C}}{2} \tag{3.115}$$

the solutions of the problem are

$$\overline{u}(t) = e^{-pt}(r_1 \cos(qt) + r_2 \sin(qt)) + \frac{D}{p^2 + q^2} \tag{3.116a}$$

and

$$\overline{v}(t) = e^{-pt} \left[\frac{(-pr_1 + qr_2 + a_1r_1)}{b_1} \cos(qt) + \frac{(-pr_2 - qr_1 + a_1r_2)}{b_1} \sin(qt) + \frac{D}{p^2 + q^2} \frac{a_1}{b_1} - \frac{c_1}{b_1} \right] \tag{3.116b}$$

where

$$r_1 = u_o - \frac{D}{p^2 + q^2}, \quad r_2 = \frac{1}{q} \left[v_o b_1 - (-p + a_1)u_o - \frac{Dp}{p^2 + q^2} + c_1 \right] \tag{3.117}$$

The following considerations about the found solutions (Equations 3.116a and b) are appropriate: This analytical solution shows that the appearance of a wind meandering is a phenomenon related to the structure of NS equations. A particular condition (in this case the equilibrium between the Coriolis and pressure forces) generates a solution that shows oscillatory characteristics. The mathematical condition (Equation 3.113b) for the meandering existence imposes that the difference between the wind component gradients is small. However, no condition is imposed on the wind speed. Thus, this latter condition could be greater than those usually found in meandering studies. However, if the wind velocity increases the Reynolds stresses cannot be disregarded anymore. Furthermore, Equations 3.106 and 3.107 suggest that an increase in the longitudinal pressure gradient (that is an imbalance between Coriolis force and the pressure gradient) causes an increase of $\partial \overline{u}/\partial x$ whereas $\partial \overline{v}/\partial y$ remains approximately constant in such a way that the difference in Equation 3.113b becomes greater than b_1b_2 (Equation 3.113b) and consequently, the analytical solution shows that, in this case, the meandering phenomenon does not appear.

The oscillation period in Equations 3.116a and b depends upon the $\partial \bar{u}/\partial y$ and $\partial \bar{v}/\partial x$ values (see Equations 3.108, 3.109a and b, and 3.112) and the Coriolis parameter. It can be seen that when $\partial \bar{u}/\partial y$ and $\partial \bar{v}/\partial x$ are both equal to zero, the present solution gives the inertial oscillation period $2\pi/f_c \approx 17\,\mathrm{h}$ (Stull, 1988). It may also be noted that with increasing wind speed, and thus increasing spatial derivates of the wind speed components, the oscillation period decreases. This means that variance associated to meandering shifts to higher frequencies in power spectra.

An interesting result is obtained by performing a scale analysis of Equations 3.116a and b. The term $(D/p^2 + q^2)$ is approximately zero, $((-pr_1 + qr_2 + a_1r_1)/b_1)$ is two orders of magnitude lower than $((-pr_2 - qr_1 + a_1r_2)/b_1)$, also r_2 is two orders of magnitude lower than r_1, a_1/b_1 and c_1/b_1 are also negligible with respect to the remaining terms, $-pr_2 + a_1r_2$ is approximately zero, and $(q/b_1) \approx 1$. Accounting for these simplifications, Equations 3.116a and b become:

$$\bar{u}(t) = r_1 e^{-pt} \cos(qt) \tag{3.118a}$$

$$\bar{v}(t) = r_1 e^{-pt} \sin(qt) \tag{3.118b}$$

These can be written in an analytical function form as

$$\bar{u}(t) = \alpha_1 e^{-(pt-iqt)} \tag{3.119}$$

Solutions (118a circles) and (118b triangles) are plotted in Figure 3.5, where they are labeled $\bar{u}_S(t)$ and $\bar{v}_S(t)$, respectively. It clearly appears that these simplified

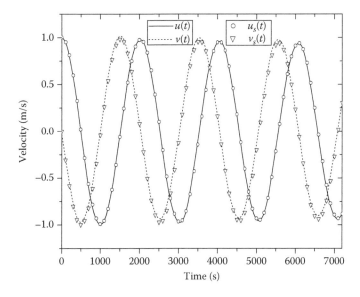

FIGURE 3.5 Time series of calculated velocities with Equations 3.116a and b and the corresponding simplified Equations 3.118a and b.

solutions are practically coincident with the complete ones (Equations 3.116a and b) and, thus, can be used in applications. In particular, if we compute the normalized autocorrelation function of Equation 3.119, we find:

$$R(\tau) = e^{-p\tau}\left(\cos(q\tau) + i\sin(q\tau)\right) \tag{3.120}$$

By taking the real part of Equation 3.120, the following EAF is obtained:

$$R(\tau) = e^{-p\tau}\cos(q\tau) \tag{3.121}$$

It is worth mentioning that the EAF presented in this last equation is exactly the one proposed by Frenkiel (1953), which was verified by Anfossi et al. (2005) as being the correct EAF for LWS conditions.

According to the analytical derivations, there seems to be no need for a specific mechanism, such as gravity waves (Olesen et al., 1984; Etling, 1990) and/or particular stability conditions, to generate horizontal meandering of a flow in a LWS condition. Furthermore, the present analysis shows that the appearance of wind meandering is a phenomenon related to the structure of the NS equations; that is, a special condition (in this case the equilibrium between Coriolis force and the pressure gradient) generates a solution that shows oscillatory characteristics. The solutions suggest that an increase in the horizontal pressure gradient (i.e., an imbalance between Coriolis force and the pressure gradient) prevents the meandering phenomenon appearance. More generally, with increasing production terms on the right-hand side of the NS equations (departure from geostrophic balance, increasing Reynolds-stress terms) meandering is being damped and turbulence begins to play a key role. Finally, on the basis of these results, a system of stochastic Langevin equations that could be used for computing dispersion in LWS could be derived.

REFERENCES

Anfossi D., Oettl, D., Degrazia, G., and Goulart, A. 2005, An analysis of sonic anemometer observations in low wind speed conditions, *Bound. Lay. Meteorol.*, 114, 179–203.

Anfossi, D., Degrazia, G., Ferrero, E., Gryning, S.E., Morselli, M.G., and Trini C. 2000, Estimation of the Lagrangian structure function constant C_0 from surface-layer wind data, *Bound. Lay. Meteorol.*, 95, 249–270.

Angell, J.K., Pack, D.H., Hoecker, W.H., and Delver, N. 1971, Lagrangian-Eulerian time-scale estimated from constant volume balloon flights past a tall tower, *Quart. J. Roy. Meteorol. Soc.*, 97, 87–92.

Arya, S.P. 1995, Modeling and parameterization of near-source diffusion in weak winds, *J. Appl. Meteorol.*, 34, 1112–1122.

Batchelor, G.K. 1949, Diffusion in a field of homogeneous turbulence, I: Eulerian analysis, *Aust. J. Sci. Res.*, 2, 437–450.

Caughey S.J. and Palmer, S.G. 1979, Some aspects of turbulence structure through the depth of the convective boundary layer, *Quart. J. Roy. Meteorol. Soc.*, 105, 811–827.

Champagne, F.H., Friche, C.A., Larve, J.C., and Wyngaard, J.C. 1977, Flux measurements, flux estimation techniques, and fine scale turbulence measurements in the unstable surface layer over land, *J. Atmos. Sci.*, 34, 515–520.

Corrsin, S. 1963, Estimates of the relations between Eulerian and Lagrangian scales in large Reynolds number turbulence, *J. Atmos. Sci.*, 20, 115–119.

Degrazia, G.A. Moraes O.L.L., and Goedert, J. 1991, Estimation of Lagrangian parameters from a diffusion experiment, *Il Nuovo Cimento C*, 14, 615–621.

Degrazia, G.A. and Anfossi, D. 1998, Estimation of the Kolmogorov constant from classical statistical diffusion theory, *Atmos. Environ.*, 32, 3611–3614.

Degrazia, G.A. and Moraes, O.L.L. 1992, A model for eddy diffusivity in a stable boundary layer, *Bound. Lay. Meteorol.*, 58, 205–214.

Degrazia, G.A., Campos Velho, H.F., and Carvalho, J.C. 1997, Nonlocal exchange coefficients for the convective boundary layer derived from spectral properties, *Beitr. Phys. Atmos.*, 70(1), 57–64.

Degrazia, G.A., Anfossi, D., Carvalho, J.C., Mangia, C., and Tirabassi, T. 1998, Turbulence parameterization for PBL dispersion models in all stability conditions, *Atmos. Environ.*, 34, 3575–3583.

Degrazia, G.A., Moreira, D.M., and Vilhena, M.T. 2001, Derivation of an eddy diffusivity depending on source distance for vertically inhomogeneous turbulence in a convective boundary layer, *J. Appl. Meteorol.*, 40, 1233–1240.

Degrazia, G.A., Goulart, A., Anfossi, D., Velho, H.F.C., Lukaszcyk, J.P., and Palandi, J.A. 2003, Model based on Heisenberg's theory for the eddy diffusivity in decaying turbulence applied to the residual layer. *Il Nuovo Cimento C*, 26(1), 39–51.

Draxler, R.R. 1976, Determination of atmospheric diffusion parameters, *Atmos. Environ.*, 10, 99–105.

Etling, D. 1990, On plume meandering under stable stratification, *Atmos. Environ.*, 8, 1979–1985.

Frenkiel, F.N. 1953, Turbulent diffusion: Mean concentration distribution in a flow field of homogeneous turbulence, *Adv. Appl. Mech.*, 3, 61–107.

Frisch, U. 1995, *Turbulence*, Cambridge University Press, Cambridge, MA, 296 pp.

Gardiner, C.W. 1983, *Handbook of Stochastic Methods*, 2nd edn., 1990, Springer-Verlag, Berlin, Germany, 442 pp.

Gifford, F.A. 1955, A simultaneous Lagrangian-Eulerian turbulence experiment, *Mon. Weath. Rev.*, 83, 293–301.

Goulart, A., Degrazia, G.A., Anfossi, D., and Acevedo, O. 2002, Modelling and eddy diffusivity for convective decaying turbulence in the residual layer, *15th Symposium on Boundary Layers and Turbulence*, American Meteorological Society, Wageningen, the Netherlands, pp. 267–268.

Goulart, A., Degrazia, G., Rizza, U., and Anfossi, D. 2003, A theoretical model for the study of convective turbulence decay and comparison with large-eddy simulation data, *Bound. Lay. Meteorol.* 107, 143–155.

Hanna, S.R. 1981, Lagrangian and Eulerian time-scale in the daytime boundary layer, *J. Appl. Meteorol.*, 20, 242–249.

Hanna, S.R. 1982, Applications in air pollution modeling. In *Atmospheric Turbulence and Air Pollution Modeling*, F.T.M. Nieuwstadt and H. von Dop, Eds., D. Reidel Publishing, Dordrecht, the Netherlands, pp. 275–310.

Hay, J.S. and Pasquill, F. 1959, Diffusion from a continuous source in relation to the spectrum and scale of turbulence. *Atmospheric Diffusion and Air Pollution*, F.N. Frenkiel and P.A. Sheppard, Eds., Acadamic Press, New York, Advances in Geophysics, 345 pp.

Hinze, J.O. 1975, *Turbulence*, McGraw-Hill, New York, 790 pp.

Højstrup, J. 1982, Velocity spectra in the unstable planetary boundary layer, *J. Atmos. Sci*, 33, 2152–2169.

Hunt, J.C.R. 1982, Diffusion in the stable boundary layer. In *Atmospheric Turbulence and Air Pollution Modeling*, F.T.M. Nieuwstadt and H. von Dop, Eds., D. Reidel Publishing, Dordrecht, the Netherlands, pp. 231–274.

Kaimal, J.C. and Finnigan, J.J. 1994, *Atmospheric Boundary Layer Flows*, Oxford University Press, Oxford, U.K., 289 pp.

Kolmogorov, A.N. 1941, Dissipation of energy in locally isotropic turbulence, *Dolk. Akad. Nauk. SSSR*, 32, 16–18 (reprinted in *Proc. Roy. Soc. Lond., A* 434, 15–17 (1991)).

Kristensen, L., Lenchow, D., Kirkegaard, P.E., and Courtney, M. 1989, The spectral velocity tensor for homogeneous boundary layer, *Bound. Lay. Meteorol.*, 47, 149–193.

Monin, A.S. and Yaglom, A.M. 1975, *Statistical Fluid Mechanics 2*, J. Lumley, Ed., MIT Press, Cambridge, MA.

Nieuwstadt, F.T.M. and Brost, R.A. 1986, The decay of convective turbulence. *J. Atmos. Sci.* 43, 532–546.

Olesen, H.R., Larsen, S.E., and Højstrup, J. 1984, Modelling velocity spectra in the lower part of the planetary boundary layer, *Bound. Lay. Meteorol.*, 29, 285–312.

Panofsky, H.A. and Dutton, J.A. 1984, *Atmospheric Turbulence*, Wiley, New York, 397 pp.

Pao, Y.H. 1965, Structure of turbulent velocity and scalar fields at large wavenumbers, *Phys. Fluid.*, 8, 1063–1075.

Pasquill, F. 1974, *Atmospheric Diffusion*, 2nd edn., Ellis Horwood Ltd., Chichester, U.K., 228 pp.

Pasquill, F. and Smith, F.B. 1983, *Atmospheric Diffusion*, Wiley & Sons, New York, 437 pp.

Saffman, P.G. 1960, On the effect of the molecular diffusivity in turbulent diffusion, *J. Fluid Mech.*, 8, 273–283.

Sorbjan, Z. 1989, *Structure of the Atmospheric Boundary Layer*, Prentice Hall, Englewood Cliffs, NJ, 317 pp.

Stull, R.B. 1988, *An Introduction to Boundary Layer Meteorology*, Kluwer Academic Publishers, Boston, MA, 666 pp.

Taylor, G.I. 1921, Diffusion by continuous movements, *Proc. Lond. Math. Soc.*, Ser. 2, 20, 196–211.

Tennekes, H. 1981, Similarity relation, scaling laws and spectral dynamics. In *Atmospheric Turbulence and Air Pollution Modeling*, F.T.M. Nieuwstadt, H. van Dop, Eds., D. Reidel Publishing, Dordrecht, the Netherlands, pp. 37–68.

Tennekes, H. and Lumley J.L. 1972, *A First Course in Turbulence*, The MIT Press, Cambridge, MA, 300 pp.

Venkatram, A. 1988, Dispersion in the stable boundary layer. In *Lectures on Air Pollution Modeling*, A. Venkatram and J.C. Wyngaard, Eds., American Meteorological Society, Boston, MA, pp. 229–258.

Wandel C.F. and Kofoed-Hansen, O. 1962, On the Eulerian–Lagrangian transform in the statistical theory of turbulence, *J. Geophys. Res.*, 67, 3089–3093.

Wyngaard, J.C. 1982, Boundary-layer modeling. In *Atmospheric Turbulence and Air Pollution Modeling*, F.T.M. Nieuwstadt and H. von Dop, Eds., D. Reidel Publishing, Dordrecht, the Netherlands, pp. 69–106.

4 Parameterization of Convective Boundary Layer Turbulence and Clouds in Atmospheric Models

Pedro M. M. Soares, João Teixeira, and Pedro M. A. Miranda

CONTENTS

4.1   INTRODUCTION

Earth's climate system is largely controlled by the transfer of properties across different interfaces. For this reason, boundary layers, which occupy a very small fraction of the atmosphere and ocean, are disproportionably important. Atmosphere-ocean and atmosphere-land fluxes of water, momentum, and energy are mediated by these boundary layers.

Planetary boundary layer (BL) fluxes occur at different scales. At very small scales, smaller than 1 mm, fluxes are due to molecular diffusion. At larger scales, we enter the domain of turbulence. The larger BL turbulent structures, which occupy its full depth up to more than 1 km, may organize themselves in BL convective systems and lead to BL convective clouds. Clouds profoundly modify the surface fluxes, interfering not only in the vertical fluxes of heat and moisture but also in the radiative fluxes, shortwave and longwave.

Clouds play an important role in the global climate. *Stratocumulus* and *cumulus*, prevalent BL clouds in subtropical latitudes, seem to have a controlling effect in the atmospheric circulation at the synoptic scale (Philander et al. 1996; Siebesma 1998; Larson et al. 1999). Any good climate or numerical weather prediction model heavily relies on one or more parameterization schemes, representing the effects of these subgrid scale clouds (Tiedtke 1987).

Due to the special nature of cloud processes, qualitatively different from other turbulent fluxes, such as those involved in dry convection over land, it is not surprising that many atmospheric models use different schemes to deal with saturated and nonsaturated BLs. However, such procedure implies the switching on and off of the different schemes, with some sort of discontinuity in model behavior. On the other hand, results of current parameterization schemes are often poor (e.g., Ayotte et al. 1996; Lenderink et al. 2004).

In this chapter, a short review of the main parameterizations for BL turbulence and clouds is presented. These are at the base of the development of a set of unified parameterizations for BL turbulence and convection, applicable to the convective boundary layer (CBL), either dry or with shallow cumulus (Siebesma and Teixeira 2000; Soares et al. 2004; Siebesma et al. 2007). The developed approaches are easily integrated in models of different scales, and could be extended to the problem of the deep convection parameterization, aiming the establishment of a unified parameterization scheme for turbulence and convection in the troposphere.

It is clear that, in the future, the computational development will allow carrying out simulations with global circulation and/or mesoscale models with much higher resolutions than is possible currently. However, it is also certain that the turbulent processes, with and without condensation, will continue outside the range of direct numerical simulation, justifying an increasing effort in the parameterization development.

4.2 THE ATMOSPHERIC BOUNDARY LAYER

The troposphere is often divided in two layers: the BL and the free atmosphere (FA). The BL corresponds to the turbulent region where the surface has a direct influence. The timescales of the adjustment of the BL to the different forces are relatively short. In this layer of the troposphere exist the great majority of living creatures and take place most of the human activities, conferring to its study an enormous importance. The transport of momentum, heat, and moisture between the atmosphere, the land, and the ocean takes place in the BL, and it directly influences the local and regional weather, and the global circulation of the atmosphere.

The understanding of the phenomenology that occurs in the atmospheric BL has relevance in many domains, including the dispersion of pollutants, the forecast of temperature, moisture, and wind near the surface, and the processes of cloud and storm formation. Some key activities are critically dependent on BL monitoring and forecast, with emphasis on aeronautics.

In anticyclonic fair weather, the solar diurnal cycle of heating and cooling of the surface determines the temporal evolution of the vertical structure of the BL. At sunrise, the land starts to get warmed; heat is transferred to the overlaying air in a heterogeneous pattern, provoking the turbulent mixing of the air. Turbulent transport depends on the difference of properties between neighboring air parcels, but is essentially controlled by the velocity field. In a stably stratified environment, such as the one that exists in the early morning surface layer and most of the time in the FA, vertical velocity is inhibited, and turbulence must work its way up into the atmosphere by progressively eroding the temperature profile through the warming of the lowest layer. Convective movements, contributing to the vertical turbulent transport of heat and other properties, will be driven by changes in the surface layer thermal structure due to the upward surface heat flux, and will extend into the upper boundary layer, and intensify, into the afternoon, leading to BL growth and to the establishment of quasi-neutral layer, often called the mixed layer, on top of a shallow unstable surface layer, and beneath a top stable inversion layer.

Turbulent structures in the BL have a spectrum of scales. In the CBL, the eddies containing more energy have a vertical dimension of the order of magnitude of the height of the BL itself. They are generally referred to as updrafts or thermals, and can reach more than 2 km in height. The thermals penetrate into the FA and carry air from this back into the BL, contributing to the its vertical growth, a process often referred to as top-entrainment. In these conditions, a top-entrainment region goes through the top BL inversion layer.

In the same case of a clear sky BL, the ground cools after sunset by the emission of longwave radiation. This cooling inhibits surface turbulence leading to the establishment of a low stable layer, which develops vertically less than the mixed layer in daytime. A remainder of the previous day mixed layer may be present, laying above the stable BL, and is then called the residual layer. In the absence of convection, wind shear causes turbulence in the night period. If the convective situation remains for successive days, the daytime growth of the BL will be enhanced by the presence of the residual layer.

The clear sky CBL was the subject of study of many field observational campaigns: Wangara (Clarke et al. 1971), HAPEX (*Hydrologic-Atmospheric Pilot EXperiment*; André et al. 1986), Phoenix CBL (Phoenix 78 *Convective Boundary Layer field experiment*; Kropfli and Hildebrand 1980; Young 1988a,b,c), etc. Detailed studies with large eddy simulation (LES) models were also carried out, including those of Moeng (1984), Schumann and Moeng (1991a,b), Sorbjan (1996a,b), and Sullivan et al. (1998), that contributed for a better understanding of the structure and dynamics of the CBL.

Great extensions of the globe, and in particular of the oceans, are covered by low clouds. Clouds modify the radiative balance of the surface and of the BL. The exchange of latent heat associated with condensation and evaporation of precipitation in the sub-cloud layers constitutes other important thermal forces. Both are essential for the dynamics of the BL and for the global circulation.

The clouds that are more frequently present in the BL include shallow *cumulus*, *stratus*, *stratocumulus*, and *nimbostratus*. *Stratus*, *stratocumulus,* and *nimbostratus* are stratiform clouds, characterized by a great horizontal extension and a relatively small vertical thickness.

Stratocumulus above the oceans are often associated with anticyclonic subsidence in the subtropical and middle latitudes, and above the land surface in the cold season. In the subtropical eastern ocean basins, the subsidence associated with the descending branches of the Hadley cell, together with the cold oceanic currents, induces the persistent presence of *stratocumulus*, for example, off the coast of California, Peru, Namibia, and Mauritania (Hanson 1991; Klein and Hartmann 1993; Ma et al. 1996). *Stratocumulus* have a great impact on climate and its variability (e.g., Philander et al. 1996; Clement and Seager 1999). The thermodynamic and turbulent structures of *stratocumulus* are known, essentially, due to some large observational campaigns, for example, FIRE (*First* ISCCP [*International Satellite Cloud Climatology Project*] *Regional Experiment*, Albrecht et al. 1988), ASTEX (*Atlantic Stratocumulus Transition EXperiment*, Albrecht et al. 1995), and other studies (Duynkerke et al. 1987, Hignett 1991, Duynkerke and Teixeira 2000). More recently, LES simulations have been contributing to the understanding of the mechanism involved in the formation and maintenance of *stratocumulus* (Moeng et al. 1996; Stevens et al. 1998; Duynkerke et al. 1999).

A CBL with shallow cumulus is present in the entire globe. On average, 12% of the oceanic surface is covered by this type of clouds, while the terrestrial surface presents a mean coverage of 5% (Duynkerke 1998). Shallow cumulus are omnipresent in the trade-wind region of oceans, where they are known as trade-wind-cumulus. They are also frequent in the middle latitudes of continents, in summer, where they often evolve into deep convection, turning into *cumulonimbus*.

Shallow *cumulus* directly influence the global circulation and the hydrologic cycle since they enhance the vertical transport of heat, moisture, and momentum, namely, in the intertropical zone of convergence, contributing to the efficiency of the transport of moisture and heat in the Hadley circulation (Tiedtke 1987; Siebesma 1998). The vertical transport of moisture associated with shallow *cumulus* tends to dry the BL, limiting the formation of stratiform clouds. Several modeling exercises showed that the distribution of precipitation and its variability in the tropics are strongly influenced by the presence of *cumulus* convection (Slingo et al. 1994; Gregory 1997). Tiedtke (1987) emphasized that the presence of shallow *cumulus* increases the surface evaporation in the ECMWF (European Center of Medium Weather Forecasts, Beljaars and Betts 1992) model, up to $50\,W\,m^{-2}$ in the subtropical regions.

Additionally, shallow *cumulus* exert an indirect influence in the BL through the modification of the radiative balance. This type of irregular clouds possess unusual radiative properties (Ackerman and Cox 1981; Marshak et al. 1995). In general, BL clouds have a radiative net surface cooling effect, since they have very high reflectivities. Studies with one-dimensional models (1D) of convection-radiation with BL clouds point out that an increase of 1% in the global cloud cover of these clouds could cool the surface, offsetting an equivalent of increase of 25% of CO_2 concentrations (Van Dorland 1999). Note, though, that such a change would be an increase in about 10% of the current ocean Cu cloud cover and 20% in the continental regions.

In spite of their importance for weather and climate shallow *cumuli* have received less attention than other BL clouds. However, some observational campaigns had been dedicated to its study, like BOMEX (the Barbados Oceanographic and Meteorological EXperiment, Kuettner and Holland 1969), ARM (Atmospheric Radiation Measurement, Brown et al. 2002) and SCMS (Small Cumulus Microphysics Study, French et al. 1999), leading to a number of relevant contributions (Warner 1977; Jonas 1990; Blyth 1993; Smith and Jonas 1995; Grinell et al. 1996; de Roode and Duynkerke 1997). On the other hand, LES models have been extensively used for the study of the BL with shallow cumulus (Sommeria 1976; Cuijpers and Duynkerke 1993; Siebesma and Cuijpers 1995; Siebesma and Holtslag 1996; Stevens et al. 2001; Brown et al. 2002; Neggers et al. 2004; Siebesma et al. 2004).

The BL with shallow cumulus presents a mean cloud cover of 10%–30%, usually associated with good weather. Shallow *cumuli* have a small vertical development, up to a maximum of 2 km, with a cloud-base between 500 m and 1.5 km. These clouds are turbulent convective structures, where the vertical velocity can reach the $5\,m\,s^{-1}$.

Over land surfaces, the BLs with shallow cumulus have a diurnal cycle, due to the large variation of the heat and moisture fluxes at the surface. A typical cycle of this type of BL is as follows: at sunrise the BL is cloud free; associated with the surface heat flux the first *cumuli* appear a few hours later, deepening with time; before sunset the *cumuli* start to dissipate, and eventually disappear.

4.2.1 STRUCTURE OF THE CONVECTIVE BL

Figure 4.1 shows a typical average profile of potential temperature, specific humidity, and wind, for a CBL. The profiles of these properties reflect the BL evolution. During the day, the surface BL presents a steep vertical gradient of all the properties, intensifying the transport of heat, humidity, and momentum between the surface and the air, that is, the potential temperature and humidity profiles are unstable. In the mixed layer, that is, in the interior of the BL, all these properties reflect the intense mixing, presenting almost constant profiles, in particular in what concerns potential temperature and wind. The specific humidity decreases slightly with height, and the temperature decreases in accordance with the vertical lapse rate of the temperature in an adiabatic process. In the upper region of the BL, the temperature shows an inversion, potential temperature increases with height, specific humidity decreases rapidly, and the wind has significant vertical shear approaching geostrophic values in the FA. The presence of strong vertical gradients at the inversion acts as a "lid" for the penetration of thermals in the free troposphere, restricting the domain of influence of turbulence.

The diurnal cycle of a CBL with cumulus (Figure 4.2) is similar to the one described above for a clear sky convective BL, taking into account the modifications associated with clouds and moistening of the BL. The surface BL is still superadiabatic, characterized by a decrease in the potential temperature and the total specific humidity with height. Above it is a mixed layer, where those properties are approximately constant with height, until the lifting condensation level, which constitutes cloud base, and the beginning of the cumulus cloud layer. This layer has the vertical extension of the ensemble of cumulus, and presents reduced but constant vertical gradients of potential temperature and moisture. The potential temperature increases and the humidity decreases. The typical vertical profile of a layer with cumulus clouds is a conditionally unstable profile: dry air parcels are stable while saturated parcels are unstable, resulting in the vertical development of clouds. The cloudy layer is topped by the inversion, where the gradients of the previous properties are very sharp.

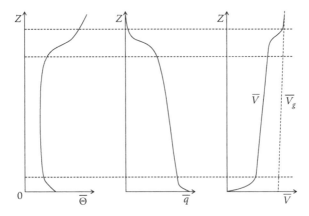

FIGURE 4.1 Average vertical profiles, for a typical dry convective boundary layer, of: potential temperature θ, specific humidity q, and wind V, V_g refers to geostrophic wind.

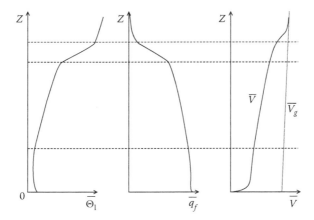

FIGURE 4.2 Average vertical profiles for a typical convective BL with shallow cumulus: liquid water potential temperature θ_l, total specific humidity q_t, and wind V. V_g refers to geostrophic wind.

4.3 PARAMETERIZATION OF TURBULENCE IN ATMOSPHERIC MODELS

The prevalence of turbulence is the main reason why fluid mechanics is, and will be, a difficult mathematical problem. Because of that, progress in this field is largely dependent on numerical simulations. These require the discretization of the system of equations, effectively reducing the number of degrees of freedom to a finite value. The discretized system represents only processes that occur in time and spatial scales greater than the chosen grid mesh, and the discretization process implies the appearance of new terms in the equations, describing the effect of the nonrepresented scales on the scales of the model. These sub-grid terms are referred to as turbulent terms.

In fact, the discretization process is inherent to the observation of the flow of a fluid. Reynolds (1895) showed that the contributions of the sub-grid-scale terms are responsible for the irregular (or turbulent) character of the flow in certain fluid regimes. In principle, the equations can be applied directly even to turbulent flows, if the resolution is enough to explicitly represent the turbulence of different scales down to the limit of the dissipative eddies. However, it is not, in general, possible to draw a model with such fine grid. Instead, the turbulence associated with scales smaller than the discretization limit has to be parameterized, through a turbulence model, where the statistical average effect of the subscale on the equations for each mean variable is represented.

The Reynolds (1895) decomposition consists in considering each atmospheric variable, ϕ, as the sum of its mean value, $\bar{\phi}$, and a perturbation, ϕ'. The mean value describes, in a numerical model, some average value in the domain represented by the grid point:

$$\phi = \bar{\phi} + \phi'. \tag{4.1}$$

$\overline{(\ )}$ may correspond to a spatial and time average in a grid:

$$\bar{\phi} = \frac{1}{VT} \int_T \int_V \phi \, dV dt. \tag{4.2}$$

Introducing the definition (Equation 4.1), the hydrostatic and Boussinesq approximations in the set of atmospheric equations and considering the properties of the averaging process leads to the Reynolds system of equations:

$$\frac{\partial \bar{u}_j}{\partial x_j} = 0, \tag{4.3}$$

$$\frac{\partial \bar{u}_i}{\partial t} = -\bar{u}_j \frac{\partial \bar{u}_i}{\partial x_j} - \frac{\partial}{\partial x_j}(\overline{u'_i u'_j}) - \frac{1}{\rho_{\text{ref}}} \frac{\partial \bar{p}}{\partial x_i} - 2\varepsilon_{ijk}\Omega_j \bar{u}_k + \delta_{i3} \frac{g}{\theta_{v\,\text{ref}}} \overline{\theta}_v + \nu \nabla^2 \bar{u}_i, \tag{4.4}$$

$$\frac{\partial \bar{\theta}_l}{\partial t} = -\bar{u}_j \frac{\partial \bar{\theta}_l}{\partial x_j} - \frac{\partial}{\partial x_j}(\overline{u'_j \theta'_l}) + \lambda_{\theta_l} \nabla^2 \bar{\theta}_l + \bar{S}_{\theta_l}, \tag{4.5}$$

$$\frac{\partial \bar{q}_t}{\partial t} = -\bar{u}_j \frac{\partial \bar{q}_t}{\partial x_j} - \frac{\partial}{\partial x_j}(\overline{u'_j q'_t}) + \lambda_{q_t} \nabla^2 \bar{q}_t + \bar{S}_{q_t}. \tag{4.6}$$

These prognostic equations contain new terms $(\partial(\overline{u'_j \theta'_l})/\partial x_j, \partial(\overline{u'_j q'_t})/\partial x_j$, and $\partial(\overline{u'_i u'_j})/\partial x_j)$ that represent divergences of turbulent fluxes. These terms result directly from the nonlinearity of the advective terms in the prognostic equations and constitute new source terms to the mean variable budgets. All the new terms are covariances. $\overline{u'_j \theta'_l}$, $\overline{u'_j q'_t}$, and $\overline{u'_i u'_j}$ are, respectively, the kinematic turbulent fluxes of heat, humidity, and linear momentum. These terms reveal that perturbations of velocity, temperature, and humidity redistribute momentum, heat, and humidity in the atmosphere.

In the atmospheric BL, the turbulent terms of the prognostic equations are some orders of magnitude greater than the molecular diffusion terms (Garratt 1992), consequently, these last terms are normally neglected in BL modeling.

4.3.1 THE TURBULENCE CLOSURE PROBLEM

The Reynolds system of equations constitutes an open system, for example, one that contains more unknowns than equations. In principle, one may deduce more equations, describing the time derivatives of the new unknowns (e.g., Mellor and Yamada 1974; André et al. 1978). However, that always leads to an even larger number of unknowns, in the form of higher statistical moments of the perturbation variables. In general, an equation of order n statistical moment always contains terms of order $n+1$. The impossibility of establishing a closed set of turbulence equations constitutes the "closure problem."

This is the reason why turbulence remains an open problem in physics research. The use of some closure is a practical necessity in any model. In the FA, one often "forgets" about this fundamental problem by just putting the turbulent terms to zero. That cannot be done in the boundary layer, because those terms are comparatively large and may be even dominant. There, models use additional empirical or semi-empirical equations to close the equation set. Those new equations need to be physically sound and must be thoroughly supported by observations.

There are different ways to solve the closure problem. In all cases, it amounts to the establishment of relations between second-order (and possibly higher) statistical moments of the perturbation variables and the mean variables (the first-order moments). These relations, together with other semi-empirical relations describing adiabatic processes, are included in the "parameterization" package of a numerical model.

4.3.2 Turbulence Parameterization and First-Order Closures

In the CBL, there are two classic approaches for the parameterization of turbulence, the ones based on local and nonlocal closures. The local approach consists of relating the unknown higher-order terms in a given point of the space and time with properties of the flow near that point. Nonlocal closures, on the other hand, allow relations between those unknowns and properties in other, possibly noncontiguous, regions of the BL.

Local closures can be established at different orders, depending on the number of statistical moments that are explicitly kept in the prognostic equations. Simplest, first-order closures keep only prognostic equations for the mean variables (first-order moments). Higher-order closures add other prognostic equations and extra variables. Local closures have proposed up to the third order (André et al. 1978), but first- and second-order schemes are, in general, considered to be good enough (and computationally expensive enough).

First-order closures are the more frequently used in atmospheric BL modeling, and are based on an analogy between the processes of molecular diffusion and turbulent mixing (Boussinesq 1877). Therefore, the turbulent flux of a property in a point of the space is considered proportional to the local gradient of that property. This approach is commonly known as theory of turbulent diffusion or diffusion-K, since the considered coefficients of proportionality are called as coefficients of turbulent mixing or eddy-diffusivity coefficients, K. The fluxes of momentum, temperature, and humidity are then described by

$$\overline{u'w'} = -K_m \frac{\partial \overline{u}}{\partial z},$$

$$\overline{v'w'} = -K_m \frac{\partial \overline{v}}{\partial z}, \tag{4.7}$$

$$\overline{w'\theta'_l} = -K_h \frac{\partial \overline{\theta}_l}{\partial z},$$

$$\overline{w'q'_t} = -K_q \frac{\partial \overline{q}_t}{\partial z} \tag{4.8}$$

where K_m, K_h, and K_q are, respectively, the turbulent eddy diffusivities of momentum, temperature, and humidity. These are, by definition, positive, since it is assumed that the transport is in the opposite sense of the gradient. The previous formula will constitute closure only if the eddy diffusivities are given as functions of the model variables.

The simplest formulation considers that turbulent eddy diffusivity coefficients are constants, but orders of magnitude greater than that associated with molecular diffusion. Prandtl (1925) proposed an improved formulation for those coefficients, using the concept of a mixing length, l, which, again, was inspired by the concept of the mean free path in kinetic theory. In a statically neutral atmosphere, a parcel displaced from its original level, z, without mixing, to the level, $z+z'$, will have a perturbation in the property, ϕ, given by (Prandtl 1925)

$$\phi' = \phi(z) - \phi(z + z') \approx -z' \frac{\partial \overline{\phi}}{\partial z}, \tag{4.9}$$

where $\overline{\phi}$ represents the initial, unperturbed, profile. Equation 4.9 assumes that ϕ is a conservative property, and it attributes the perturbation field ϕ' to the effects of advection. To compute the covariances in Equations 4.7 and 4.8, one further needs to compute the wind field perturbations u', v', and w'. Equation 4.9 can be also applied to the horizontal wind field, assuming that we have an initial profile with wind shear. Then $u' = -z'(\partial \overline{u}/\partial z)$, and an equivalent relation for v'. In a horizontally homogeneous BL, one necessarily has $\overline{w} = 0$, implying that w' must be computed in a different way. One may show (Stull 1988) that it may be defined as $w' = -cu'$, leading to $w' = cz' |\partial \overline{u}/\partial z|$. Then, for a generic property ϕ, one has

$$\overline{w'\phi'} = -c\,\overline{z'^2}\, \frac{\partial \overline{\phi}}{\partial z} \left| \frac{\partial \overline{u}}{\partial z} \right|. \tag{4.10}$$

The root mean square of $\overline{z'^2}$ is a measure of the mean distance that the parcel can travel while keeping its properties, leading to the definition of a mixing length, l, by $l^2 = c\overline{z'^2}$. Thus, the Prandtl formulation gives

$$K = l^2 \left| \frac{\partial \overline{u}}{\partial z} \right|, \tag{4.11}$$

where l represents the average dimension of the turbulent eddies. K is, therefore, proportional to l^2 and to the wind shear.

Different first-order theories developed by Prandtl, Taylor, and von Karman are presented by Monin and Yaglom (1971). Those approaches showed some skill when tested with measurements in wind tunnels and laboratorial tanks, and even with some atmospheric observations, that at the time were limited to the surface layer. Consequently, they have been widely used.

The K-diffusion theories are essentially theories of small eddies that represent well the turbulent mixing in neutral and steady BLs (Stull 1984, 1993). These are appropriate when the mixing length is smaller than the resolution of the model, and belong to the inertial subrange of the spectra of energy of the turbulence. However, they give a deficient representation of the nonlocal turbulent fluxes associated with thermals, of vertical extension larger than the vertical resolution of the BL model.

Lilly (1962) considered a formulation to take into account the effect of buoyancy, proposing that eddy diffusivities were made dependent on the Richardson number (4.13). The Prandtl approach can be derived from a steady-state equation of $\bar{e}$, equaling the production by shear with the dissipation. If buoyancy is included in the production term, it becomes

$$K = l^2 \left| \frac{\partial \bar{u}}{\partial z} \right| F(Ri), \tag{4.12}$$

where Ri is the gradient Richardson number. Ri is based on the K-diffusion approach, more precisely in Equations 4.7 and 4.8 allowing to write,

$$Ri = \frac{\dfrac{g}{\theta_v} \dfrac{\partial \overline{\theta_v}}{\partial z}}{\left[\left(\dfrac{\partial \bar{u}}{\partial z} \right)^2 + \left(\dfrac{\partial \bar{v}}{\partial z} \right)^2 \right]}, \tag{4.13}$$

To solve Equation 4.12 one has to specify l and $F(Ri)$. Usually, it is assumed that $l \rightarrow kz$ in the surface layer and $F(Ri) \rightarrow 1$ in a neutral stratification, where k is the von Karman constant.

Both Expressions 4.11 and 4.12 depend critically on l. On the basis of these observations, Blackadar (1962) built an empirical formulation for l:

$$\frac{1}{l} = \frac{1}{kz} + \frac{1}{\lambda}, \tag{4.14}$$

where λ is an adjustable parameter representing an asymptotic mixing length. This expression consists in an interpolation between the two limits: $l \rightarrow kz$ when $z \rightarrow 0$ and $l \rightarrow \lambda$ when $z \rightarrow +\infty$. Other expressions for the CBL are based on empirical prescription of profiles for K, making use of empirical expressions (Holtslag and Moeng 1991). Often, these expressions depend on the computation of the BL height, z_i, and other scale variables of the surface layer.

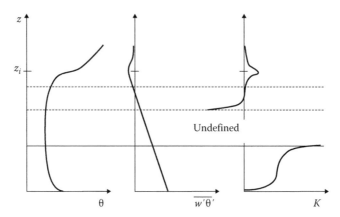

FIGURE 4.3 Typical profiles of a convective BL (built from observations) of potential temperature θ, turbulent vertical flux of potential temperature $\overline{w'\theta'}$, and the corresponding (computed) ED coefficient profile, K.

The technological advances in the measurement instruments of the turbulence allowed the monitorization of the BL interior (e.g., with airborne instruments) disclosing the fragilities of those approaches. Figure 4.3 shows a typical vertical profile of potential temperature of $\overline{w'\theta'}$ and turbulent diffusivity coefficient in a CBL. This profile illustrates the conceptual fragilities of the proposed closure.

The problem with local first-order closures is that the only way to get a finite turbulent flux in a region of no vertical gradients, as observed in the mixed layer, would be to have infinite values for the eddy diffusivity K. On the other hand, the observation of an upward flux in the slightly stable region in the upper half of the mixed layer would imply a negative eddy diffusivity. Both results contradict the basic assumptions of K-diffusion, implying that the process is not, in general, an analog of molecular diffusion. One may argue that the observed profiles can only be understood as a result of the global BL response to the surface fluxes, indicating that what is wrong in the proposed first-order closure is its local nature, that is, the fact that it is unaware of what is happening "far away" in the fluid (at the surface and/or at the inversion).

Alternatively, one may think that what is missing in the previous theories are extra variables and equations, describing the dynamics of the turbulent fields, or, in other words, that the problem is in the low order of the approximation. Both lines of development have been tried in the past decades, with various degrees of success.

4.3.3 Turbulent Kinetic Energy and Higher-Order Closures

Even when it is not explicitly required for the turbulence closure, it is important to understand the equation of the turbulent kinetic energy (TKE), one of the equations that may be derived in a higher-order closure approximation. TKE is a measure of the intensity of turbulence. When computed per unit mass, $\overline{e}$, it is equal to half of the sum of the velocity variances:

$$\bar{e} = \frac{1}{2}\overline{u_i'^2} = \frac{1}{2}(\overline{u'^2 + v'^2 + w'^2}). \tag{4.15}$$

A prognostic equation for $\bar{e}$ can be easily derived (e.g., Stull 1988):

$$\frac{\partial \bar{e}}{\partial t} + \bar{u}_j \frac{\partial \bar{e}}{\partial x_j} = -\overline{u_i'u_j'}\frac{\partial \bar{u}_i}{\partial x_j} + \delta_{i3}\frac{g}{\theta_v}\overline{u_i'\theta_v'} - \frac{\partial(\overline{u_j'e})}{\partial x_j} - \frac{1}{\rho}\frac{\partial(\overline{u_i'p'})}{\partial x_i} - \varepsilon, \tag{4.16}$$

where, from left to right, the terms are the tendency of $\bar{e}$, advection of $\bar{e}$ by the mean wind, production of $\bar{e}$ by wind shear, generation/destruction of $\bar{e}$ by buoyancy, turbulent transport of $\bar{e}$, pressure-correlation term, and dissipation, $\varepsilon = \nu\overline{(\partial u_i'/\partial x_j)^2}$. If horizontal homogeneity is considered, Equation 4.16 becomes

$$\frac{\partial \bar{e}}{\partial t} = -\overline{u'w'}\frac{\partial \bar{u}}{\partial z} - \overline{v'w'}\frac{\partial \bar{v}}{\partial z} + \frac{g}{\theta_v}\overline{w'\theta_v'} - \frac{\partial}{\partial z}\left(\overline{w'e} + \frac{\overline{w'p'}}{\rho}\right) - \varepsilon. \tag{4.17}$$

This equation emphasizes the relative importance of the mechanical and thermal source-sink effects on turbulence. It can be used in a model as a prognostic equation for turbulence intensity, giving a clear physical basis for some closure assumptions.

One may use Equation 4.17 as a new model equation, to add to the prognostic equations for the mean variables, using first-order closures for the different fluxes, as in the previous section, but using $\bar{e}$ to compute the eddy diffusivities. This closure, which includes only one extra prognostic equation for a second-order moment $(\bar{e})$, is the simplest higher-order closure, and is often called a 1.5-order closure.

Equation 4.17 includes two extra turbulent terms $(\overline{w'e}, \overline{w'p'})$. Although it is hard to fully justify, one generally represents these terms with a "joint" K-diffusion approach:

$$\overline{w'e} + \frac{\overline{p'w'}}{\rho} = -K_e\frac{\partial \bar{e}}{\partial z}. \tag{4.18}$$

However, there are important arguments against Equation 4.18, particularly in what concerns the treatment of the pressure-correlation term, which may be an important source of TKE (Hogstrom 1990) not described by that expression. Also, observations made in 1968 in Kansas (Wyngaard 1998) showed that Equation 4.18 is not always valid.

Putting aside those problems, one may use the TKE prognostic equation with Equation 4.18 and the other K-diffusion relations (Equations 4.7 and 4.8), provided that the eddy diffusivities and the dissipation are given. Accepting that those terms are explicit algebraic functions of the TKE, which is the simplest possible set of relations, leads by dimensional analysis to

$$\varepsilon = \frac{\bar{e}^{3/2}}{\Lambda_1}, \tag{4.19}$$

and

$$K_{m,h,e} = \Lambda_{m,h,e} \, \overline{e}^{\frac{1}{2}},$$ (4.20)

where different Λ are length scales associated with dissipation and mixing of the different properties (all other). The closure is complete with the establishment of functional relations for those different length scales.

In general, the performance of K-diffusion schemes in the CBL is very much dependent on the mixing length in the region of the inversion. One advantage of schemes that incorporate the TKE equation is that they are less sensitive, since they have to satisfy the equation of TKE. In the upper half of a CBL, the gradients of the horizontal components of the wind are practically zero, and the TKE balance is due mainly to buoyancy, transport, and dissipation. In the surface layer, the vertical transport of TKE is negative and considerable, changing the signal in the interior of the BL, which implies a positive transport of TKE for the BL upper region. When the surface heat flux is ascending, the surface layer is unstable.

Higher-order closures, incorporating prognostic equations for part or all second-order moments and even for third-order moments, may be easily deduced, eventually leading to a large set of simultaneous equations (e.g., Mellor and Yamada 1974; André et al. 1978). Some of these sets have been successfully tested in idealized cases, but they have not been generally incorporated in numerical weather prediction models, even at mesoscale, due to their high computational cost.

4.3.4 NONLOCAL APPROACHES

Nonlocal approaches were inspired by the observation of many thermals that make an ascent almost without lateral mixing, transporting air at significant distances in the boundary layer, a picture of the flow that is consistent with the analysis of cloud images (Lenschow and Stephens 1980; Agee 1984; Stull and Driedonks 1987; Ebert et al. 1989). Because turbulent terms arise in the Reynolds equations as a consequence of the nonlinearity of advection, it is easy to argue that at least part of its effect in the mean flow should not behave as if it were an extra diffusion term. As a consequence, nonlocal closures relate the turbulent terms with variables known in all the BL, and not just in the vicinity of the grid point.

The simplest nonlocal approach, frequently used in atmospheric models, consists in the *ad hoc* introduction of a "counter-gradient" term in the balance equations (Deardorff 1966), permitting upward heat flow in the upper half of the mixed layer, and in the inclusion of a "mass-flux" (MF) term in the parameterization of the transport in cloud updrafts (Betts 1973). Other, more complex, nonlocal closures include the transilient turbulent theory (Stull 1984) and the spectral diffusive theory (Berkowicz and Prahm 1979).

4.3.4.1 Counter-Gradient

The inconsistency of K-diffusion theory with observations was made clear in the convective BL, when an upward heat flux is maintained through a neutral or slightly stable mixed layer. In his pioneer study, Deardorff (1966) verified the existence of a counter-gradient turbulent flows with a superadiabatic surface layer, and suggested as solution the introduction of a modified vertical gradient:

$$\overline{w'\theta'} = -K_h \frac{\partial \overline{\theta}_c}{\partial z} = -K_h \left(\frac{\partial \overline{\theta}}{\partial z} - \gamma_c \right), \tag{4.21}$$

where $\gamma_c \approx 0.65 \times 10^{-3}\,\mathrm{km^{-1}}$ is a new counter-gradient term, ensuring an upward heat flux in weak stratification. While γ_c was initially introduced as an empirical correction, it has been shown that Equation 4.21 is consistent with a simplification of the prognostic equation of $\overline{w'\theta'}$, where the counter-gradient is controlled by $\gamma_c = (g/\overline{\theta}_v)\, \overline{\theta'\theta'_v}/\overline{w'^2}$ (Garratt 1992).

Holtslag and Moeng (1991), Holtslag and Boville (1993), and Holtslag et al. (1995) developed the counter-gradient theory, appealing to results from Hojtrup (1982). These authors presented different options to formulate the turbulent diffusivities and the counter-gradient term. The closure of Holtslag and Boville (1993) was implemented in the NCAR Community Climate Model (Willimason et al. 1987), with the turbulent diffusivity for the heat given by (Troen and Mahrt 1986; Holtslag et al. 1990):

$$K_h = kw_t z \left(1 - \frac{z}{z_i} \right)^2, \tag{4.22}$$

where w_t is a turbulent velocity scale and the counter-gradient term is given by

$$\gamma_c = a_\gamma \frac{w_* \overline{(w'\theta')}_s}{w_m^2\, z_i}, \tag{4.23}$$

where w_m is another velocity scale, $a_\gamma = 7.2$. $w_* = (z_i \overline{(w'\theta')}_s\, g/\overline{\theta}_v)^{1/3}$ is the convective vertical velocity scale, z_i is the CBL inversion height, and $\overline{(w'\theta')}_s$ is the surface flux of virtual potential temperature. Supported by CBL observations in Holland, Holtslag et al. (1995) evaluated the introduction of this term in the K-diffusion parameterization of Louis et al. (1982) showing the improvements of this approach.

4.3.4.2 Mass-Flux

The MF approach was inspired by observational evidence that the vertical transport of properties in a shallow cumulus is mainly done by in-cloud updrafts (e.g., Warner 1970, 1977). In spite of the small horizontal area occupied by those updrafts, they seem to contribute disproportionably to the vertical transport, once they are associated with large perturbations of vertical velocity and of the thermodynamical properties.

If one considers the probability density function (PDF) of the specific humidity of a horizontal domain of a shallow cumulus BL (Figure 4.4), the MF approach

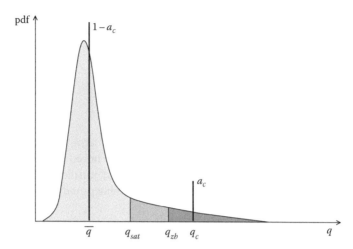

FIGURE 4.4 The PDF of a shallow cumulus BL. q_{sat} is the saturation-specific humidity and q_{zb} is the specific humidity at the level where buoyancy is zero in saturated conditions. a_c corresponds to the horizontal area with saturated air and positive buoyancy of the PDF, $1-a_c$ corresponds to the remaining area. The height of the two picks is equal to those areas. To represent this PDF by the two picks corresponds to the MF approach.

results from the association of all the vertical transport in two distinct areas of this PDF: the saturated area of uprising motion, with positive buoyancy, a_c and the remaining area, $1-a_c$. Thus, the horizontal domain is divided accordingly, and considers the average properties of each one of these areas, resulting in two peaks in the distribution function, whose intensity is equal to each one of the areas (Ooyama 1971; Betts 1973; Yanai et al. 1973). This simplification neglects some of the smaller-scale variability (Wang and Stevens 2000) since it is based on the average properties of these two sub-domains. This simplification is known in literature as "top-hat" since properties of the horizontal domain have only two possible values.

The MF schemes have been used with great success in the representation of cumulus convection (Betts 1973; Arakawa and Schubert 1974; Tiedtke 1989). That has led to its implementation in many numerical weather prediction models and to the development of different methodologies. The results of several numerical and observational studies have contributed to that development (e.g., Randall et al. 1992). The MF approach was extensively analyzed using LES simulations, for different types of BL: with *stratus* (Schumann and Moeng 1991a), *stratocumulus* (Randall et al. 1992), and shallow *cumulus* (Siebesma and Cuijpers 1995; Siebesma and Holstlag 1996; Brown et al. 2002; Siebesma et al. 2004).

Recently, this type of parameterization was also applied to the clear sky BL (Businger and Oncley 1990; Wang and Albrecht 1990; Randall et al. 1992; Wyngaard and Moeng 1992), since observations of this type of BL also point out to the prominent role played by (dry) updrafts in the vertical transport of properties (Lenschow and Stephens 1980; Lenschow et al. 1980; Nicholls 1989). Beyond the vast number of applications of this type of approaches, it is important to mention that the MF approach is suitable for the representation of the transport of active chemical tracers (Chatfield and Brost 1987).

Originally, the MF approach was based on the decomposition of the atmosphere in regions of ascending motion, in the interior of clouds, and regions in its environment. But, if it is also intended to describe the clear sky BL, this criterion obviously needs to be adjusted. A more general criterion of decomposition may use the signal of vertical velocity, not forgetting that more restrictive criteria could be applied. Therefore, the MF approach consists in the decomposition of the atmosphere in updraft regions (u) and the surrounding environment (e). Considering an horizontal domain of the atmosphere with the area $A = L_x L_y$, A_u is the area with upward motion and A_e is the surrounding area ($A = A_u + A_e$), and one can compute the mean value of a variable in the domain, $\bar{\phi}(z)$, and in each of the complementary regions, respectively $\bar{\phi}^u(z)$ and $\bar{\phi}^e(z)$:

$$\bar{\phi}(z) \equiv \frac{1}{A} \int_0^{L_x} \int_0^{L_y} \phi(x, y, z) \, dx \, dy,$$

$$\bar{\phi}^u(z) = \phi_u \equiv \frac{1}{A_u} \iint_u \phi(x, y, z) \, dx \, dy, \qquad (4.24)$$

$$\bar{\phi}^e(z) = \phi_e \equiv \frac{1}{A_e} \iint_e \phi(x, y, z) \, dx \, dy.$$

Normalizing the areas, with $a_u = (A_u/A)$ as the fraction of the domain covered by updrafts, the mean ϕ for all the domain is given by,

$$\bar{\phi} = a_u \phi_u + (1 - a_u) \phi_e. \qquad (4.25)$$

The turbulent vertical fluxes for the domain and for the partial domains can be easily deduced using the Reynolds averaging, yielding

$$\overline{w'\phi'} = \overline{w\phi} - \bar{w}\bar{\phi},$$

$$\overline{w'\phi'}^u = \overline{w\phi}^u - w_u \phi_u, \qquad (4.26)$$

$$\overline{w'\phi'}^e = \overline{w\phi}^e - w_e \phi_e,$$

and knowing that $\bar{w} = a_u w_u + (1 - a_u) w_e$ and $\overline{w\phi} = a_u \overline{w\phi}^u + (1 - a_u) \overline{w\phi}^e$, and Equation 4.25, the following relations between fluxes are obtained:

$$\overline{w'\phi'} = a_u \overline{w'\phi'}^u + (1 - a_u) \overline{w'\phi'}^e + a_u (w_u - \bar{w})(\phi_u - \phi_e). \qquad (4.27)$$

The third term of the r.h.s. expresses the updraft contribution to the vertical turbulent transport of ϕ, and is the MF term. In the context of shallow convection that is considered the more significant component of the turbulent transport (Ooyama

1971; Betts 1973; Yanai et al. 1973). Siebesma et al. (2004) showed, through LES simulations, that in the BOMEX case, corresponding to a shallow cumulus BL above the ocean, more than 80% of the total flux can be attributed to the MF term. Similar results have been used to justify the approximation:

$$\overline{w'\varphi'} \cong a_u(w_u - \overline{w})(\varphi_u - \varphi_e) \equiv M(\varphi_u - \varphi_e), \tag{4.28}$$

where M is the MF coefficient associated with the updrafts: $M = a_u(w_u - w_e)$. This result is fundamental in this approach, stating that the vertical flux of a property is proportional to the difference between the mean value of that property in the updraft regions and its mean value in the surrounding areas. Neglecting the first two terms of Equation 4.27 corresponds to consider $a_u \ll 1$ and that the turbulence in the surrounding environment contributes scarcely to the total flux. In the same context, it is acceptable to consider $w_e \cong 0$, resulting in

$$M \cong a_u w_u. \tag{4.29}$$

The validity of Equation 4.28 was extensively studied for different types of BL. It is important to mention the work of Wyngaard and Moeng (1992), where analytical and LES simulation results were compared. These authors divide the atmosphere in regions of updrafts and downdrafts (d) and assume that w and ϕ obey a Gaussian PDF, allowing to write

$$\overline{w'\phi'} = b_{ud}\sigma_w(\phi_u - \phi_d), \tag{4.30}$$

where

$b_{ud} = (\sqrt{2\pi}/4) = 0.627$
σ_w is the vertical velocity standard deviation

Admitting that vertical velocity also obeys to a Gaussian distribution, $M = (\sigma_w/\sqrt{2\pi})$, this equation can be written as a function of

$$\overline{w'\phi'} = v_u M(\phi_u - \phi_d), \tag{4.31}$$

where $v_u = (2\pi/4) \approx 1.57$, implying that in Equation 4.27, for a Gaussian PDF, approximately 60% ($\approx 1/v_u$) of the total flux is described by the MF term. Note that in Equation 4.28 v_u is implicitly equal to 1.

Equation 4.28 constitutes the fundamental expression of the MF approach and is presented as an alternative way to parameterize the turbulent fluxes and to close the Reynolds system of Equations 4.3 through 4.6. For that, one needs to know the profiles of M, ϕ_u, and ϕ_e. The computation of those profiles is not trivial. A moving air parcel in the BL is not an isolated system; whenever it ascends or descends it exchanges properties with the surrounding air modifying its own properties in the mixing process.

The evolution of the variable ϕ of an updraft horizontal section, with area a_u, that is, ϕ_u, has to reflect the lateral mixing process. Following Siebesma (1996), starting from the prognostic equation of ϕ:

$$\frac{\partial \phi}{\partial t} + \nabla_h \cdot \vec{v}_H \phi + \frac{\partial (w\phi)}{\partial z} = F, \tag{4.32}$$

where $\vec{v}_H$ is the horizontal velocity vector and F gathers all the external forces, and integrating ϕ in the A_u area, dividing by A, applying the Leibnitz and the divergence theorems leads to

$$\frac{\partial a_u \phi_u}{\partial t} + \frac{1}{A} \oint_{\text{fronteira}} \vec{n} \cdot (\vec{v} - \vec{v}_f) \, \phi dl + \frac{\partial a_u \overline{w\phi}^u}{\partial z} = a_u F_u, \tag{4.33}$$

where
  $\vec{n}$ is a unit vector perpendicular to the boundary that separates the updraft from the surroundings
  $\vec{v}$ is the velocity vector
  $\vec{v}_f$ is the boundary velocity

Similarly, the continuity Equation 4.3, or Equation 4.32 assuming $\phi = $ cte, integrated in the same horizontal section of an updraft can be written

$$\frac{\partial a_u}{\partial t} + \frac{1}{A} \oint_{\text{fronteira}} \vec{n} \cdot (\vec{v} - \vec{v}_f) \, dl + \frac{\partial a_u w_u}{\partial z} = 0. \tag{4.34}$$

The Equation 4.34 corresponds to the mass balance of air crossing the boundary, that is, constitutes the net result of lateral mixing. If one defines E (entrainment) as the lateral mixing associated with the air mass from the surroundings that enters in the ascent, and D (detrainment) the term representing the process of exporting updraft air, $D-E$ represents the net result of both processes, therefore

$$D - E = \frac{1}{A} \oint_{\text{fronteira}} \vec{n} \cdot (\vec{v} - \vec{v}_f) \, dl, \tag{4.35}$$

where E is given by

$$E = -\frac{1}{A} \oint_{\vec{n} \cdot (\vec{v} - \vec{v}_f) < 0} \vec{n} \cdot (\vec{v} - \vec{v}_f) \, dl, \tag{4.36}$$

and D is written as

$$D = \frac{1}{A} \oint_{\vec{n} \cdot (\vec{v} - \vec{v}_f) > 0} \vec{n} \cdot (\vec{v} - \vec{v}_f) \, dl, \tag{4.37}$$

Finally, from Equation 4.34:

$$\frac{\partial a_u}{\partial t} + (D - E) + \frac{\partial a_u w_u}{\partial z} = 0. \tag{4.38}$$

Considering the following simplifications for the lateral mixing that exchanges average properties:

$$\frac{1}{A}\oint_{\vec{n}\cdot(\vec{v}-\vec{v}_f)>0} \vec{n}\cdot(\vec{v}-\vec{v}_f)\,\phi\,dl \approx \frac{\phi_u}{A}\oint_{\vec{n}\cdot(\vec{v}-\vec{v}_f)>0}\vec{n}\cdot(\vec{v}-\vec{v}_f)\,dl = D\phi_u,$$

$$\frac{1}{A}\oint_{\vec{n}\cdot(\vec{v}-\vec{v}_f)<0} \vec{n}\cdot(\vec{v}-\vec{v}_f)\,\phi\,dl \approx \frac{\phi_e}{A}\oint_{\vec{n}\cdot(\vec{v}-\vec{v}_f)<0}\vec{n}\cdot(\vec{v}-\vec{v}_f)\,dl = -E\phi_e, \quad (4.39)$$

Equation 4.33 is simplified to

$$\frac{\partial a_u\phi_u}{\partial t} - E\phi_e + D\phi_u + \frac{\partial a_u\overline{w\phi}^u}{\partial z} = a_u F_u. \quad (4.40)$$

Analogously, for the surrounding area $(1-a_u)$, the following equation can be deduced:

$$\frac{\partial(1-a_u)\phi_e}{\partial t} - E\phi_e + D\phi_u + \frac{\partial(1-a_u)\overline{w\phi}^e}{\partial z} = (1-a_u)F_e. \quad (4.41)$$

The system of Equations 4.38, 4.40, and 4.41 establishes the mass balance in the context of the MF approach.

In the CBL, other decompositions can be formulated. For the shallow cumulus BL, the better established decomposition (Siebesma and Cuijpers 1995; Siebesma and Holtslag 1996) takes as criterion of the domain for the turbulent vertical flux the positive velocities ($w>0$) with positive buoyancy and liquid water, usually called cloud core (c). Accordingly, it is acceptable to consider (Tiedtke 1989; Siebesma and Holtslag 1996) that: the cumulus ensemble is in steady state, ($\partial\theta_c/\partial t=0$); $v_u=1$, that is, Equation 4.28 is valid; the cloud core fractional area is much smaller than one, $a_u \ll 1$, consequently $\phi_e \approx \overline{\phi}$.

Jointly with Equation 4.29, the previous simplifications allow to obtain a simplified system for Equations 4.38, 4.40, and 4.41:

$$\frac{\partial M_c}{\partial z} = E - D, \quad (4.42)$$

$$\frac{\partial M_c\phi_c}{\partial z} = E\overline{\phi} - D\phi_c, \quad (4.43)$$

$$\frac{\partial\overline{\phi}}{\partial t} = -\frac{\partial M_c(\phi_c-\overline{\phi})}{\partial z} + F. \quad (4.44)$$

Expression 4.40 is equivalent to Equations 4.5 or 4.6, in the MF approach, where the source terms are gathered in F, with the exception of the term of vertical divergence of the turbulent flux (first term in the r.h.s. member). Thus, the tendency of a property $\bar{\phi}$, that can be any meteorological variable in a grid point of a numerical weather prediction model, depends on knowing M_c and ϕ_c.

4.3.5 TRANSFER OF PROPERTIES AT INTERFACES

4.3.5.1 Lateral Mixing

Entrainment and detrainment were previously defined. They can be both scaled by the total MF (Tiedtke 1989):

$$E = \varepsilon M_c,$$

$$D = \delta M_c, \tag{4.45}$$

defining ε and δ, respectively, the fractional rate of lateral mixing of entrainment and detrainment, and can be interpreted as inverse mixing length scales (m^{-1}).

The studies about lateral mixing processes in isolated thermals and plumes (Squires and Turner 1962; Simpson and Wiggert 1969; Simpson 1971) suggested a lateral mixing rate, ε, inversely proportional to the plume radius:

$$\varepsilon \approx \frac{\vartheta}{\text{radius}} \tag{4.46}$$

with $\vartheta \approx 0.2$. If one considers a typical cumulus radius, say 500 m, the mixing rate has the order of magnitude 10^{-4} m^{-1}. This approximation was widely used for representing the lateral mixing associated with an ensemble of *cumulus*, namely, in the MF scheme of the ECMWF model (Tiedtke 1989), where these rates were assumed as $\varepsilon = \delta = 3 \times 10^{-4}$ m^{-1}.

Siebesma and Cuijpers (1995) constructed a LES case study, from observations of the BOMEX field campaign, with the objective of better understanding the processes of lateral mixing that occurs in an ensemble of shallow cumulus. These authors concluded that the lateral mixing rates would have to be 10 times higher than those that were until then considered:

$$\varepsilon \approx 2 \times 10^{-3} \text{ m}^{-1}; \quad \delta \approx 3 \times 10^{-3} \text{ m}^{-1}. \tag{4.47}$$

These values were tested in a 1D model, improving the results for the thermodynamic properties of the cumulus ensemble (Siebesma and Holtslag 1996).

It is worthwhile to mention that this description of the lateral mixing refers to a *cumulus* ensemble and not to an individual *cumulus*. Unlike a dry thermal, where the lateral mixing seems to be the dominant mixing process, in cumulus this process has less importance when compared with the cloud-top entrainment (Stull 1988). Hence, to assume a constant lateral mixing with height is based on the fact that in a cumulus ensemble there are cumulus with the tops at different heights, where the mixing process is more relevant, and in the overall justify the constancy of the mixing rate. Siebesma (1998) suggested that the lateral mixing rate of a cumulus ensemble may be represented by the vertical distance to the cloud base, that is, $\varepsilon \approx 1/(z - z_b)$. On the other hand, through a scale analysis, Nordeng (1994) proposed that $\varepsilon \propto 1/w_c$. Other suggestions were developed to represent the lateral mixing (Grant and Brown 1999; Lin 1999; Gregory 2001), but this issue is still in debate.

Equations 4.42 and 4.45 allowed to obtain an equation that when vertically integrated can describe the MF in the cumulus layer:

$$\frac{1}{M_c} \frac{\partial M_c}{\partial z} = \varepsilon - \delta. \tag{4.48}$$

4.3.5.2 Cloud-Top Entrainment

Cloud-top entrainment is a very important physical process in the dynamics of the cloudy BL. The process of mixing in the top of cumulus consists in the incorporation of neighboring unsaturated air, resulting in the cooling of the cloud due to droplet evaporation. This process also occurs in stratocumulus, where it is even more relevant, and it is known by its relation with the onset of cloud-top entrainment instability. The region of the cloud top becomes, by this process, denser than the surrounding air, presenting negative buoyancy and consequently inducing descending air motion, leading to vertical mixing with cloudy air throughout its passage (Paluch 1979, Pontikis et al. 1987).

Blyth and Latham (1985) and Jensen et al. (1985) analyzed radiosonde data of a cumulus BL to identify the origin of the air inside the clouds. From profiles of conserved variables, Betts (1985) observed that properties of the air inside *cumulus* vary almost linearly in the vertical, between the base and the top. This distribution can be calculated by a linear combination of the properties observed in the vertical extremities of the cloud, suggesting that the air of the cloud comes from the top and the base of the cloud. This interpretation attributes a lesser importance to the process of lateral mixing in *cumulus*.

4.3.5.3 BL Top-Entrainment

The BL top-entrainment process is related to the penetration of free-atmosphere air into the CBL, enhancing the growth of the BL and playing a key role in its structure.

The mixing of warmer air of the upper stable layer into the colder BL requires the presence of a downward heat flux in the region of the inversion. Dryer and warmer air is less dense. Therefore, the negative heat flux is associated with energy consumption, presumably kept by turbulence and with a consequence in the TKE balance.

The BL top-entrainment can be understood as a conversion of TKE into potential energy, since warmer and dryer air (less dense) is incorporated in the BL. This conversion reflects the balance that is established between the turbulence and the entrainment: more TKE leads to more entrainment, but the bigger the entrainment the less TKE is available.

The source of this process is the surface buoyancy flux $\overline{(w'\theta'_v)}_s$. The process of top-entrainment also contributes, in general, for the dilution of clouds, once it mixes dry and warm air into the BL (Randall 1984). It is not possible to represent this process explicitly in large-scale and mesoscale models, where it needs to be parameterized. The K-diffusion schemes, in general, underestimate the effect of top-entrainment (Ayotte et al. 1996), and there is not yet a scheme that represents well this process for different types of BL. Additionally, the representation of this process depends highly on the resolution of the model.

4.4   EDDY DIFFUSIVITY/MASS-FLUX PARAMETERIZATION IN THE CBL

In GCMs, the sub-grid turbulent mixing is parameterized using different schemes. When cloud formation exists in the CLC, these models use, in general, an alternative parameterization for the vertical transport in the cloud layer: the MF approach, whereas the sub-cloud layer is parameterized by K-diffusion. This discontinuity seems to contribute for weak results in shallow cumulus simulations (Lenderink et al. 2004). Some MF parameterizations have been applied to the dry BL (Wang and Albrecht 1990; Randall et al. 1992), pointing out that the MF schemes are adjusted for the parameterization of the turbulent mixing due to thermals. These facts associated with the discontinuity in the treatment of the different scales of turbulent mixing suggest the development of a unifying scheme, for the representation of the turbulent mixing of heat and moisture in the BL with and without clouds.

   This section introduces the eddy diffusivity/mass-flux (EDMF) parameterization. This parameterization resulted from an original idea of Siebesma and Teixeira (2000) and was developed by Soares et al. (2004) and Siebesma et al. (2007). These different developments aimed at the implementation of EDMF in different scale models, from GCMs to LAMs. The EDMF scheme includes both a turbulent diffusion approach and a MF contribution. This combination is based on the division of the CBL turbulent mixing between the mixing done by thermals and the mixing due to smaller eddies. The nonlocal transport associated with thermals, structurally nonsymmetrical, is represented by a MF term. The local mixing due to smaller eddies is described by the diffusion contribution. As mentioned above, the need of nonlocal contribution is found to be crucial in the description of the CBL evolution.

4.4.1   The Eddy Diffusivity/Mass-Flux Scheme

The EDMF scheme is based on the decomposition of mixing scales, responsible for the subgrid turbulent transport (Figure 4.5).

   Considering an horizontal slab of the CBL divided into an area with strong thermals, with a fixed fractional area a_u, and a complementary environment, the turbulent flux of a moist conserved variable ϕ can be decomposed into three terms:

$$\overline{w'\phi'} = a_u \overline{w'\phi'}^u + (1-a_u)\overline{w'\phi}^e + a_u(w_u - w_e)(\phi_u - \phi_e), \qquad (4.49)$$

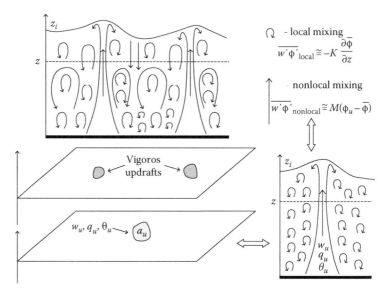

FIGURE 4.5 Schematic of the different eddy scales in the convective BL and conceptualization of the EDMF scheme.

where

　　u is related to the updrafts, and

　　e refers to the surrounding environment

The first term in the r.h.s. is associated with the in-cloud turbulence. The second term in the r.h.s. describes the turbulent mixing in the surroundings, and the third term expresses the updrafts contribution to the vertical transport of ϕ. This last term is, in the context of the MF approach, considered to be the most significant to the turbulent transport (Ooyama 1971; Betts 1973; Yanai et al. 1973).

　　If one considers Equation 4.49 and the following simplifications: (1) the area fraction occupied by thermals is very small ($a_u \ll 1$), implies that the first term on r.h.s. member can be neglected and $\phi_e \cong \bar{\phi}$; (2) $w_e \cong 0$; and (3) the environment turbulence (second term in the r.h.s.) can be described by a K-diffusion approach, and not neglected like in MF schemes. The subgrid turbulent flux is then represented by the sum of the diffusion and MF contributions:

$$\overline{w'\varphi'} \cong -K\frac{\partial\bar{\phi}}{\partial z} + M(\phi_u - \bar{\phi}), \qquad (4.50)$$

where $M = a_u w_u$ is the MF coefficient associated with the strong thermals. This expression requires the knowledge of the eddy diffusivity (ED), K; the MF coefficient, M; and the properties of the stronger thermals ensemble, ϕ_u in the BL.

　　Those functionals have been characterized by a set of LES simulations inspired in the dry BL case of Nieuwstadt et al. (1992), done with the KNMI LES model (Cuijpers and Duynkerke 1993) where a large set of diagnostics were coded to compute the properties of the BL and of the different coherent structures present in the domain. The updrafts were defined as the grid points containing a vertical velocity greater than a certain threshold.

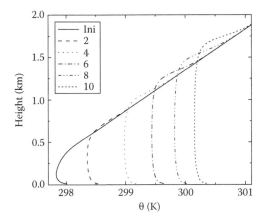

FIGURE 4.6 Temporal evolution of the potential temperature. Hourly average of the vertical profiles of a LES run.

Figure 4.6 presents the temporal evolution of the mean potential temperature in the LES simulation. A typical structure of a growing CBL can be observed. The average profile of potential temperature corresponds to an unstable surface BL, a region of a quasi-neutral mixed layer in the inner BL and a strong inversion at the top.

In what concerns the most vigorous updrafts (Figure 4.7), it can be seen in the surface BL that they present an excess of potential temperature of the order of 0.2 K, relative to the average domain value, then a monotonous decrease in potential

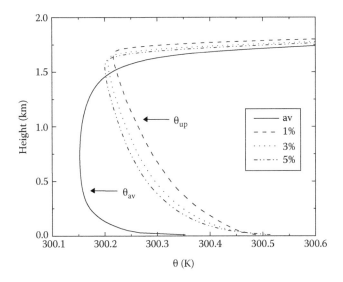

FIGURE 4.7 Vertical profiles of potential temperature: (av) horizontal domain average, 1%, 3%, and 5% average of the grid points with stronger vertical motion correspondent to the percentiles 0.01, 0.03, and 0.05, respectively. Results of the fourth hour of the LES simulation.

temperature until the inversion region, where they converge to the values of the mean environment. This decrease is related to the lateral mixing with the surrounding air that penetrates in the ascendants. Globally, the vertical evolution of this excess of the potential temperature of thermals relative to the average seems to be able to be described by an expression not too complex. However, in the inversion region it is clear that different processes, linked to top-entrainment and the penetration of thermals in the FA, come into play.

4.4.2 Mass-Flux Contribution

The contribution of MF to the total turbulent flux depends on the product of the MF coefficient, M, proportional to the thermals vertical velocity, by the difference between the ensemble thermals' properties and the horizontal domain average $(\phi_u - \overline{\phi})$. Hence, it is necessary to build an updraft model representing the properties of the updrafts ensemble responsible for the nonlocal transport, ϕ_u.

4.4.2.1 Updraft Model

The proposed updraft model follows the methodology of Betts (1973) to describe cumulus convection. Consider an air parcel moving upward in the BL with lateral mixing. From relations (Equations 4.43 and 4.48) for an updraft u, one can obtain the equation to determine its vertical structure, given by

$$\frac{\partial \phi_u}{\partial z} = -\varepsilon(\phi_u - \overline{\phi}),$$
(4.51)

where
   ε is the lateral mixing rate
   ϕ_u and $\overline{\phi}$ are, respectively, a generic property, of thermals and its horizontal domain average

This parcel model is designed to establish updraft's properties: the liquid water potential temperature, $\theta_{l\,u}$; the total specific humidity, $q_{t\,u}$; the vertical velocity, w_u; and the BL height, z_i.

Equation 4.51 needs a bottom boundary condition, that is, the initial value, to allow its vertical integration. To the initialization of the parcel one has to estimate its virtual potential temperature excess relatively to the surroundings, $\Delta\theta_{v\,u}$:

$$\Delta\theta_{vu} = \theta_{vu}(z) - \overline{\theta}_v(z) \approx f((\overline{w'\theta'_v})_s...)$$
(4.52)

It is assumed that the excess is directly related to the surface BL variability, expressed in function of the excess of $\Delta\theta_{lu}$ and Δq_{tu}. Considering a constant sensible heat flux and knowing that $\theta_{l*} = -(\overline{w'\theta'_l})_s/u_*$, $q_{t*} = -(\overline{w'q'_t})_s/u_*$, one can write

$$(\overline{w'\theta'_v})_s = -u_*\theta_* \approx b_\sigma \sigma_{\theta_v} \sigma_w,$$
(4.53)

where b_σ is a constant, σ_{θ_v}, σ_w are, respectively, the standard deviation of θ_v and of the vertical velocity, w. Siebesma and Teixeira (2000) admitted that $\Delta\theta_v \approx c\sigma_{\theta_v}$, justifying to scale $\Delta\theta_{vu}$ with the ratio between $\overline{(w'\theta'_v)}_s$ and σ_w, for any level z_1 of the surface BL:

$$\theta_{vu}(z_1) = \overline{\theta}_v(z_1) + b_i \frac{\overline{(w'\theta'_v)}_s}{\sigma_w(z_1)}. \tag{4.54}$$

The coefficient value b_i of Equation 4.54 was adjusted, through diagnostics of LES results, to 0.3. Similar expressions could be applied for the other variables, like θ_{lu} and q_{tu}. This initialization requires the knowledge of σ_w.

The vertical velocity of the thermals ensemble, w_u, is then computed through a modified version of the Simpson and Wiggert (1969) equation, adding a buoyancy term $B = g(\theta_{vu} - \overline{\phi}_v)/\overline{\phi}_v$:

$$w_u \frac{\partial w_u}{\partial z} = -\varepsilon b w_u^2 + aB, \tag{4.55}$$

where ε represents, as before, the lateral mixing rate. The presence of the coefficients a and b is discussed in several papers (e.g., Siebesma et al. 2004) and is to account approximately for the effects of the pressure perturbations and sub-plume turbulence. The values of these coefficients, still in discussion, were diagnosed from LES runs and were considered as: $a = 2.0$ and $b = 1.0$. The BL height, z_i, corresponds to the level where w_u is nil.

Following Siebesma and Cuijpers (1995), the lateral mixing rates for the dry BL were diagnosed. Expression 4.51 was applied to potential temperature, to compute the vertical profiles of the lateral mixing for each of the three referred decompositions, corresponding to the fractions 1%, 3%, and 5%. The obtained profiles can be observed in Figure 4.8, suggesting a strong relation between the lateral mixing rate, the height, and the BL height z_i, despite some dispersion. The same type of scaling was, previously, suggested by Siebesma (1998) for the lateral mixing of cumulus cores, proposing that the rate should be inversely proportional to the distance to the cloud base ($\varepsilon \approx 1/z - z_b$).

The LES results, depicted in Figure 4.8, allowed to propose an empirical expression for the solid curve, to compute the rate of lateral mixing, present in Equations 4.51 and 4.55, given by

$$\varepsilon = c_\varepsilon \left(\frac{1}{z} + \frac{1}{z_i - z} \right), \tag{4.56}$$

where $c_\varepsilon \approx 0.4$.

The last expression showed some unwanted dependency on the vertical resolution, thus a modified expression was proposed to deal with coarser resolutions:

$$\varepsilon = \max \left\{ 0, \min \left[c_\varepsilon \frac{1}{\Delta z}, c_\varepsilon \left(\frac{1}{z} + \frac{1}{z_i - z} \right) \right] \right\}. \tag{4.57}$$

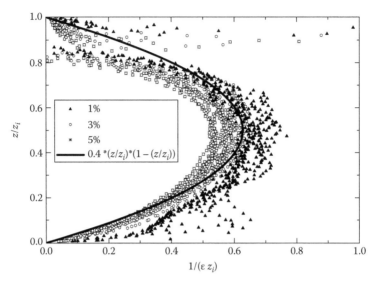

FIGURE 4.8 Triangles, circles, and stars represent the lateral mixing of thermals of fractions, 1%, 3%, and 5%, respectively. The solid line represents a best fitting function.

4.4.2.2 The Mass-Flux Profile

Figure 4.9 shows the hourly average profiles of the vertical velocity for different fractions of the strongest ascents, and the standard deviation of the vertical velocity, diagnosed by LES. The profiles support the possibility of scaling the MF coefficient

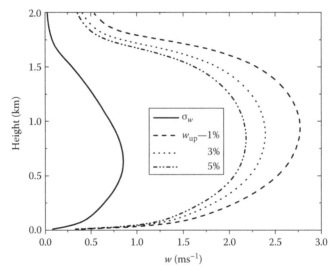

FIGURE 4.9 Vertical profile of the vertical velocity for fractions of the strongest thermals and standard deviation of the vertical velocity. Hourly average of LES runs.

of thermals, $M = a_u w_u$, with the vertical velocity standard deviation. Therefore, it was assumed

$$M \approx c_\sigma \sigma_w, \tag{4.58}$$

where $c_\sigma = 0.5$, making the M formulation dependent on the standard deviation of the vertical velocity σ_w, like in the parcel initialization.

As previously mentioned, this scheme was developed to be applied in global models, namely, the ECMWF model, hence the use of empirical expressions is appealing due to its computational economy. Accordingly, the standard deviation is computed using the empirical expression, derived from observations, tank measurements, and LES simulations (Holtslag and Moeng 1991):

$$\frac{\sigma_w}{w_*} \cong 1.26 \left[\left(\left(\frac{u_*}{w_*} \right)^3 + 0.6 \frac{z}{z_i} \right)^{\frac{1}{3}} \left(1 - \frac{z}{z_i} \right)^{\frac{1}{2}} \right], \tag{4.59}$$

where w_* is the convective velocity scale, given by

$$w_* = (g\beta \overline{(w'\theta'_v)}_s z_i)^{\frac{1}{3}}. \tag{4.60}$$

To illustrate the quality of this expression for the convective BL, Figure 4.10 shows the profiles obtained with the Expression 4.59 for three LES runs and the corresponding σ_w diagnosed profiles.

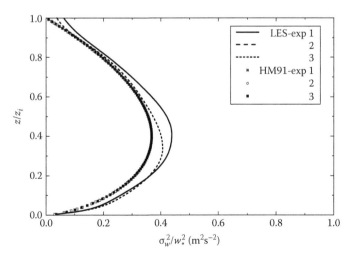

FIGURE 4.10 Vertical profiles of the vertical velocity variance scaled by the square of the convective vertical velocity, given by LES results and the Expressions 4.59 of Holtslag and Moeng (1991) (HM91).

4.4.3 The K-Diffusion Contribution

To complete the EDMF formulation, one needs to specify the eddy diffusivities for heat and moisture. Troen and Mahrt (1986) and Holtslag et al. (1995) proposed the prescription of these profiles is a simple but robust way:

$$K_h = ku_*\phi_{h0}^{-1}z\left(1 - \frac{z}{z_i}\right)^2,$$

(4.61)

where ϕ_{h0} is a stability function given by

$$\phi_{h0} = \left(1 - 39\frac{z}{L_{MO}}\right)^{-\frac{1}{3}}.$$

(4.62)

These profiles have three important properties: (1) they obey to the similarity theory in surface BL, (2) are nill in the inversion, and (3) its nondimensional maximum value is $(K_{max}/w_* z_i) \cong 0.1$.

4.4.4 EDMF-EMP Results

The parameterizations described in the previous subsection may be called EDMF-EMP (*Eddy Diffusivity/Mass-Flux EMPirirical formulation scheme*) due to the empirical base of its formulation. The following results correspond to an idealized case of a dry convective BL. This case is based on a LES model intercomparison study (Nieuwstadt et al. 1992), and the LES results presented here were obtained with the KNMI LES model (Cuijpers and Duynkerke 1993).

The surface force in this case corresponds to the prescription of a constant sensible heat flux, $\overline{w'\theta'}_s = 6.10^{-2}\,\mathrm{K\,m\,s^{-1}}$. The surface variables are $\theta_s = 300\,\mathrm{K}$ and $p_s = 1000\,\mathrm{hPa}$. The initial profile of potential temperature is displayed in Figure 4.11. In the initial state, the mean wind is negligible $(u, v) = (0.01, 0)\mathrm{ms^{-1}}$, allowing for a small but nonzero turbulent flux. Two vertical resolutions were used: 20 m (h) and the one of the ECMWF model vertical grid (lr), having approximately 300 m in the inner BL. The temporal evolution of potential temperature (hourly average) is presented in Figure 4.11. The BL growth and structure are clearly well captured.

The comparison of those profiles against LES, for the 5th and 10th hour, shows the ability of the scheme to represent the vertical structure of the convective BL. The high-resolution results are, in fact, very good. The EDMF scheme is capable of reproducing the properties of the three BL regions: the unstable surface BL, the quasineutral mixed layer, and the slightly stable upper mixed layer. These properties appear worst described by the coarser resolution simulation. Nevertheless, the temporal evolution and the vertical structure remain essentially correct (Figure 4.12).

The linear vertical profiles of the vertical flux of potential temperature, presented in Figure 4.13, are also correctly captured by EDMF-EMP. The flux in the inversion region, responsible for the BL growth, shows similar values to those of LES results,

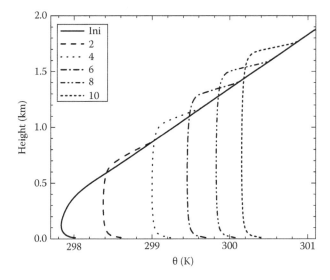

FIGURE 4.11 Temporal evolution of the potential temperature. Hourly mean profiles, with the EDMF-EMP scheme with a vertical resolution of 20 m.

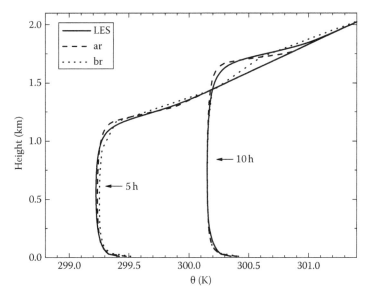

FIGURE 4.12 Vertical profiles of the potential temperature. Hourly average of the 5th and 10th hour. Results of the new scheme, (hr) 20 m resolution and (cr) ecmwf-40 levels resolution, and LES model.

in agreement with the theory of Driedonks (1982). However, in both resolutions, there is a slight underestimation of these fluxes.

Figure 4.14 depicts the total flux of potential temperature, and the values of the ED and MF terms contributing to the vertical flux of potential temperature. It can be

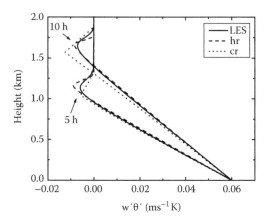

FIGURE 4.13 Vertical profiles of the vertical flux of potential temperature. Hourly average of the 5th and 10th hour. Results of the new scheme, (hr) 20 m resolution and (cr) ecmwf-40 levels resolution, and LES model.

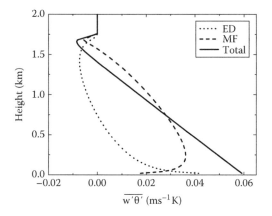

FIGURE 4.14 The ED and MF contributions for the vertical flux of potential temperature. Hourly average of the 10th hour.

seen that the MF term gives the leading contribution to the heat flux, except in the lower part of the BL. In particular, in the inversion layer this dominance assumes an important role, since it enhances considerably the buoyancy flux associated with the top-entrainment process, crucial to the BL growth. LES results (Sullivan et al. 1998) confirm the important role of thermals in top-entrainment.

In spite of the simplicity of the parcel model, in what concerns its basic equation, and of the lateral mixing formulation, the vertical profiles of vertical velocity computed with Equation 4.55 are quite close to those diagnosed through LES simulations as fractions of the stronger ascents (Figure 4.15).

4.4.4.1 Top-Entrainment

The top-entrainment is a typical interface process, present in all geophysical flows. Although crucial to the BL growth, it is still not a well-understood process and is often

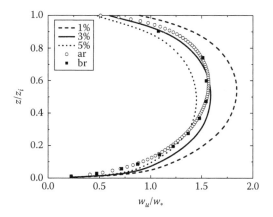

FIGURE 4.15   Vertical profiles of updraft vertical velocity. Hourly average results of the EDMF-EMP scheme: (hr) 20 m resolution and (cr) ecmwf-40 resolution; and LES diagnostics for three different fractions of the most vigorous thermals.

poorly represented in atmospheric models (Ayotte et al. 1996). In order to organize the study of the top-entrainment process, one may define a rate of top-entrainment as the difference between the time rate of BL height growth and the subsidence velocity at the inversion, w_i, expressed by the top-entrainment velocity, w_{ent}:

$$w_{ent} = \frac{dz_i}{dt} - w_i. \tag{4.63}$$

The first conceptual model to compute the top-entrainment rate of the CBL was suggested by Lilly (1968). This model is known as *zeroth-order jump model* and represents the rate at which the mixed layer air penetrates into the upper stable layer (see Figure 4.16a). For this, consider the inversion thickness δz null, z_i is the inversion height (where $\overline{(w'\theta'_v)}_{z_i}$ is minimum), the lapse rate $\gamma = \partial \overline{\phi}_v / \partial z$, and the inversion intensity is given by virtual potential temperature discontinuity at the inversion, $\Delta \overline{\phi}_{vi}$, leading to

$$\left(\frac{dz_i}{dt} - w_i \right) \Delta \overline{\theta}_{vi} = w_{ent} \Delta \overline{\theta}_{vi} = -\overline{(w'\theta'_v)}_{z_i}. \tag{4.64}$$

One of the better known approximations for the dry CBL considers the turbulent flux of virtual potential temperature at the top of the BL, $\overline{(w'\theta'_v)}_{z_i}$, as a fixed fraction, $A_{w\theta}$, of the surface flux, $\overline{(w'\theta'_v)}_s$ (Betts 1973; Carson 1973; Tennekes 1973), that is,

$$\overline{(w'\theta'_v)}_{z_i} = \overline{(w'\theta'_v)}_{min} = A_{w\theta} \overline{(w'\theta'_v)}_s. \tag{4.65}$$

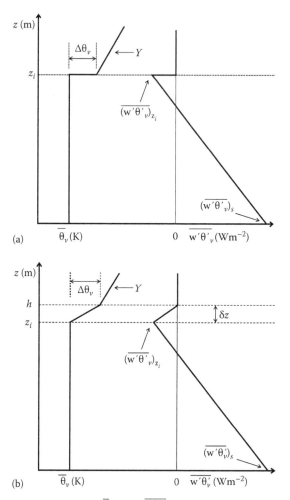

FIGURE 4.16 Vertical profiles of $\overline{\phi}_v$ and $(\overline{w'\theta'_v})_s$ in jump models: (a) zeroth-order and (b) first-order. $\gamma = \partial\overline{\phi}_v/\partial z$ in the inversion region, z_i is the inversion height, $\Delta\overline{\phi}_{vi}$ is the intensity of the inversion, and δz is the inversion thickness.

Frequently, this fraction is taken to be constant and equal to 0.2 (e.g., Stull 1976). This assumption was the basis of a parameterization used in several models. In these parameterizations, the ED in the CBL, K_{topo}, is set to satisfy that ratio between the fluxes, that is,

$$K_{topo} = -0.2(\overline{w'\theta'_v})_s \frac{\Delta z}{\Delta\theta_v}. \qquad (4.66)$$

Such approximation was implemented in the ECMWF model (Beljaars and Betts 1992).

Nevertheless, in agreement with Sullivan et al. (1998), this constant ratio between fluxes, given by Equation 4.65, is not general. If one considers Figure 4.17, presenting the temporal evolution of $\left|\overline{(w'\theta'_v)}_{min}\big/\overline{(w'\theta'_{vs})}\right|$ in a LES simulation, it can be seen that this ratio is not constant, varying approximately between 0.05 and 0.25.

Deardorff et al. (1980) carried out a number of laboratory tank experiments of thermally forced convection, aiming to investigate the relation between the top-entrainment rate and Ri, suggesting the following expression:

$$\frac{w_{ent}}{w_*} = \frac{A_{Ri}}{Ri},\qquad(4.67)$$

where $0.1 < A_{Ri} < 0.2$, Ri corresponds to a bulk value in the BL, given by $Ri = (g/\theta_{v0})\Delta\overline{\theta}_{vi}z_i/w_*^2$. This relation reveals that the mixing rate depends on the BL turbulent state and on the difference of virtual potential temperature across the interface. Turner (1973) had already suggested this type of dependency applying dimensional analysis. However, later studies did not achieve general expressions that could deal with the variability of the coefficient A_{Ri}. This inability was linked with the $\delta z = 0$ hypothesis. Additionally, it was verified that A_{Ri} varies with wind shear (Moeng and Sullivan 1994). Consequently, higher-order conceptual models were developed.

The first-order-jump model considers a finite thickness for the inversion layer, $\delta z \neq 0$ (Figure 4.16b). The height where $\overline{(w'\theta'_v)}$ is minimum is still z_i, but the height where $\overline{(w'\theta'_v)}$ is zero is $h = z_i + \delta z$. Defining $\Delta\overline{\theta}_{vi} = (\overline{\theta}_{vh} - \overline{\theta}_{vz_i})$ and $\hat{\theta} = (\theta_h - \theta_{z_i})/2$, Betts (1974) showed that:

$$w_{ent}\Delta\overline{\theta}_{vi} = -\overline{(w'\theta'_v)}_{z_i} + \delta z\frac{\partial\hat{\theta}}{\partial t}.\qquad(4.68)$$

This expression whenever $\delta z = 0$ reduces to Equation 4.64. Betts (1974), Deardorff (1979), and van Zanten (2000) developed different models based on this last Equation 4.68 without the time derivative.

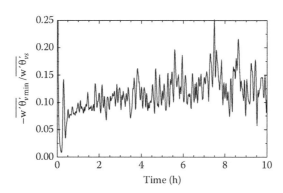

FIGURE 4.17 Temporal evolution of the ratio between the fluxes of virtual potential temperature at the BL inversion and surface. LES results.

Sullivan et al. (1998) used LES simulations to address the problems of top-entrainment and the structure of different CBLs, characterized by distinct Richardson numbers, Ri. From this study, it is relevant to emphasize two conclusions for the interval $13.6 \leq Ri \leq 43.8$: top-entrainment is mostly due to thermals and the normalized top-entrainment rate (w_{ent}/w_*) varies inversely with Ri. Dividing Equation 4.68 by w_*, simplifying the expression, and defining $A_{w\theta} = -\overline{(w'\theta'_v)}_{z_i} / \overline{(w'\theta'_v)}_s$ and $A_{\delta z} = (\delta z / \overline{(w'\theta'_v)}_s)\, (\partial\hat{\theta}/\partial t)$, one gets

$$\frac{w_{ent}}{w_*} = \frac{1}{Ri}(A_{w\theta} + A_{\delta z}). \tag{4.69}$$

Sullivan et al. (1998) considering $\left(\partial\hat{\theta}/\partial t = \partial\langle\theta\rangle_{z_i}^h / \partial t\right)$ verified the consistency of this model, stressing the need for two contributions, $A_{w\theta}$ and $A_{\delta z}$ for A_{Ri}. $A_{w\theta}$ is related to the contribution of the top buoyancy flux and $A_{\delta z}$ is associated with the inversion thickness.

Similarly, using results from LES experiments, one may diagnose the ratio w_{ent}/w_*, and the terms $A_{w\theta}/Ri$ (Figure 4.18a) and $A_{\delta z}/Ri$ (Figure 4.18b) separately, and their sum (Figure 4.18c), verifying the validity of Equation 4.69. Figures 4.18a and b show that both ratios are greater than w_{ent}/w_*, emphasizing the insufficiency of the simpler approach, supported in the representation of the top-entrainment by the ratio between minimum buoyancy flux and the surface flux. Therefore, top-entrainment parameterizations should include a nonzero inversion thickness.

The requirement of the two previous contributions for the top-entrainment rate allows establishing a parallelism with the need of the two contributions, K-diffusion and MF, for the total vertical flux in the region of the inversion. Actually, the MF contribution is preponderant in the inversion region of finite thickness, where $(\phi_u - \bar{\phi}) < 0$, taking into account the effect of the penetration of thermals into the free-atmosphere. If the inversion thickness was zero, then $A_{Ri} = A_{w\theta} = \text{cons}$ and the approximation Equation 4.66 would be sufficient to parameterize the top-entrainment. In this validation experiment, like in Sullivan et al. (1998), the term $A_{\delta z}$ contributes more than $A_{w\theta}$ to the value of A_{Ri}, suggesting that top-entrainment is not simply controlled by surface fluxes, but requires the representation of the more vigorous mixing at the inversion done by the overshooting of thermals.

4.4.4.2 Comparison with Other Approaches

The previous results illustrate the potential of the EDMF-EMP scheme. However, it is important to compare it with other schemes, namely, K-diffusion (Holtslag 1998) and K-diffusion with counter-gradient term (Holtslag and Moeng 1991). The first scheme corresponds to consider the turbulent diffusivities described by the Expressions 4.61 and 4.62. The second parameterization is modified with the introduction of a counter-gradient term:

$$\overline{w'\phi'} = -K\frac{\partial\bar{\phi}}{\partial z} + K\gamma_c, \tag{4.70}$$

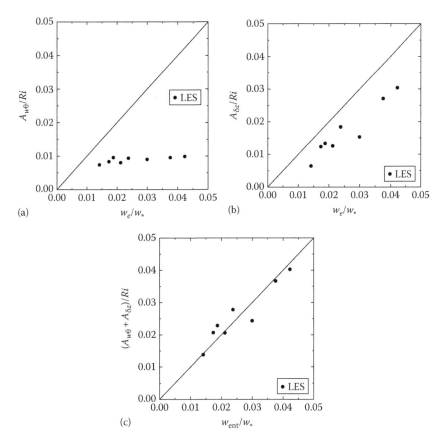

FIGURE 4.18 Comparison between the first-order-jump model and the LES results for the top-entrainment rate: (a) contribution of the minimum buoyancy flux $A_{w\theta}/Ri$, (b) contribution of the inversion thickness $A_{\delta z}/Ri$, and (c) total $(A_{w\theta} + A_{\delta z})/Ri$.

where the counter-gradient term is given by $K\gamma_c$. This approximation also considers the turbulent diffusivities given by Equations 4.61 and 4.62. The counter-gradient term obeys to

$$\gamma_c = a_\gamma \frac{w_*}{\sigma_w^2 \, z_i} \overline{(w'\theta')}_s, \tag{4.71}$$

where $a \cong 2$.

Simulations with those two schemes, considering the same initial conditions of the previous section, were made to compare its results. Figure 4.19 shows the vertical profiles of potential temperature resulting from the three different schemes and LES (fifth hour average). From the inspection of Figure 4.19, it can be seen that EDMF is the scheme showing the best agreement in all the BL sublayers.

The K-diffusion parameterization presents a typical unstable profile. The scheme with counter-gradient does not have this problem. However, the BL growth is rather

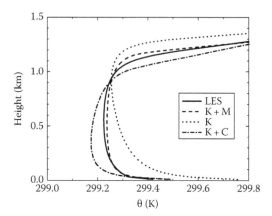

FIGURE 4.19 Vertical profiles of potential temperature from the three different parameterizations. Fifth hour average for the schemes: (K+M) EDMF-EMP, (K) K-diffusion, (K+C) K-diffusion with counter-gradient, and the LES model.

underestimated, producing insufficient top-entrainment, and the inner layer is less well mixed.

The analysis of the vertical profiles of the two contributions (K-diffusion and counter-gradient) to the total flux of potential temperature reveals the reason for the weak BL growth result of the K+C scheme (Figure 4.19). In fact, the nonlocal transport, that is, the counter-gradient term is always positive, unlike the MF contribution in the EDMF-EMP. Thus, the buoyancy flux associated with top-entrainment is reduced. In this example, the two contributions in the inversion region are almost symmetrical, provoking the inexistence of top-entrainment and the lack of BL growth. By adding nonlocal transport in the upper half BL, the counter-gradient inhibits the top-entrainment (Figure 4.20).

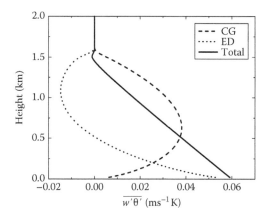

FIGURE 4.20 The eED and *Counter-gradient* (CG) contributions for the vertical flux of potential temperature. Hourly average of the 10th hour.

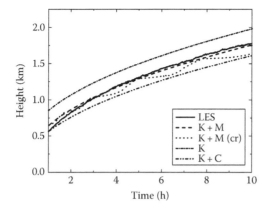

FIGURE 4.21   Temporal evolution of the BL height. Results from the three different schemes: (K + M) EDMF-EMP (cr - ecmwf-40 resolution), (K) *K*-diffusion, (K + C) *K*-diffusion with counter-gradient term, and LES model.

The temporal evolution of the BL height (Figure 4.21) computed with the three different parameterizations leads to similar conclusions. The EDMF-EMP scheme (K + M) presents the best description of this parameter in the BL, revealing the good representation of the top-entrainment effect. The *K*-diffusion scheme is too aggressive; the prescribed eddy diffusivities overestimate this process and the BL growths 200 m plus that in the LES, 10 h after the simulation starts. On the other hand, the *K*-diffusion scheme with counter-gradient (K + C) underestimates, at the same time, also by 200 m, the BL height. Even the coarser resolution results of the EDMF-EMF are quite reasonable and better than any of the other alternative schemes at a 20 m resolution. This is an important result due to its possible implementation in GCMs.

4.5   BOUNDARY LAYER CLOUD PARAMETERIZATIONS

4.5.1   INTRODUCTION

A longstanding problem in weather and climate prediction is how to estimate cloud fraction and liquid/ice water in a partially cloudy grid square. In the first attempts of climate simulation with general circulation models, climatological clouds were prescribed. But early in the 1960s relative humidity–based cloud fraction diagnostic schemes were already in use. There are today three major types of cloud parameterizations used in climate or weather prediction models: diagnostic cloud parameterizations (e.g., Slingo 1987), where the cloud fraction is diagnosed as a function of relative humidity and some other parameters; PDF-based parameterizations (e.g., Mellor 1977; Sommeria and Deardorff 1977) where the cloud fraction and liquid water are diagnosed based on assumed probability distributions for the sub-grid variability of the thermodynamic variables; and prognostic cloud parameterizations (e.g., Tiedtke 1993), where the mean liquid/ice water content and sometimes cloud fraction can be determined prognostically. In this chapter, these three different types of cloud schemes are described.

The Slingo (1987) scheme is based on diagnostic relations between cloud fraction and relative humidity. One of its main objectives was to be able to simulate stratocumulus clouds realistically. To be able to do that, the cloud fraction is also related diagnostically to the strength of the BL inversion. This scheme is described in Section 4.5.2.

The PDF-based parameterizations (e.g., Mellor 1977; Sommeria and Deardorff 1977) were not initially developed to be used in global models of the atmosphere and its initial applications were cloud-resolving models of the atmospheric boundary layer, followed by 1D BL models. Only later was this method applied to large-scale models. This scheme is described in Section 4.5.3.

Only about a decade ago did prognostic cloud schemes started to be implemented in global models. Until 1991, prognostic cloud schemes treated cloud fraction by means of diagnostic relations (e.g., Sundqvist 1988). In Tiedtke (1991) a prognostic equation for cloud fraction was proposed, but in this work only convective clouds were considered. In Tiedtke (1993), a more complete parameterization with a prognostic cloud fraction equation was presented. This scheme is described in detail in Section 4.5.4.

The prognostic cloud schemes based on the Sundqvist approach (e.g., Sundqvist 1988) have a prognostic variable for liquid/ice water only. These schemes are not described here because they were not aimed at parameterizing BL clouds. Also, Sundqvist's approach of having a prognostic equation for the liquid/ice water content and a diagnostic equation for the cloud fraction is believed to be less consistent.

To finalize this section, simplified versions of the PDF-based and the prognostic cloud parameterizations are discussed. In the following section, only the parts of the cloud parameterizations that are relevant for the BL are discussed.

4.5.2 THE SLINGO DIAGNOSTIC CLOUD SCHEME

In this section, the scheme introduced by Slingo (1987) is described in the way it was operational in the ECMWF model until 1995. This cloud scheme allows for four cloud types: convective and three layer clouds (high, middle, and low level). Clouds are computed at every model level using different relations depending on where they are located in the vertical.

The convective clouds are computed whenever the convection scheme is active and can fill any number of model layers. The convective cloud depth is determined by the convection scheme. The convective clouds C_c are determined from the convective precipitation rate P (mm day^{-1}), $C_c = 0.4 \min (0.8, 0.125 \log_e (P) - 1.5)$. The cloud base and top heights are also derived from the convection scheme.

Two types of BL clouds are parameterized: clouds associated with extratropical fronts and tropical disturbances and clouds that occur in relatively quiescent conditions and are directly associated with the boundary layer. The first class of clouds are diagnosed from relative humidity and vertical velocity:

$$C_L = \frac{\omega}{\omega_c} \left\{ \max \left[\frac{(RH - C_c) - 0.7}{0.3}, 0 \right] \right\}^2 \tag{4.72}$$

where the critical velocity ω_c is -0.1 Pa s^{-1}. $\omega/\omega c$ has to be always less or equal to 1 and if there is subsidence ($\omega > 0$) this cloud fraction is zero.

The second class of BL clouds is usually associated with subsidence and BL inversions. This is the part of the scheme that parameterizes stratocumulus. These clouds are parameterized using:

$$C_L = -10 \left(\frac{\Delta\theta}{\Delta p} \right) - 0.9 \tag{4.73}$$

where $\Delta\theta/\Delta p$ is the lapse rate K Pa^{-1} in the most stable layer below $\eta = 0.8$. An additional dependence on relative humidity at the base of the inversion can be introduced to prevent cloud forming under dry inversions such as those over deserts and the winter poles:

$$C_L = 0 \quad \text{if } RH_{base} < 0.5$$

$$C_L = C_L \left(1 - \frac{0.7 - RH_{base}}{0.2} \right) \quad \text{for } 0.5 < RH_{base} < 0.7 \tag{4.74}$$

$$C_L = C_L \quad \text{otherwise}$$

A supersaturation of 5% within the stratiform clouds and a constant liquid water content of 10^{-4} kg kg^{-1} for the convective clouds is typically assumed.

One of the main goals of the Slingo scheme was to be able to realistically simulate stratocumulus clouds. The assumption that the cloud cover in the stratocumulus areas is related to the strength of the inversion was based on observations (Slingo 1980, 1987). The association of subtropical cloud amount and inversion strength is supported by observational data at least since Von Ficker's (1936) observational campaigns in the Atlantic. Klein and Hartmann (1993) discuss the seasonal cycle of stratocumulus and this type of parameterization. On the basis of extensive climatological observations, they conclude that in all regions they have considered, except for the Arctic, the season of maximum stratus corresponds to the season of greatest lower troposphere static stability.

These observations give a good support for the Slingo parameterization. However, there is the potential for a strong positive feedback in this scheme, since the cloud-top radiative cooling strengthens the inversion, which increases the cloud amount and so on. This feedback was actually observed and is discussed in Slingo (1987). In any case, the Slingo parameterization produced, probably for the first time in global atmospheric models, a realistic simulation of stratocumulus cloud cover.

4.5.3   PDF-BASED CLOUD PARAMETERIZATIONS AND A SIMPLE ITERATIVE VERSION

PDF-based cloud schemes basically assume a statistical description of the sub-grid scale condensation processes. Traditionally, in cloud-resolving models (CRMs), condensation processes are modeled using an "all-or-nothing" approach, where condensation only occurs when the mean specific humidity reaches its saturation value. However, if the mean specific humidity is close to saturation, regions of saturation can probably be found along with regions of subsaturation. In theory, it can be assumed that the cloud fraction and the mean liquid water content can be defined as

$$\bar{l} = \int\int lG(q_t,\theta_l)dq_t d\theta_l$$

$$a = \int\int H(q_t - q_s)G(q_t,\theta_l)dq_t d\theta_l \tag{4.75}$$

where
$G(q_t, \theta_l)$ is a PDF
$H(x)$ is the Heaviside function
a is the cloud cover
l is the liquid water content
q_t is the total specific water
θ_l is the liquid water potential temperature
The overbar represents the mean

To calculate the values of cloud fraction and mean liquid/ice water, it is then necessary to compute the integrals from the prognostic variables: total specific water and liquid water potential temperature. These variables are used because they are conserved during condensation processes (e.g., Deardorff 1976). In their original paper, Sommeria and Deardorff (1977) propose a scheme that needs the evaluation of double integrals. Mellor (1977) showed that the problem can be reduced to the integration over one variable. After some algebra (e.g., Mellor 1977), the integrals can be written as a function of a variable s, which is in the phase space (q_t,θ_l) the coordinate along the transect perpendicular to the tangent of the Clausius–Clapeyron equation (see Mellor (1977) for details).

Different types of distributions have been proposed for $F(s)$. Initially, Sommeria and Deardorff (1977) and Mellor (1977) suggested a Gaussian distribution:

$$F(s) = \frac{1}{\sqrt{2\pi}}\exp\left(-\frac{s^2}{2}\right) \tag{4.76}$$

With a Gaussian PDF, it is possible to obtain simple relations for the integrals that give the cloud fraction and the mean liquid water as functions of the ratio between the moisture deficit (the difference between the mean total water and the saturation-specific humidity) and the variance of s, that is related to the variance of the total water content and liquid water potential temperature.

Assuming, for simplicity, that using a Gaussian distribution is acceptable and taking into account that the mean values are produced by the model, what remains to be determined is the variance. To estimate temperature and water variances, prognostic equations can be obtained using Reynolds decomposition and averaging, with three important terms in the r.h.s.: a term representing the production of variance, a second term representing the transport, and a third term the dissipation of variance.

A simplification of these prognostic equations is often used as a first approximation, where steady state is assumed and the third-order transport term is neglected. This simplification allows to go from a partial differential equation to an algebraic equation in determining the variance as a function of the sub-grid vertical flux and the vertical gradient of the mean temperature or water variable.

The major issue with PDF-based methods is how to determine the characteristics of the PDF itself. As briefly discussed above, in terms of the model, it is possible (even if not always feasible), using Reynolds decomposition and averaging, to obtain equations for the variance and for the skewness of the PDF. What is still unclear is how the PDFs look like for different physical situations and different types of clouds.

Aircraft observations (e.g., Larson et al. 2001) and LES models (e.g., Cuijpers and Bechtold 1994) support the idea that for most stratus clouds, a Gaussian PDF is a realistic approximation. However, for cumulus clouds, the skewness of the PDF may well play an important role in determining the cloud properties. For example, Bougeault (1981) analyzed the results produced by a 3D LES model, in order to produce a PDF that fitted the data better for trade cumulus situations. The suggested PDF was

$$F(s) = H(s+1)\exp[-(s+1)] \tag{4.77}$$

PDF-based cloud parameterizations are based on reasonably solid physical and mathematical concepts and as such offer promise for more realistic cloud predictions in weather and climate prediction models. However, for many years since the earlier studies that were discussed above, PDF-based cloud parameterizations were not implemented in global climate or weather prediction models. Notable exceptions include very much simplified versions by Smith (1990), who suggests a simpler triangular distribution, LeTreut and Li (1988), and LeTreut (1989). Recently, there has been a revival of these ideas as the studies from Bony and Emanuel (2001), Tompkins (2002), and Chaboureau and Bechtold (2002) illustrate.

In what follows, we describe in some more detail the approach taken by Teixeira and Hogan (2002) in adapting, simplifying, and implementing in a global model the PDF-based cloud parameterization suggested by the LES studies of Cuijpers and Bechtold (1994) (hereafter CB95). In CB95, it is suggested that cloud fraction can be diagnosed as

$$a = 0.5 + \alpha \arctan(\gamma Q) \tag{4.78}$$

where $\alpha = 1/\pi$, $\gamma = 1.55$, $Q = q_t - q_s/\sigma$. In Teixeira and Hogan (2002), hereafter TH02, it is assumed that the standard deviation σ can be determined as $\sigma = \lambda\sqrt{q_t'q_t'} = \eta\sqrt{\theta'\theta'}$, where λ and η are constants of proportionality and θ is the potential temperature. The variance of potential temperature is obtained from the steady-state version (neglecting the transport term) of the prognostic variance equation (e.g., Stull 1988) assuming an eddy-diffusivity approach for the flux. Assuming that the turbulent diffusion coefficient is simplified to $k = (1-a)k_d + ak_c$, where k_d and k_c are constants and are respectively the turbulent diffusion coefficients for the dry and the cloudy parts, and that $q_t = al_c + q = a\beta q_s + q$, the following iterative equation to diagnose cloud fraction is obtained:

$$a_{i+1} = 0.5 + \alpha \arctan\left(\gamma \frac{q_s(\beta a_i - (1-RH))}{B\sqrt{(1-a_i)k_d + a_i k_c \left|\frac{\partial \theta}{\partial z}\right|}} \right) \qquad (4.79)$$

where B is a constant.

This iterative equation converges rapidly, normally in less than 10 iterations, for realistic values of the terms in the r.h.s. (see TH02). TH02 shows that versions of this approach can lead to successful simulations of stratocumulus cloud cover in a global model with significant impacts on the ocean surface solar radiation budget.

4.5.4 A PROGNOSTIC CLOUD FRACTION PARAMETERIZATION AND A STEADY-STATE VERSION FOR BOUNDARY LAYER CUMULUS

Prognostic cloud schemes started to be implemented in global climate and weather prediction models only recently and many of them treat cloud fraction by means of diagnostic relations (e.g., Sundqvist 1988). In Tiedtke (1991, 1993), a new approach with prognostic equations for both cloud fraction and mean liquid/ice water content was presented. In practice, this parameterization was inspired by the wish to couple the sources of liquid water and cloud fraction to the moist convection scheme.

A more complete description of the entire parameterization is given in Tiedtke (1993). The scheme consists of two prognostic equations:

$$\frac{\partial \overline{l}}{\partial t} = A_l + CV_l + BL_l + C - E - P - ENT$$

$$\frac{\partial a}{\partial t} = A_a + CV_a + BL_a + CL_a - D$$

where
A_l and A_a represent the advection of liquid/ice water and cloud fraction
CV_l and CV_a represent the sources of liquid water/ice and cloud fraction from the moist convection processes
BL_l and BL_a represent the source terms due to BL turbulence
C is the condensation/sublimation rate
E is the evaporation rate
P is the conversion of liquid/ice water into precipitation
ENT is the destruction of liquid/ice water due to cloud-top entrainment
C_a is the generation of cloud fraction by stratiform condensation
D is the dissipation of cloud fraction by turbulent erosion

The prognostic cloud scheme allows the ice and the liquid water to coexist (mixed phase) based on a simple quadratic temperature dependence between 0°C and –23°C.

Here, we concentrate on the usefulness of this parameterization for BL shallow cumulus situations. It has been shown (Teixeira and Miranda 2001) that, for cumulus, the prognostic cloud fraction equation of Tiedtke (1993) is typically dominated by the production of clouds by detrainment from shallow convection and the erosion of clouds by turbulent mixing with the environment. This relationship can be written approximately as

$$\frac{\partial a}{\partial t} = D(1-a) - \frac{a}{l_c} K(q_s - q) \qquad (4.81)$$

where
D is the detrainment rate
l_c is the in-cloud liquid water content
q_s is the saturation specific humidity
K is an erosion coefficient

Moreover, in typical shallow convection situations, the two terms in the r.h.s. are of similar magnitude and the cloud fraction tendency is negligible in comparison. In this case, a simple diagnostic relation (Teixeira 2002, TH02) to calculate cloud fraction as a function of relative humidity, detrainment, and erosion can be written as

$$a = \frac{D}{D + \dfrac{K}{\beta}(1 - \mathrm{RH})} \qquad (4.82)$$

where RH is the relative humidity, and we assume $l_c = \beta q_s$ (e.g., ECMWF 1991).

In TH02, this approach was implemented in a global atmospheric model and was shown to produce realistic results for shallow cumulus cloud cover. At each time-step, the shallow cumulus cloud parameterization uses information from the convection parameterization to set the height of the cloud base and cloud top. In TH02, it was decided to use constant values of D, as opposed to Tiedtke (1993) where only the erosion coefficient is taken as constant, because it was found (Teixeira 2001) that it is more important to maintain a realistic balance between the creation of clouds due to detrainment and destruction due to the erosion, than to use the value of detrainment estimated by the convection scheme at each time-step. In this way, the clouds are coupled to the convection scheme but are independent of the details of the convection parameterization. The results from Tiedtke (1993) and TH02 are promising in terms of the applicability of this particular type of parameterization to moist shallow convection clouds.

4.6   EDMF PARAMETERIZATION OF SHALLOW CUMULUS

4.6.1   INTRODUCTION

The development of a scheme for the dry and the shallow cumulus BLs was carried out in Soares et al. (2004). The scheme was implemented in the MesoNH research model (Lafore et al. 1998), taking advantage of the eddy-diffusivity turbulence closure scheme based on the TKE budget equation of Cuxart et al. (2000) (CBR hereafter). The MesoNH model has a convection scheme based on the Kain and Fritsch (1993) bulk MF convection parameterization for deep and shallow convection (Bechtold et al. 2001).

The shallow cumulus representation is an extension of the dry EDMF scheme, when an ascending parcel condensates. Therefore, the updraft model may also be used as a trigger function for the initiation of deep convective clouds.

The updraft originates at the surface layer and ascends entraining environmental air, up to the level where its vertical velocity vanishes. During the ascent, the occurrence of oversaturation is checked, in which case condensation takes place in the entraining rising plume. The condensation test used is based on the algorithm proposed by Davies and Jones (1983). Cloud base height is in that case defined as the height of the lifting condensation level, while cloud-top height is defined as the level above the lifting condensation level at which the vertical velocity vanishes according to Equation 4.55. The cloud model aims to represent an ensemble of convective shallow *cumulus* clouds.

The EDMF scheme proposed by Soares et al. (2004) showed some improvements in the description of the dry BL when compared with the version presented in Section 4.4.4 (EDMF-EMP). The main modifications in the EDMF formulation for the dry BL are related to the parcel initialization and the MF coefficient.

Once it is aimed to describe both the dry and shallow BLs, the rising parcel is designed to determine the updraft liquid water potential temperature, total specific humidity and vertical velocity, and BL height. In order to initialize the parcel, it is necessary to estimate its excess virtual potential temperature, which is a function of the surface layer variability. Soares et al. (2002), in close agreement with Troen and Mahrt (1986), showed that the excess scales very well with the ratio between the surface heat flux and $e^{1/2}$, for any level k in the surface layer:

$$\theta_{vu}(z_k) = \overline{\theta_v}(z_k) + \beta \frac{\overline{(w'\theta_v')}_s}{e^{1/2}(z_k)}, \qquad (4.83)$$

where the value of the coefficient β was adjusted to 0.3. Similar equations apply to other properties such as θ_{lu} and q_{tu}. This initialization requires the solution of the TKE prognostic equation and is rather insensitive to the initialization level k.

If Equation 4.55 is a reasonable approximation to the ensemble vertical velocity, it may be directly used to compute M from its definition, once a_u has been determined. It was proposed that $a_u = 0.1$, in agreement with the diagnosed value of the average horizontal area containing buoyant updrafts (Lenschow and Stephens 1980, van Ulden and Siebesma 1997). This way the parameterization is able to represent the overshooting process. Instead of considering z_i as the level where the buoyancy is zero, the vertical velocity equation is used to determine the level where w_u vanishes.

Compared to the updraft model presented before, this latter model has some advantages: (1) it is less sensitive to the initialization level, z_k, or to model resolution; (2) the initialization is directly based on the TKE budget equation, as the eddy-diffusivity formulation; (3) the MF profile avoids the use of an empirical formula for σ_w (Holtslag and Moeng 1991) and requires almost no tuning; and, additionally; and (4) the MF profile can be directly interpreted in the light of the standard conceptual picture of turbulence in the CBL.

For the shallow cumulus BL, Soares et al. (2004) followed the bulk mass flux scheme, where the mass flux vertical profile is given by the cloud-core continuity Equation 4.48 where M_c is the cloudy updraft MF. To integrate it, one needs boundary conditions for M_c (the cloud base MF) and the fractional entrainment and detrainment rates. The thermodynamic structure of the cloud is given by Equation 4.51.

The cloud base closure needs particular care since it prescribes the initial cloudy plume properties and represents the ventilation of the subcloud layer (Tiedtke et al. 1988). Betts (1976) introduced a closure of the MF at cloud base to describe the coupling between the two layers. Neggers et al. (2004) examined three different closures for the MF at the cloud base for a diurnal cycle of shallow cumulus convection. They concluded that the convective sub-cloud velocity scale closure (Grant 2001) captures the coupling between the two layers at cloud base, reproducing the timing of both the maximum and the final decrease in the cloud base MF in LES results. This type of closure relies on the relationship between the cloud base MF and the TKE in the subcloud layer. For these reasons, the cloud base mass flux was taken as the product of the core fraction by the updraft vertical velocity ($M_c = a_{co} w_u$). The core fraction was estimated as 10% of the cloud fraction, $a_{co} = 0.1 a_c$, where the cloud fraction a_c is given by the sub-grid condensation scheme, all values computed at cloud base. The values of entrainment and detrainment rates chosen were Equation 4.47.

The diagnostic of cloud cover and cloud water mixing ratio is a crucial component of any NWP or mesoscale model due to its potential impacts on the radiation budget. MesoNH has a statistical sub-grid condensation scheme, based on the distributions of the grid scale values of θ_l and q_t, and their variances, which are supplied by the general turbulence scheme (Sommeria and Deardorff 1977; Cuijpers and Bechtold 1994). Consistent with the EDMF, the evaluation of those variances have both the ED and MF contributions. Following Lenderink and Siebesma (2000), the variance of a conserved variable ϕ was computed according to

$$\overline{\phi'^2} \cong 2\tau_\phi K \left(\frac{\partial \overline{\phi}}{\partial z} \right)^2 - 2\tau_\phi M (\phi_u - \overline{\phi}) \frac{\partial \overline{\phi}}{\partial z} \tag{4.84}$$

where the two terms in the r.h.s. account, respectively, for the ED and MF contributions, assuming for simplicity that $\tau_\phi = 600\,\text{s}$ is a typical eddy-turnover time (Cheinet and Teixeira 2003).

4.6.2 EDMF RESULTS

4.6.2.1 Dry BL

The idealized case presented here differs slightly from the previous case, once humidity is taken into account. The surface force in this case corresponds to the prescription of constant latent and sensible heat fluxes, $\overline{w'q'}_s = 2.5 \times 10^{-5}\,\mathrm{m\,s^{-1}}$ and $\overline{w'\theta'}_s = 6.10^{-2}\,\mathrm{K\,m\,s^{-1}}$, respectively. The surface variables are $\theta_s = 300\,\mathrm{K}$, $q_s = 5\,\mathrm{g\,kg^{-1}}$ and $p_s = 1000\,\mathrm{hPa}$. The initial profiles of potential temperature and humidity can be summarized:

$$\theta = 300\,\mathrm{K}, \quad \partial q_t/\partial t = -3.7 \times 10^{-4}\,\mathrm{km^{-1}} \quad 0 < z < 1350\,\mathrm{m}$$

$$\partial\theta/\partial z = 2\,\mathrm{K\,km^{-1}}, \quad \partial q_t/\partial t = -9.4 \times 10^{-4}\,\mathrm{km^{-1}} \quad z > 1350\,\mathrm{m}$$

Simulations were done with the one-dimensional version of MesoNH, where the EDMF scheme was implemented. A constant vertical resolution of 20 m (as in the LES simulations) and a 60 s time step were used.

In general, the results were better than those produced with the EDMF version presented before. The results revealed a significant improvement of the MesoNH model with the EDMF closure, especially in what concerns the evolution of the top inversion, but also in the vertical structure of the mixed layer. The EDMF scheme showed improvements in BL features, associated with a closer agreement of the over-all profiles. Most important is the fact that EDMF is capable of reproducing a slightly stable upper mixed layer and a much better BL growth. The improvement of the rep-resentation of the average properties of the BL is, obviously, related to a better repre-sentation of the vertical fluxes. The time evolution of the vertical fluxes of potential temperature (not shown) and total specific humidity (Figure 4.22) illustrates this enhanced transport. In particular, the heat flux shows a realistic linear profile with a top-entrainment ratio $\left|\min(\overline{w'\theta'})/(\overline{w'\theta'})_s\right|$ of 0.17, on spot with LES results. In what concerns the humidity fluxes, there is again a clear improvement in the profiles, associated with a better BL growth, but some discrepancies remain, and they are responsible for the moister BL found in both 1D simulations.

In this version, the MF contribution for the potential temperature is enhanced. Figure 4.23 shows the values of the ED and the MF terms contributing to the poten-tial temperature vertical flux. It can be seen that the MF term gives the leading

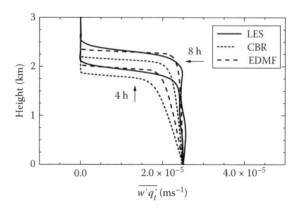

FIGURE 4.22 Hourly averaged specific humidity flux profiles. Results from the EDMF scheme, Cuxart et al. (2000) (CBR) scheme, and KNMI LES.

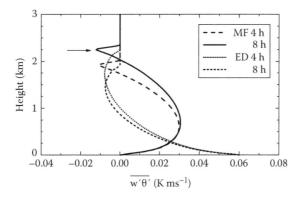

FIGURE 4.23 The ED and MF contributions to the potential temperature vertical flux. Hourly average results from the new scheme at fourth and eighth hour.

contribution to the heat flux, except in the lower part of the BL. This is in agreement with the LES study performed by Ebert et al. (1989) where it is concluded that mass and heat transport spectra showed the relatively minor contribution made by small-size eddies as compared with medium and large thermals. On the other hand, the MF term dominates over the eddy-diffusivity term in the upper half of the mixed layer, implying an upward (counter-gradient) heat flux across that region. In the inversion, the MF term is also dominant and the fact that it goes to zero slightly above the eddy-diffusivity profile is an indication of overshooting of thermals.

Figure 4.24 shows the BL height evolution, defined as the level of minimum buoyancy flux, for both the LES and CBR scheme, and the height where the vertical velocity vanishes, in the EDMF scheme. Clearly, results of the EDMF scheme are in a better agreement with LES results than with the CBR scheme.

4.6.2.2 Shallow Cumulus BL

The results presented here concern to a case based on an idealization of observations made in the framework of the ARM experiment. This case corresponds to a diurnal

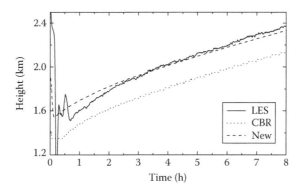

FIGURE 4.24 Time evolution of the PBL height. Hourly average results from the EDMF scheme, CBR scheme, and KNMI LES.

cycle of cumulus clouds over land. Lenderink et al. (2004) presented a single-column intercomparison study based on this case. This intercomparison was based on several single column versions of (semi)-operational models. The MesoNH was one of the models included, using its standard options, referred above. Like the majority of the other models, MesoNH produced too large values of cloud cover, and unlike the other models too small values of cloud liquid water.

In this case, surface latent and sensible heat fluxes were prescribed, with values close to zero in early morning and the evening, and a maximum at midday of 500 and 140 Wm^{-2}, respectively. The initial profiles of potential temperature and humidity are displayed in Figure 4.27. The mean wind initial vertical profiles correspond to $(u, v) = (10, 0)$ ms^{-1}.

Figures 4.25 and 4.26 show the time series of cloud properties. The EDMF scheme captures well the diurnal cycle of shallow convection. Both the timing of the onset and disappearance of the cumulus clouds are well described. The cloud cover and liquid water path (LWP), given by the sub-grid condensation scheme, are in a rather good agreement with LES results, revealing both the proper time evolution and order of magnitude. The maximum of cloud cover (Figure 4.25a) around 0.2 is very close to the LES simulation result, and is reached after 18 UTC (12 LT) only slightly later than in the LES. The cloud base height also agrees well with LES results. The diagnostic of cloud base height given by the updraft condensation level ("new up" in Figure 4.25b) indicates the onset of cumulus clouds at 15.30 UTC (9.30 LT), above 700 m. The cloud base goes up in time, attaining a maximum of around 1200 m after 24.00 UTC (18 LT).

Some of the models that participated in the intercomparison study of Lenderink et al. (2004) were able to reproduce the cumulus onset time. However, most models were unable to correctly predict the dissipation of clouds by the end of the day. With the EDMF scheme, the cumulus field dissolves at approximately the correct time, and shows a low level of intermittency. In its standard version, MesoNH had a rather high cloud fraction (around 50%) and almost no liquid water during the simulation. These features appear much improved mainly due to a better estimate of the variance, associated with the contribution of the MF term in Equation 4.84.

Figure 4.25c shows that the cloud scheme is able to reproduce the height of maximum cloud cover given by the LES model. The cloud-top height (Figure 4.25d) is underestimated, probably due to the over simplistic constant entrainment profile used. The LWP results in Figure 4.26a show a realistic evolution but with a slight underestimation by the EDMF model. The cloud base MF time series shown in Figure 4.26b resembles the one diagnosed from LES by Neggers et al. (2004).

The vertical profiles of potential temperature and humidity are shown in Figure 4.27. The subcloud layer potential temperature evolution is very much realistic. The specific humidity is too high in the sub-cloud layer due to insufficient transport into the cloud layer. This deficiency is consistent with the cloud fraction and liquid water profiles displayed in Figure 4.28; they tend to have the right order of magnitude but they lack sufficient vertical extension. This is also reflected in the potential temperature profile in the cloud layer (Figure 4.27a). The discrepancy again is the result of insufficient turbulent mixing in the cloud layer.

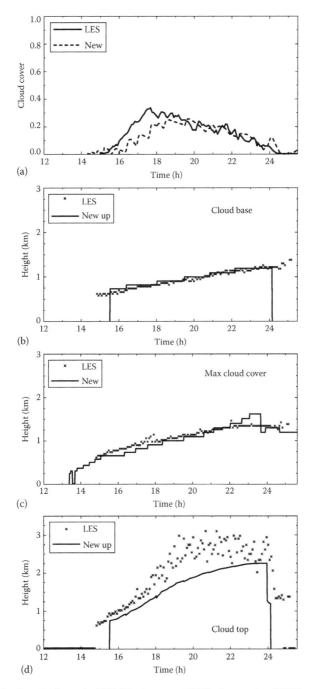

FIGURE 4.25 Results from the EDMF scheme and LES, time series (UTC) of (a) the cloud cover, (b) the height of cloud base, (c) the height of maximum cloud cover, and (d) the height of cloud top.

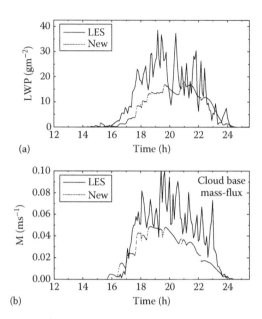

(a)

(b)

FIGURE 4.26 Time series (UTC) of (a) the cloud LWP and (b) the cloud base mass-flux. EDMF scheme and LES results.

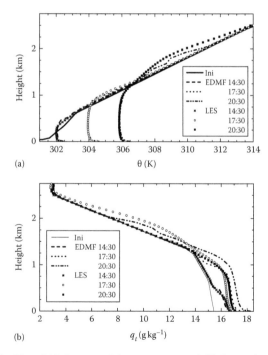

(a)

(b)

FIGURE 4.27 Profiles of (a) the potential temperature and (b) the total specific humidity, at 14th, 16th, 18th, 20th, and 22nd hour UTC. Hourly average results from the EDMF scheme and KNMI LES.

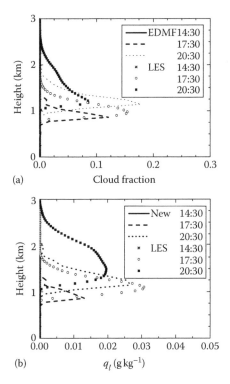

(a)

(b)

FIGURE 4.28 Profiles of (a) the cloud fraction and (b) the liquid water at 14th, 16th, 18th, 20th, and 22nd hour UTC. Hourly average results from the EDMF scheme and KNMI LES.

4.7   FINAL REMARKS

In this chapter, a review of some of the most common methods for parameterizing turbulent and convective mixing in the atmospheric BL is presented, together with a short review on the different types of cloud parameterizations utilized in global climate and weather prediction models. In particular, it is shown in detail that simple combinations of ED and MF closures can realistically describe the turbulent transport in the CBL.

The main advantage of this eddy diffusivity/mass-flux (EDMF) parameterization is that it allows for the unification of parameterizations for the clear CBL and the moist shallow cumulus boundary layer. No switching to a separate convection scheme is then necessary, until deep convection is triggered. This approach has the conceptual advantage that the whole CBL is parameterized by a single scheme that has local (diffusive) and nonlocal (MF) closures in both the cloud and sub-cloud layers.

With EDMF, the clear CBL is rather well represented, with a good prediction of both the time evolution and the vertical development of the mixed layer. The potential temperature profiles clearly show the typical mixed layer shape: an unstable surface layer, followed by a well-mixed mid-BL and a slightly stable layer below the inversion. The improved behavior of the CBL structure is associated with a better representation of counter-gradient fluxes and top-entrainment, including the effect of thermals overshooting. Note that recent ED schemes were developed based on the TKE prognostic equation and a definition of the mixing length that involves the square root of TKE and a timescale (Cheinet and Teixeira 2003; Teixeira and Cheinet 2004). This type of parameterization can produce a realistic top-entrainment, but the instability in the lower BL (although small) still persists.

The analysis of the performance of the EDMF scheme for the cumulus diurnal cycle also shows good results. The onset, dissipation time, and cloud cover of cumulus clouds are all well captured. The same applies to the vertically integrated quantities, namely, the LWP. Nevertheless, there are still some unsolved problems in the cloud layer properties, requiring further research. While the EDMF scheme seems to lead to realistic potential temperature profiles, there remain some discrepancies in the mean humidity profile and significant problems in the momentum profiles. As the momentum balance was not modified in the EMDF scheme presented here, it may well be that a nonlocal effect needs to be added to those equations.

It is worthwhile to emphasize that a rather simple entraining parcel model is used to diagnose the different properties of the ensemble of updrafts, including the mean profile of vertical velocity and the condensation level. This was found to provide the necessary information for the computation of the MF contribution. The cloud base MF closure allows for a direct link between the convective state of the sub-cloud layer and the cloud layer. This is based on the concept that clouds are actually the visible extensions of thermals that penetrate into the stable layer.

Finally, an important operational advantage of the EDMF scheme is associated with a significant reduction in model computational costs in moist conditions. In the dry BL case, the inclusion of the EDMF scheme in the MesoNH model implied an increase in its computational cost of about 7%, when compared to the standard version of the same model with a TKE scheme. However, in the shallow moist convection case, where the standard MesoNH also uses the shallow convection scheme of Bechtold et al. (2001), EDMF leads to a substantial reduction in computational cost to 30% of its standard value. This important reduction leaves room for further improvements of the code.

As mentioned, EDMF approaches represent a step toward the unification of turbulence and convection parameterizations in weather and climate prediction models. Hopefully, it will also allow for a better understanding of the physical processes associated with the CBL.

ACKNOWLEDGMENTS

The authors are grateful to the editors for the opportunity to write this chapter and for their encouragement in all the writing process. This work has been supported by Fundação para a Ciência e Tecnologia (FCT) under project REWRITE, PTDC/CLI/73814/2006, cofinanced by the EU under Program FEDER.

REFERENCES

Ackerman, S. A. and S. K. Cox. 1981. Aircraft observations of the shortwave fractional absorptance of non-homogeneous clouds. *J. Appl. Meteorol.* 20:1510–1515.

Agee, E. M. 1984. Observations from space and thermal convection: A historical perspective. *Bull. Amer. Meteorol. Soc.* 65:938–949.

Albrecht, B. A., C. S. Bretherton, D. Johnson, W. H. Schubert, and A. S. Frisch. 1995. The Atlantic stratocumulus transition experiment—ASTEX. *Bull. Am. Meteorol. Soc.* 76:889–903.

Albrecht, B. A., D. A. Randall, and S. Nicholls. 1988. Observations of marine stratocumulus clouds during FIRE. *Bull. Amer. Meteorol. Soc.* 69:618–626.

André, J. C., G. De Moor, P. Lacarrère. G. Thérry, and R. du Vachat. 1978. Modeling the 24-hour evolution of the mean and turbulent structures of the planetary boundary layer. *J. Atmos. Sci.* 35:1861–1883.

André, J.-C., J.-P. Goutorbe, and A. Perrier. 1986. HAPEX-MOBILHY, a hydrologic atmospheric pilot experiment for the study of water budget and evaporation flux at the climate scale. *Bull. Am. Meteorol. Soc.* 67:138–144.

Arakawa, A. and W. H. Schubert. 1974. Interaction of a cumulus cloud ensemble with the largescale environment. Part I. *J. Atmos. Sci.* 31:674–701.

Ayotte, K. M., P. P. Sullivan, A. Andren, C. S. Doney, A. A. M. Holtslag, W. G. Large, J. C. McWilliams et al. 1996. An evaluation of neutral and convective planetary boundary layer parameterisations relative to large eddy simulations. *Bound. Layer Meteorol.* 79:131–175.

Bechtold, P., E. Bazile, F. Guichard, P. Mascart, and E. Richard. 2001. A mass flux convection scheme for regional and global models. *Q. J. R. Meteorol. Soc.* 127:869–886.

Beljaars, A. and A. Betts. 1992. Validation of the boundary layer representation in the ECMWF model. In *Proceedings ECMWF Seminar on the Validation of Models over Europe.* ECMWF, Reading, U.K., pp. 159–196.

Berkowicz, R. and L. P. Prahm. 1979. Generalization of K-theory for turbulent diffusion. Part 1: Spectral turbulent diffusivity concept. *J. Appl. Meteorol.* 18:266–272.

Betts, A. K. 1973. Non-precipitating cumulus convection and its parameterization. *Q. J. R. Meteorol. Soc.* 99:178–196.

Betts, A. K. 1974. Reply to comment on the paper Non-precipitating cumulus convection and its parameterization. *Q. J. R. Meteorol. Soc.* 100:469–471.

Betts, A. K. 1976. Modeling subcloud layer structure and interaction with a shallow cumulus layer. *J. Atmos. Sci.* 33:2363–2382.

Betts, A. K. 1985. Mixing line analysis of clouds and cloudy boundary layers. *J. Atmos. Sci.* 42:2751–2763.

Blackadar, A. K. 1962. The vertical distribution of wind and turbulent exchange in a neutral atmosphere. *J. Geophys. Res.*, 67:3095–3102.

Blyth, A. M. 1993. Entrainment in cumulus clouds. *J. Appl. Meteorol.* 32:626–641.

Blyth, A. M. and J. Latham. 1985. An airborne study of vertical structure and microphysical variability within small cumulus. *Q. J. R. Meteorol. Soc.* 111:773–792.

Bony, S. and K. A. Emanuel. 2001. A parameterization of the cloudiness associated with cumulus convection; evaluation using TOGA COARE data. *J. Atmos. Sci.* 58:3158–3183.

Bougeault, P. 1981. Modeling the trade-wind cumulus boundary layer. Part I: Testing the ensemble cloud relations against numerical data. *J. Atmos. Sci.* 38:2414–2428.

Boussinesq, J. 1877. Essai sur la théorie des eaux courantes. *Mem. Academie de Sci. de Paris.* 23:1–680.

Brown, A. R., R. T. Cederwall, A. Chlond, P. G. Duynkerke, J. C. Golaz, J. M. Khairoutdinov, D. C. Lewellen, et al. 2002. Large eddy simulation of the diurnal cycle of shallow cumulus convection over land. *Q. J. R. Meteorol. Soc.* 128:1075–1094.

Businger, J. A. and S. P. Oncley. 1990. Flux measurement and conditional sampling. *J. Atmos. Oceanic Tech.* 7:349–352.

Carson, D. J. 1973. The development of a dry inversion-capped convectively unstable boundary layer. *Q. J. R. Meteo. Soc.* 99:450–467.

Chaboureau, J. P. and P. Bechtold. 2002. A simple cloud parameterization derived from cloud resolving model data: Diagnostic and prognostic applications. *J. Atmos. Sci.* 59:2362–2372.

Chatfield, R. B. and R. B. Brost. 1987. A two-stream model of the vertical transport of trace species in the convective boundary layer. *J. Geophys. Res.* 92:13262–13276.

Cheinet, S. and J. Teixeira. 2003. A simple formulation for the eddy-diffusivity parameterization of cloud-topped boundary layers. *Geophys. Res. Lett.* 30:18–19.

Clarke, R. H., A. J. Dyer, R. R. Brook, D. G. Reid, and A. J. Toup, 1971: *The Wangara Experiment: Boundary Layer Data.* Technical paper No. 19, Division of Meteorological Physics, CSIRO, Australia, 21 pp.

Clement, A. and R. Seager. 1999. Climate and the tropical oceans. *J. Climate.* 12:3383–3401.

Cuijpers, J. W. M. and P. Bechtold. 1994. A scale and skewness independent parameterisation of cloud water related variables. *J. Atmos. Sci.* "Notes and Correspondances."

Cuijpers, J. W. M. and P. G. Duynkerke. 1993. Large-eddy simulation of trade wind cumulus clouds. *J. Atmos. Sci.* 50:3894–3908.

Cuxart, J., P. Bougeault, and J. L. Redelsperger. 2000. A turbulence scheme allowing for mesoscale and large eddy simulations. *Q. J. R. Meteorol. Soc.* 126:1–30.

Davies-Jones, R. P. 1983. An accurate theoretical approximation for adiabatic condensation temperature. *Mon. Wea. Rev.* 111:1119–1121.

De Roode, S. R. and P. G. Duynkerke. 1997. Observed Lagrangian transition of stratocumulus into cumulus during astex: Mean state and turbulence structure. *J. Atmos. Sci.* 54:2157–2173.

Deardorff, J. W. 1966. The counter-gradient heat flux in the lower atmosphere and in the laboratory. *J. Atmos. Sci.* 23:503–506.

Deardorff, J. W. 1976. Usefulness of liquid-water potential temperature in a shallow-cloud model. *J. Appl. Meteorol.* 15:98–102.

Deardorff, J. W. 1977. A parameterization of ground-surface moisture content for use in atmospheric prediction models. *J. Appl. Meteorol.,* 16:1182–1185.

Deardorff, J. W. 1979. Prediction of convective mixed-layer entrainment for realistic capping inversion structure, *J. Atmos. Sci.,* 36:424–436.

Deardorff, J. W., G. E. Willis and B. H. Stockton. 1980. Laboratory studies of the entrainment zone of a convectively mixed layer. *J. Fluid Mech.* 100:41–64.

Driedonks, A. G. M. 1982. Models and observations of the growth of the atmospheric boundary layer. *Bound. Layer Meteorol.* 23:283–306.

Duynkerke, P. G. 1998. Modelling of atmospheric boundary layers. In *Clear and Cloudy Boundary Layers,* eds. A.A.M. Holtslag and P.G. Duynkerke. Royal Netherlands Academy of Arts and Sciences, Amsterdam, the Netherlands, pp. 85–110.

Duynkerke, P. G. and A. Driedonks. 1987. A model for the turbulent structure of the stratocumulus-topped atmospheric boundary layer. *J. Atmos. Sci.* 44:43–64.

Duynkerke, P. G. and J. Teixeira. 2000. Comparison of the ECMWF Re-analysis with FIRE I observations: Diurnal variation of marine stratocumulus. *J. Climate.* 14:1466–1478.

Duynkerke, P. G., P. Jonker, A. Chlond, M. C. van Zanten, J. Cuxart, P. Clark, E. Sanchez, G. Martin, G. Lenderink, and J. Teixeira. 1999. Intercomparison of three- and one-dimensional model simulations and aircraft observations of stratocumulus. *Bound. Lay Meteorol.* 92:453–487.

Ebert, E. E., U. Schumann. and R. B. Stull. 1989. Nonlocal turbulent mixing in the convective boundary layer evaluated from large eddy simulation. *J. Atmos. Sci.* 46:2178–2207.

French, J. R., G. Vali, and R. D. Kelly. 1999. Evolution of small cumulus clouds in Florida: observations of pulsating growth. *Atmos. Res.* 52:143–165.

Garratt, J. R. 1992. *The Atmospheric Boundary Layer.* Cambridge University Press, Cambridge, U.K.

Grant, A. L. M. 2001. Cloudbase fluxes in the cumulus-capped boundary layer. *Q. J. R. Meteorol. Soc.* 127:407–422.

Grant, A. L. M. and A. R. Brown. 1999. A similarity hypothesis for shallow cumulus transports. *Q. J. R. Meteorol. Soc.* 125:1913–1936.

Gregory, D. 1997. Sensitivity of general circulation model performance to convective param-
    eterization. In *The Physics and Parameterization of Moist Atmospheric Convection*, ed.
    R. K. Smith, NATO ASI Series C. 505. Kluwer Academic Publishers, Dordrecht, the
    Netherlands.

Gregory, D. 2001. Estimation of entrainment rate in simple models of convective clouds.
    Q. J. R. Meteorol. Soc. 127:53–72.

Grinell, S. A., C. S. Bretherton, D. E. Stevens, and A. M. Fraser. 1996. Vertical mass flux
    calculations in Hawaiian trade cumulus clouds from dual-Doppler radar. *J. Atmos. Sci.*
    53:1870–1886.

Hanson, H. 1991. Marine stratocumulus climatologies. *Int. J. Climatol.* 11:147–164.

Hignett, P. 1991. Observations of the diurnal variation in a cloud-capped marine boundary
    layer. *J. Atmos. Sci.* 48:1474–1482.

Högström, U. 1990. Analysis of turbulence structures in the surface layer with a modified
    similarity formulation for near neutral conditions. *J. Atmos. Sci.* 47:1949–1972.

Hojstrup, J. 1982. Velocity spectra in the unstable planetary boundary layer. *J. Atmos. Sci.*, 46,
    2178–2207.

Holtslag, A. A. M. 1998. Modelling of atmospheric boundary layers. In *Clear and Cloudy
    Boundary Layers*, eds. A.A.M. Holtslag and P.G. Duynkerke. Royal Netherlands
    Academy of Arts and Sciences, Amsterdam, the Netherlands, pp. 85–110.

Holtslag, A. A. M. and B. A. Boville. 1993. Local versus nonlocal boundary-layer diffusion in
    a global climate model. *J. Climate.* 6:1825–1842.

Holtslag, A. A. M. and C.-H. Moeng. 1991. Eddy diffusivity and countergradient transport in
    the convective atmospheric boundary layer. *J. Atmos. Sci.* 48:1640–1698.

Holtslag, A. A. M., E. I. F. de Bruijn, and H.-L. Pan. 1990. A high resolution air mass transfor-
    mation model for short-range weather forecasting. *Mon. Wea. Rev.* 92:1561–1575.

Holtslag, A. A. M., E. van Meijaard, and W. C. de Rooy. 1995. A comparison of boundary-layer
    diffusion schemes in unstable conditions over land. *Bound. Layer Meteorol.* 76:69–95.

Jensen, J. B., P. H. Austin, M. B. Baker, and A. M. Blyth. 1985. Turbulent mixing, spectral
    evolution and dynamics in a warm cumulus clouds. *J. Atmos. Sci.* 42:173–192.

Jonas, P. R. 1990. Observations of cumulus cloud entrainment. *Atmos. Res.* 25:105–127.

Kain, J. S. and J. M. Fritsch. 1993. Convective parameterization for mesoscale models: The
    Kain and Fritsch scheme. *Meteorol. Monogr.* 46:165–170.

Klein, S. and D. Hartmann. 1993. The seasonal cycle of low stratiform clouds. *J. Climate.*
    6:1587–1606.

Kropfli, R. A. and P. H. Hildebrand. 1980. Three-dimensional wind measurement in the optically
    clear planetary boundary layer with dual-Doppler radar. *Radio Sci.* 15(2):283–296.

Kuettner, J. P. and J. Holland. 1969. The BOMEX project. *Bull. Amer. Meteorol. Soc.*
    50:394–402.

Lafore, J.-P., J. Stein, N. Asensio, P. Bougeault, V. Ducrocq, J. Duron, C. Fischer, et al. 1998.
    The Meso-NH atmospheric simulation system. Part 1: Adiabatic formulation and con-
    trol simulations. *Ann. Geophys.* 16:90–109.

Larson, K., D. L. Hartmann, and S. A. Klein. 1999. On the role of clouds, water vapor, circula-
    tion and boundary layer structure on the sensitivity of the tropical climate. *J. Climate.*
    12:2359–2374.

Larson V. E., R. Wood, P. R. Field, J.-C. Golaz, T. H. Vonder Haar and W. R. Cotton. 2001.
    Small-scale and mesoscale variability of scalars in cloudy boundary layers: One-
    dimensional probability density functions. *J. Atmos. Sci.*, 58, N14, 1978–1994.

Lenderink, G. and A. P. Siebesma. 2000. Combining the mass flux approach with a statis-
    tical cloud scheme, In *Proceedings of the 14th Symposium on Boundary Layers and
    Turbulence.* Aspen, pp. 66–69.

Lenderink, G., A. P. Siebesma, S. Cheinet, S. Irons, C. Jones, P. Marquet, F. Muller,
    D. Olmeda, E. Sanchez, and P. M. M. Soares. 2004. The diurnal cycle of shallow Cumulus

clouds over land: A single column model intercomparison study. *Q. J. R. Meteorol. Soc.* 130:3339–3364.

Lenschow, D. H. and P. L. Stephens. 1980. The role of thermals in the convective boundary layer. *Bound. Lay. Meteorol.* 19:509–532.

Lenschow, D. H., J. C. Wyngaard, and W. T. Pennell. 1980. Mean-field and second-moment budgets in a baroclinic, convective boundary layer. *J. Atmos. Sci.* 37:1313–1326.

LeTreut, H. 1989. *First Studies with a Prognostic Cloud Generation Scheme.* ECMWF Tech. Memo. 155, ECMWF. Reading. U.K., 32 pp.

LeTreut, H. and Z. X. Li. 1988. Using Meteosat to validate a prognostic cloud generation scheme. *Atmos. Res.* 21:273–292.

Lilly, D. K. 1962. On the numerical simulation of buoyant convection. *Tellus.* XIV:148–172.

Lilly, D. K. 1968. Models of cloud-topped mixed layers under a strong inversion. *Q. J. R. Meteorol. Soc.* 94:292–309.

Lin, C. 1999. Some bulk properties of cumulus ensembles simulated by a cloud-resolving model. Part II: Entrainment profiles. *J. Atmos. Sci.* 56:3376–1151.

Louis, J. F., M. Tiedtke, and J. F. Geleyn. 1982. A short history of the PBL parameterization at ECMWF. In *Proceedings of the ECMWF Workshop on Boundary-Layer Parameterization.* ECMWF, Reading, U.K., pp. 59–79.

Ma, C.-C., C. R. Mechoso, A.W. Robertson, and A. Arakawa. 1996. Peruvian stratus clouds and the tropical Pacific circulation: A coupled ocean-atmosphere GCM study. *J. Climate,* 9:1635–1646.

Marshak, A., A. Davis, and W. Wiscombe. 1995. Radiation smoothing in fractal clouds. *J. Geophys. Res.* 100:26247–26261.

Mellor, G. L. 1977. The Gaussian cloud model relations. *J. Atmos. Sci.* 34:356–358.

Mellor, G. L. and T. Yamada. 1974. A hierarchy of turbulence closure models for planetary boundary layers. *J. Atmos. Sci.* 31:1791–1806.

Moeng, C.-H. 1984. A large-eddy simulation model for the study of the planetary boundary-layer turbulence. *J. Atmos. Sci.* 41:2052–2062.

Moeng, C.-H. and P. P. Sullivan. 1994. A comparison of shear and buoyancy driven planetary-boundary-layer flows. *J. Atmos. Sci.* 51:999–1022.

Moeng, C.-H., W. R. Cotton, C. S. Bretherton, A. Chlond, M. H. Khairoutdinov, S. Krueger, W. S. Lewellen, et al. 1996. Simulation of a stratocumulus-topped PBL: Intercomparison among different numerical codes. *Bull. Am. Meteorol. Soc.* 77:261–278.

Monin, A. S. and A. M. Yaglom. 1971. *Statistical Fluid Mechanics—Mechanics of Turbulence,* vol. I. MIT Press, Cambridge, MA.

Neggers, R. A. J., A. P. Siebesma, G. Lenderink, and A. A. M. Holtslag. 2004. An evaluation of mass flux closures for diurnal cycles of shallow cumulus convection. *Mon. Wea. Rev.* 132:2525–2538.

Nicholls, S. 1989. The structure of radiatively driven convection in stratocumulus. *Q. J. R. Meteorol. Soc.* 115:487–511.

Nieuwstadt, F. T. M., P. J. Mason, C.-H Moeng, and U. Schumann. 1992. Large eddy simulation of the convective boundary layer: A comparison of four codes, In *Turbulent Shear Flows 8.* Springer-Verlag, Berlin, Germany, pp. 343–367.

Nordeng, T. E. 1994. Extended versions of the convective parameterization scheme at ECMWF and their impact on the mean and transient activity of the model in tropics. Tech. Memo. No 206. ECMWF. U.K.

Ooyama, V. K. 1971. A theory on parameterization of cumulus convection. *J. Meteorol. Soc. Jpn.* 49:744–756.

Paluch, I. R. 1979. The entrainment mechanism in Colorado cumuli. *J. Atmos. Sci.* 36:2467–2478.

Philander, S. G. H., D. Gu, D. Halpern, G. Lambert, N.-C. Lau, T. Li, and R. C. Pawcanowski. 1996. Why the ITCZ is mostly north of the equator. *J. Climate.* 9:2958–2972.

Pontikis, C. A., A. Rigaud, and E. Hicks. 1987. Entrainment and mixing as related to the micro-physical properties of shallow warm cumulus clouds. *J. Atmos. Sci.* 44:2150–2165.

Prandtl, L. 1925. Bericht uber untersuchungen zur ausgebildeten turbulenz. *Zs. Angew. Math. Mech.*, 5, 136–139.

Randall, D. A. 1984. Stratocumulus cloud deepening through entrainment. *Tellus.* 36:446–457.

Randall, D. A., Q. Shao, and C.-H. Moeng. 1992. A second-order bulk boundary-layer model. *J. Atmos. Sci.* 49:1903–1923.

Reynolds, O. 1895. On the dynamical theory of incompressible viscous fluids and the determination of the criterion. *Phil. Trans. R. Soc. Lon. A* 186:123–164.

Schumann, U. and C.-H. Moeng. 1991a. Plume fluxes in clear and cloudy convective boundary layers. *J. Atmos. Sci.* 48:1746–1757.

Schumann, U. and C.-H. Moeng. 1991b. Plume budgets in clear and cloudy convective bound-ary layers. *J. Atmos. Sci.* 48:1758–1770.

Siebesma, A. P. 1996. On the mass flux approach for atmospheric convection. In *ECMWF Workshop Proceedings New Insights and Approaches to Convective Parameterization.* ECMWF. Reading, U.K., pp. 25–57.

Siebesma, A. P. 1998. Shallow cumulus convection. In *Buoyant Convection in Geophysical Flows*, eds. E. J. Plate et al. Kluwer Academic Publishers, Dordrecht, the Netherlands, pp. 441–486.

Siebesma, A. P. and J. W. M. Cuijpers. 1995. Evaluation of parametric assumptions for shallow cumulus convection. *J. Atmos. Sci.* 52:650–666.

Siebesma, A. P. and A. A. M. Holtslag. 1996. Model impacts of entrainment and detrainment rates in shallow cumulus convection. *J. Atmos. Sci.* 53:2354–2364.

Siebesma, A. P. and J. Teixeira. 2000. An advection-diffusion scheme for the convective boundary layer, description and 1D results. In *Proceedings of 14th Symposium on Boundary Layers and Turbulence*, Aspen, CO, pp. 133–136. American Meteorological Society.

Siebesma, A. P., C. S. Bretherton, A. Brown, A. Chlond, J. Cuxart, P. G. Duynkerke, H. Jiang, et al. 2003. A large eddy simulation intercomparison study of shallow cumulus convection. http://www.knmi.nl/~siebesma/PUBS/siebesma_03_jas.pdf, *J. Atmos. Sci.* 60:1201–1219.

Siebesma, A.P., P. M. M. Soares, and J. Teixeira. 2007. A combined eddy diffusivity mass-flux approach for the convective boundary layer. *J. Atmos. Sci.* 64:1230–1248.

Simpson, J. 1971. On cumulus entrainment and one-dimensional models. *J. Atmos. Sci.* 28:449–455.

Simpson, J. and V. Wiggert. 1969. Models of precipitating cumulus towers. *Mon. Wea. Rev.* 97:471–489.

Slingo, J. M. 1980. A cloud parameterization scheme derived from GATE data for use with a numerical model. *Q. J. R. Meteorol. Soc.* 106:747–770.

Slingo, J. M. 1987. The development and verification of a cloud prediction scheme in the ECMWF model. *Q. J. R. Meteorol. Soc.* 113:899–927.

Slingo, J. M., M. Blackburn, A. Betts, R. Brugge, B. J. Hoskins, M. J. Miller, L. Steenman-Clark, and J. Thurburn. 1994. Mean Climate and transience in the tropics of the UGAMP GCM: Sensitivity to convective parameterization. *Q. J. R. Meteorol. Soc.* 120:881–922.

Smith, R. N. B. 1990. A scheme for predicting layer clouds and their water content in a general circulation model. *Quart. J. Roy. Meteor. Soc.*, 116:435–460.

Smith, S. A. and P. R. Jonas. 1995. Observations of the turbulent fluxes in fields of cumulus clouds. *Q. J. R. Meteorol. Soc.* 121:1185–1208.

Soares, P. M. M., A. P. Siebesma, and J. Teixeira. 2002. The role of entrainment in the mass-flux/K-diffusion parameterization of the convective boundary layer. In *Proceedings of 3° Encontro Luso-Espanhol de Meteorologia*, Évora, Portugal, pp. 177–182.

Soares, P. M. M., P. M. A. Miranda, A. P. Siebesma, and J. Teixeira. 2004. An eddy-diffusivity/ mass-flux turbulence closure for dry and shallow cumulus convection. *Q. J. R. Meteorol. Soc.* 130:3365–3384.

Sommeria, G., 1976. Three-dimensional simulation of turbulent processes in an undisturbed trade-wind boundary layer. *J. Atmos. Sci.* 33:216–241.

Sommeria, G. and J. W. Deardorff. 1977. Subgrid-scale condensation in models of nonprecipitating clouds. *J. Atmos. Sci.* 34:344–355.

Sorbjan, Z. 1996a. Numerical study of penetrative and "solid lid" nonpenetrative convective boundary layers. *J. Atmos. Sci.* 53:101–112.

Sorbjan, Z. 1996b. Effects caused by varying the strength of the capping inversion based on a large eddy simulation model of the shear-free convective boundary layer. *J. Atmos. Sci.* 53:2015–2024.

Squires, P. and J. S. Turner. 1962. An entraining jet model for cumulonimbus updrafts. *Tellus.* 14:422–434.

Sullivan, P. P., C.-H. Moeng, B. Stevens, D. H. Lenschow, and S. D. Mayor. 1998. Structure of the entrainment zone capping the convective atmospheric boundary layer. *J. Atmos. Sci.* 55:3042–3064.

Sundqvist, H. 1988. Parameterization of condensation and associated clouds in models for weather prediction and general circulation simulation. *Physically Based Modelling and Simulation of Climate and Climate Change,* ed. M. E. Schlesinger. Kluwer, Dordrecht, the Netherlands, pp. 433–461.

Stevens, B., W. R. Cotton, G. Feingold, and C.-H. Moeng. 1998. Large-eddy simulations of strongly precipitating, shallow, stratocumulus-topped boundary layers. *J. Atmos. Sci.* 55:3616–3638.

Stevens, B., A. S. Ackerman, B. A. Albrecht, A. R. Brown, A. Chlond, J. Cuxart, P. G. Duynkerke, D. C. et al. 2001. Simulations of tradewind cumuli under a strong inversion. *J. Atmos. Sci.* 58:1870–1891.

Stull, R. B. 1976. Mixed layer depth model based on turbulent energetics. *J. Atmos. Sci.* 33:1260–1267.

Stull, R. B. 1984. Transilient turbulence theory. Part 1: The concept of eddy mixing across finite distances. *J. Atmos. Sci.* 41:3351–3367.

Stull, R. B. 1988. *An Introduction to Boundary Layer Meteorology.* Kluwer Academic Publishers, Boston, MA, 666pp.

Stull, R. B. 1993. Review of transilient turbulence theory and non-local mixing. *Bound. Layer Meteorol.* 62:21–96.

Stull, R. B. and A. G. M. Driedonks. 1987. Applications of the transilient turbulence parameterization to atmospheric boundary-layer simulations. *Bound. Layer Meteorol.* 40:209–239.

Teixeira, J. 2001. Cloud fraction and relative humidity in a prognostic cloud fraction scheme. *Monthly Weather Review*, 129, 1750–1753.

Teixeira, J. and S. Cheinet. 2002. Eddy-diffusivity and convective boundary layers: A new mixing length formulation. In *15th Symposium on Boundary Layer and Turbulence*, Wageningen, NL, pp. 55–58. American Meteorological Society.

Teixeira, J. and S. Cheinet. 2004. A simple mixing length formulation for the eddy-diffusivity parameterization of dry convection. *Bound. Layer Meteorol.* 110:435–453.

Teixeira, J. and T. F. Hogan. 2002. Boundary layer clouds in a global atmospheric model: Simple cloud cover parameterization. *J. Climate.* 15:1261–1276.

Teixeira, J. and P. M. A. Miranda. 2001. Fog prediction at Lisbon airport using a one-dimensional boundary layer model. *Meteorological Applications*, 8:497–505.

Tennekes, H. 1973. Model for dynamics of inversion above a convective boundary-layer. *J. Atmos. Sci.* 30:558–567.

Tiedtke, M. 1987. The parameterization of moist processes. Part 2: The parameterization of cumulus convection. *ECMWF Lecture Series*. 56pp.

Tiedtke, M. 1989. A comprehensive mass flux scheme for cumulus parameterization in larges-cale models. *Mon. Wea. Rev.* 177:1779–1800.

Tiedtke, M. 1991. Aspects of cumulus parameterization. *Proceedings of the ECMWF Seminar on Tropical Extra-Tropical Interactions*, Reading, U.K., ECMWF, pp. 441–466.

Tiedtke, M. 1993. Representation of clouds in large-scale models. *Mon. Wea. Rev.* 121:3040–3061.

Tiedtke, M., W. A. Heckley, and J. Slingo. 1988. Tropical forecasting at ECMWF: On the influence of physical parameterization on the mean structure of forecasts and analyses. *Q. J. R. Meteorol. Soc.* 114:639–664.

Tompkins, A. M. 2002. A prognostic parameterization for the subgrid-scale variability of water vapour and clouds in large-scale models and its use to diagnose cloud cover. *J. Atmos. Sci.* 59:1917–1942.

Troen, I. and L. Mahrt. 1986. A simple model of the atmospheric boundary layer: sensitivity to surface evaporation. *Bound. Layer Meteorol.* 37:129–148.

Turner, J. S. 1973. *Buoyancy Effects in Fluids*. Cambridge University Press, Cambridge, MA.

Van Dorland, R. 1999. Radiation and climate: From radiative transfer modelling to global temperature response. PhD thesis. University of Utrecht, the Netherlands.

VonFicker, H. 1936. *Die Passatinversion*. Veröffentlichungen des Meteorologischen Institutes der Universität Berlin, Berlin, Germany, 33pp.

van Ulden, A. P. and A. P. Siebesma. 1997. A model for strong updrafts in the convective boundary layer. In *Proceedings of the 12th Symposium on Boundary Layers and Turbulence*. Vancouver BC, pp. 258–260.

van Zanten, M. 2000. Entrainment processes in stratocumulus. PhD thesis. University of Utrecht, the Netherlands, 139pp.

Wang, S. and B. A. Albrecht. 1990. A mean-gradient model of the dry convective boundary layer. *J. Atmos. Sci.* 47:126–138.

Wang, S. and B. Stevens. 2000. On top-hat representations of turbulence statistics in cloud-topped boundary layers. *J. Atmos. Sci.* 57:423–441.

Warner, J. 1970. On steady-state one dimensional models of cumulus convection. *J. Atmos. Sci.* 27:1035–1040.

Warner, J. 1977. Time variation of updrafts and water content in small cumulus clouds. *J. Atmos. Sci.* 34:1306–1312.

Willimason, D. L., J. T. Kiehl, V. Ramanathan, R. E. Dockinson, and J. J. Hack. 1987. *Description of NCAR Community Climate Model (CCM1)*, NCAR Tech. Note, NCAR/TN-285+STR.

Wyngaard, J. C. 1998. Convection viewed from a turbulence perspective. In *Buoyant Convection in Geophysical Flows*, eds. E. J. Plate et al., Kluwer Academic Publishers, Dordrecht, the Netherlands, pp. 23–39.

Wyngaard, J. C. and C.-H. Moeng. 1992. Parameterizing turbulent diffusion through the joint probability density. *Bound. Layer Meteorol.* 60:1630–1649.

Yanai, M., S. Esbensen, and J. Chu. 1973. Determination of bulk properties of tropical cloud clusters from large-scale heat and moisture budgets. *J. Atmos. Sci.* 30:611–627.

Young, G. 1988a. Turbulence structure of the convective boundary layer. Part I: Variability of normalized statistics. *J. Atmos. Sci.* 45:719–726.

Young, G. 1988b. Turbulence structure of the convective boundary layer. Part II: Phoenix 78 aircraft observations of thermals and their environment. *J. Atmos. Sci.* 45:727–735.

Young, G. 1988c. Turbulence structure of the convective boundary layer. Part III: The vertical velocity budgets of thermals and their environment. *J. Atmos. Sci.* 45:2039–2049.

5 Mathematical Air Pollution Models: Eulerian Models

Tiziano Tirabassi

CONTENTS

5.1 INTRODUCTION

The management and safeguard of air quality presupposes a knowledge of the state of the environment. Such knowledge involves both cognitive and interpretative aspects. Monitoring networks and measurements in general, together with an inventory of emission sources, are of fundamental importance for the construction of the cognitive picture, but not for the interpretative one. In fact, air quality control requires

131

interpretative tools that are able to extrapolate in space and time the values measured by analytical instrumentation at field sites, while environmental improvement can only be obtained by means of a systematic planning of reduction of emissions, and, therefore, by employing instruments (such as mathematical models of atmospheric dispersion) capable of linking the causes (sources) of pollution with the respective effects (pollutant concentrations).

The processes governing the transport and diffusion of pollutants are numerous and of such complexity that it would be impossible to describe them without the use of mathematical models. Such models therefore constitute an indispensable technical instrument of air quality management. Mathematical models are in fact able to

- Describe and interpret experimental data
- Control air quality in either real or deferred time
- Monitor accidental emissions and assess risk areas
- Identify pollutant sources
- Evaluate the contribution of a single source to pollutant loading
- Assist in territorial management and planning

There exist innumerable, sometimes very diverse, mathematical models of atmospheric pollutant diffusion that may be utilized for the aforementioned purposes. In fact, the phenomenon of turbulent diffusion in the atmosphere has no single formulation, in the sense that no one approach has yet been proposed that is able to explain all of the observed phenomena.

5.2   OPERATIVE CHARACTERISTICS OF MATHEMATICAL MODELS

The choice of model is closely linked to the problem being confronted and to the meteoclimatic and orographic characteristics of the site under consideration. Available models can be subdivided on the basis of source characteristics:

- Point source
- Linear, area, and volume sources

or on the basis of orography:

- Flat terrain
- Complex terrain

They can also be classified on the basis of the size of the field they are describing:

- Short distance (distance from source less than 30–50 km)
- Mesoscale models (describing concentration fields of the order of hundreds of kilometers)
- Continental or planetary circulation models

Finally, models can be classified on the basis of the time resolution of the concentrations produced:

- Episodic models (temporal resolution of less than an hour)
- Short-time models (temporal resolutions greater than or equal to an hour, and less than or equal to 24 h)
- Climatological models (with resolution greater than 24 h, generally seasonal or annual)

5.3 THEORETICAL CHARACTERISTICS OF MATHEMATICAL MODELS

The theoretical approach to the problem essentially assumes four basic forms. In the *K* approach, diffusion is considered, at a fixed point in space, proportional to the local gradient of the concentration of the diffused material. Consequently, it is fundamentally Eulerian since it considers the motion of fluid within a spatially fixed system of reference. Such models are most suited to confronting complex problems, for example, the dispersion of pollutants over complex terrain or the diffusion of noninert pollutants. They are based on the numerical resolution, on a fixed spatial-temporal grid, of the equation of the mass conservation of the pollutant chemical species.

Among the Eulerian models, box models constitute the most simple mathematical approach since they neglect the spatial structure of phenomena. They assume that the pollutants are uniformly distributed within a parallelepiped (box). From the theoretical viewpoint, this is equivalent to assuming infinite diffusion coefficients that provoke an instantaneous propagation of the pollutant within the considered box. The pollutant present in the box originates from internal sources or from external contributions transported by the wind or flows across the summit due to variations in the height of the box itself, which generally coincides with the height of the mixing layer.

Lagrangian models differ from Eulerian ones in adopting a system of reference that follows atmospheric motions. Initially, the term Lagrangian was used only to refer to the box or moving box models that followed the mean wind trajectory. Currently, this class includes all models that decompose the pollutant cloud into discrete "elements," such as segments, puffs, or computer particles. In particle models, pollutant dispersion is simulated through the motion of computer particles whose trajectories allow the calculation of the concentration field of the emitted substance. The underlying hypothesis is that the combination of the trajectories of such particles to simulate the paths of the air particles situated, at the initial moment, in the same position. The motion of the particles can be reproduced both in a deterministic and in a stochastic way. Gaussian models are theoretically based upon an exact, but not realistic, solution of the equation of transport and diffusion in the atmosphere, in cases where both wind and turbulent diffusion coefficients are constant with height. The solution is forced to represent real situations by means of empirical parameters, referred to as "sigmas." They can be either stationary (the time-independent plume models) or time-dependent (puff models). The name given to these models is derived from the fact that the pollutant distribution, both vertical and transverse to wind direction, is described by the famous curve discovered by the physicist-mathematician Gauss. The various versions of Gaussian models essentially differ in the techniques utilized to calculate the "sigmas" as a function of atmospheric stability and the downwind distance from the emission source. Two basic techniques can be identified as serving this purpose: the first employs adimensional functions built on the basis of available measurements of turbulent intensity and the second adopts semiempirical functions for "sigmas" built for each stability class with which atmospheric turbulence has been schematized.

Analytic models can be considered an intermediate stage between K and Gaussian models. They conserve the simplicity of the latter, in that the concentration field is described by a simple formula, but, at the same time, they are also able to confront, in a theoretically correct way, situations in which the wind and turbulent diffusion coefficient vary with height.

5.3.1 Eulerian Approach: K Models

Eulerian models are the most suitable for tackling problems of greater complexity, for example, the dispersion of pollutants over complex terrain or the diffusion of noninert pollutants. They are based on the resolution, on a fixed spatial-temporal grid, of the equation of mass conservation of the pollutant chemical species, expressed in terms of concentration $c(x, y, z, t)$ (Zannetti, 1990):

$$\frac{\partial c}{\partial t} = -\mathbf{u} \cdot \nabla c + D\nabla^2 c + S \tag{5.1}$$

where

$\mathbf{u}$ is the wind speed vector of the components u, v, w
$D\nabla^2 c$ is the molecular diffusion term (generally neglected), with D the molecular diffusion coefficient
S is the term referring to the source, measuring the emission intensity and representing the pollutant removal kinetic
∇ is the gradient operator
∇^2 is the Laplacian

In order to resolve Equation 5.1, it is necessary to know the wind field $\mathbf{u}$, something that is not possible since it is extremely variable in space and time, from the scale of centimeters to kilometers. Consequently, wind is divided into two parts:

$\bar{\mathbf{u}}$: The so-called ensemble average
$\mathbf{u}'$: The turbulent fluctuations of wind at mean nil

Thereupon the wind speed is expressed as the sum of the two components, mean and turbulent:

$$\mathbf{u} = \bar{\mathbf{u}} + \mathbf{u}' \tag{5.2a}$$

The same considerations can be made for c. Therefore:

$$c = \bar{c} + c' \tag{5.2b}$$

The ensemble average refers to the mean value obtained by the repetition of many experiments in the same meteorological and emission conditions.

The new **u** and c are introduced into Equation 5.1; after several calculations and hypothesizing a wind with divergence nil, the following is obtained:

$$\frac{\partial \bar{c}}{\partial t} = -\bar{\mathbf{u}} \cdot \nabla \bar{c} - \nabla \cdot \overline{c'\mathbf{u}'} + D\nabla^2 \bar{c} + \bar{S} \tag{5.3}$$

This equation includes some new variables (those with an apex) whose values are unknown. The appearance of new terms in equations for mean quantities leads to a number of unknowns greater than the number of equations. Thus, the system of equations is not closed and is therefore irresolvable. To close it, in fact, new equations of variance and covariance (second-order moments) would be required, but this would only shift the problem to a higher order since it would yield further unknown quantities that are third-order moments. Now, if it were decided to find equations for the third-order moments, this would yield unknowns of a higher order, that is, fourth-order moments, requiring the introduction of new equations. Iterating the procedure, the conclusion would be reached that the number of unknowns is always greater than the number of equations. A solution to this problem consists of utilizing only a finite number of equations, relative to a certain number of unknowns, parameterizing the remaining ones in terms of known quantities.

The most classic and widely used approach to obviate this problem is the parameterization of second-order moments, assuming a hypothetical analogy between molecular diffusion and the turbulent transfers. Such approach is referred to as the *K*-theory or flux-gradient theory, as it assumes that the flow of a given field is proportional to the gradient of an appropriate mean variable. This is a first-order closure of the set of equations under examination, since it conserves the equations relative to the first moments and parameterizes the second moments:

$$\overline{c'\mathbf{u}'} = -K\nabla \bar{c} \tag{5.4}$$

where K is the eddy diffusivity coefficient.

The simplicity of the *K*-theory of turbulent diffusion has led to its widespread use as the mathematical basis for simulating urban, photochemical pollution. However, *K*-closure has its own limits. In contrast to molecular diffusion, turbulent diffusion is scale-dependent. This means that the rate of diffusion of a cloud of material generally depends on the cloud dimensions and the intensity of turbulence. As the cloud grows, larger eddies are incorporated in the expansion process, so that a progressively larger fraction of turbulent kinetic energy is available for the cloud expansion. However, eddies much larger than the cloud itself are relatively unimportant in its expansion. Thus, the gradient-transfer theory works well when the dimension of dispersed material is much larger than the size of turbulent eddies involved in the diffusion process, that is, for ground-level emissions and for large travel times. Strictly speaking, one should introduce a diffusion coefficient function not only of atmospheric stability and emission height but also of the travel time or distance from source. However, such time-dependence makes it difficult to treat the diffusion equation in a fixed-coordinate system where multiple sources have to be treated simultaneously. Otherwise, one should limit the application of the gradient theory to large travel times (Pasquill and Smith, 1983). A further problem is that the down-gradient transport hypothesis is inconsistent with observed features of turbulent diffusion in the upper portion of the mixed layer (ML), where counter-gradient material fluxes are known to occur (Deardoff and Willis, 1975).

In addition, unlike molecular diffusion, turbulent diffusion is not a property of fluids, but of the turbulence itself or of flows, and it may vary greatly from one flow to another and from one region to another of the same flow. The above relations are essentially based only on a qualitative analogy between molecular and turbulent diffusion. For the first-order closure to be realistic, the mean concentration field must have a much larger timescale than that of turbulent transport.

Despite these well-known limits, the K-closure is widely used in several atmospheric conditions, because it describes the diffusive transport in an Eulerian framework, where almost all measurements are Eulerian in character. It produces results that agree with experimental data as well as any more complex model, and it is not as computationally expensive as higher-order closures.

The reliability of the K-approach strongly depends on the way the eddy diffusivity is determined on the basis of the turbulence structure of the PBL, and on the model's ability to reproduce experimental diffusion data. A great variety of formulations exist (Ulke, 2000). Most of them are based on similarity theory, and give different results for the same atmospheric stability, as well as discontinuities and jumps at the transition between different stability regimes of the PBL.

The tensor K (3×3) of turbulent diffusion, whose elements can be extrapolated from experimental measurements, is introduced in Equation 5.3. Then, by also applying the following approximations:

- The K tensor is diagonal.
- The molecular diffusion is negligible.
- c represents the concentration of a nonreactive pollutant (thus $\overline{S} = S$).

Equation 5.3 can be written in the form:

$$\frac{\partial \overline{c}}{\partial t} = -\overline{\mathbf{u}} \cdot \nabla \overline{c} + \nabla \cdot K \nabla \overline{c} + S \qquad (5.5)$$

Equation 5.5 can be integrated (analytically or numerically) if input data for u, K, and S are provided, together with the initial and boundary conditions for $\overline{c}$.

Eulerian models and K models mainly differ in the functions utilized for the K coefficients and the techniques used for the integration of Equation 5.5.

Equation 5.5 can be resolved in two ways:

1. With analytic methods, obtaining exact solutions
2. With numerical methods, obtaining approximate solutions

5.3.2 Analytical Solutions

Analytical solutions of equations are of fundamental importance in understanding and describing physical phenomena. Analytical solutions (as opposed to numerical ones) explicitly take into account all the parameters of a problem, so that their influence can be reliably investigated, and it is easy to obtain the asymptotic behavior of the solution, which is usually difficult to generate through numerical calculations.

There are analytical solutions of the two-dimensional advection–diffusion equation (Tirabassi, 1989, 2003):

$$u\frac{\partial C}{\partial x} = \frac{\partial}{\partial z}\left(K_z\frac{\partial C}{\partial z}\right) + S \tag{5.6}$$

where
  u is mean velocity (the wind is assumed along x-axis, while z is the height)
  C is the mean concentration
  S is the source term
  K_z is the vertical eddy exchange coefficient

Moreover, as usual, the along-wind diffusion was neglected because it was considered little in respect to the advection. Recently, a steady-state mathematical model for dispersion of contaminants in low winds was formulated by taking into account the longitudinal diffusion in the advection–diffusion equation (Moreira et al., 2005a).

Unfortunately, no general solution is known for equations describing the atmospheric transport and dispersion of air pollution. There are some specific solutions, the best-known being the so-called Gaussian solution, which does not, however, realistically describe the concentrations of pollutants in the air; in fact, the models based on it (so-called Gaussian models) use empirical parameters of dispersion in order to force the Gaussian solution to represent the actual concentration field. However, there are models based on non-Gaussian analytical solutions.

Roberts (1923) presented a bidimensional solution, for ground-level sources only, in cases where both the wind speed and vertical diffusion coefficients follow power laws as a function of height, that is,

$$u = u_1\,(z/z_1)^\alpha \tag{5.7a}$$

$$K_z = K_1\,(z/z_1)^\beta \tag{5.7b}$$

where z_1 is the height where u_1 and K_1 are evaluated.

Rounds (1955) obtained a bidimensional solution valid for elevated sources, but only for linear profiles of K_z. Smith (1957a) resolved the bidimensional equation of transport and diffusion with u and K_z power functions of height with the exponents of these functions following the conjugate law of Schmidt (i.e., "wind exponent" = $1 -$ "K_z exponent").

Smith (1957b) also presented a solution in the case of constant u, but K_z following:

$$K_z = K_0 z^a (H - z)^b \tag{5.8}$$

where K_0 is a constant and a and b can be

$a \geq 0$ and $b = 0$

$a = 0$ and $b > 0$ for $0 \leq z \leq H$

$a = 1$ and $b > 0$ for $0 \leq z \leq H$

$a = 1$ and $b = 0$ for $0 \leq z \leq H/2$; $a = 0$ and $b = 1$ for $H/2 \leq z \leq H$

where H is the height of the atmospheric boundary layer.

Scriven and Fisher (1975) proposed a solution with constant u and K_z as

$$K_z \equiv z \quad \text{for } 0 \leq z \leq z_s \tag{5.9a}$$

$$K_z = K_z(z_1) \quad \text{for } z_s < z \leq H \tag{5.9b}$$

where z_s is a predetermined height (generally, the height of the surface layer). This solution allows (as boundary conditions) a net flow of material toward the ground:

$$K_z \frac{\partial C}{\partial z} = V_g C \tag{5.10}$$

where V_g is the deposition velocity. The Scriven and Fisher solution has been used in the United Kingdom for long-range transport of pollutant. In Fisher (1975), the deposition of sulfur over the United Kingdom, Sweden, and the rest of Europe was compared, and it was found that the British contribution to deposition over rural Sweden was about one half of the Swedish contribution.

Yeh and Huang (1975) and Berlyand (1975) published bidimensional solutions for elevated sources with u and K_z following power profiles, but for an unbound atmosphere, that is,

$$K_z \frac{\partial C}{\partial z} = 0 \quad \text{at } z = \infty \tag{5.11}$$

Demuth (1978) put forward a solution with the same conditions, but for a vertically limited boundary layer, that is,

$$K_z \frac{\partial C}{\partial z} = 0 \quad \text{at } z = H \tag{5.12}$$

The solutions of Yeh and Huang, Berlyand, and Demuth are used in KAPPAG air pollution model (Tagliazucca et al., 1985; Tirabassi et al., 1986; Tirabassi 1989).

By applying the Monin–Obukhov similarity theory to diffusion, van Ulden (1978) derived a solution for vertical diffusion from continuous sources near the ground only with the assumption that u and K_z follow power profiles. His results are similar to that of Roberts', but he provided a model for non–ground-level sources, but applicable to sources within the surface layer. SPM (Tirabassi and Rizza, 1995) is a model that utilizes the solution proposed by van Ulden.

Nieuwstadt (1980) presented a solution, which was a particular case of Smith's (1975b) solution noted above. Subsequently, Nieuwstadt and de Haan (1981) extended that solution to the case of a growing boundary layer height. Catalano (1982), in turn, extended the latter solution to the case of nonzero mean vertical wind profiles. Lin and Hildemann (1997) extended the solution of Demuth (1978) with boundary conditions suitable for simulating dry deposition to the ground.

Recently, Brown et al. (1997) derived equations for point source releases for the first four moments of the vertical concentration distribution and the magnitude and downwind location of the maximum ground concentration from the solution of Yeh and Huang (1975).

Finally, Moreira et al. (2005b) found a general two-dimensional steady-state solution for any profiles of wind and eddy coefficient diffusions.

5.3.3 Numerical Solutions

Among the techniques used to resolve Equation 5.5, the following should be mentioned:

* Finite difference method
* Finite elements method
* Finite volume method
* Spectral methods
* The method of confined elements

The finite difference method is the most simple technique and was the first to be used. The approximation of the finite differences of the advection term $\bar{\mathbf{u}} \cdot \nabla \bar{c}$, is, however, always associated with an error that artificially increases diffusion in the final results of the simulated concentrations. Several techniques have been developed with the aim of reducing this error, and notwithstanding the limitations posed by this model, it remains one of the most important and widely adopted methods of simulation.

Unlike analytic approaches, numerical techniques allow, from the theoretical viewpoint, the use of any function for $K(x, y, z, t)$.

5.3.4 Eddy Diffusivity

The literature proposes several expressions for K_z, which is generally a function of height z (for instance, Pleim and Chang, 1992). For instance, an approach for estimating the eddy diffusivity K and dispersion parameters as functions of eddy scale size in the PBL and relative amount of turbulent energy has been recently proposed by Degrazia and Moraes (1992) and Degrazia et al. (1997, 2000). Making use of Taylor's statistical theory (Taylor, 1921), the Hay and Pasquill working approximation of the relationship between Lagrangian and Eulerian turbulence spectra (Hay and Pasquill, 1959), and a model for Eulerian spectra, such approach relates plume dispersion in a boundary layer mainly to the turbulent eddies acting in the different stability regimes of the boundary layer (Pasquill and Smith, 1983). Bearing the K-theory limitations in mind, the main idea of the said approach is to obtain an eddy diffusivity scheme for practical applications in air pollution modeling, which reveals the essential features of turbulent diffusion, but which as far as possible preserves the simplicity and flexibility of the K-theory formulation. Degrazia et al. (1997, 2000) propose the vertical profiles of diffusion coefficients obtained by means of spectral techniques.

Several difficulties arise in the evaluation of the transversal turbulent diffusion. It is often, and not always correctly, hypothesized that,

$$K_h \cong K_y \tag{5.13}$$

where K_y is the transversal turbulent diffusion (supposing the wind is blowing in the direction of x-axis).

Another problem in the simulation of horizontal dispersion is the numerical error associated with the advection of the pollutant due to part of the mean wind. Such error is linked to the size of the spatial grid used to schematize the diffusion field, and can turn out to be greater than parameter K_h itself.

In the past, the most complex simulation techniques made widespread use of dynamic grid models, in particular with the application of the numerical method of finite differences and K closure. However, several major limitations of such applications have come to light:

1. The numerical approximation of the advection term often produces a fictitious diffusion.
2. K closure is a fundamentally incorrect approximation in strong turbulence conditions.
3. Since the concentrations are calculated as spatial means within three-dimensional cells of the grid, it is difficult to compare them with measurements carried out at single points in space.
4. It is difficult to link K eddy coefficients with experimental measurements in the atmosphere.
5. A correct application of K closure requires that the grid dimensions be smaller than those of the pollutant cloud, a condition that is difficult to satisfy the proximity to the source.

5.3.5 BOX MODELS

Of the Eulerian models, box models constitute the most simple mathematical approach, since they ignore the spatial structure of phenomena. They assume that the pollutants are uniformly distributed within a parallelepiped. From a theoretical point of view, this is equivalent to assuming infinite diffusion coefficients, which provoke an instantaneous propagation of the pollutant introduced into the box under consideration. The pollutant present in the box originates from internal sources or from external contributions transported by wind or flows through the summit as a consequence of variations in height of the box itself, which generally coincides with the ML height.

In mathematical terms, the continuity equation that describes the aforementioned phenomenon is formulated as

$$X \cdot h \frac{\partial C}{\partial t} = X \cdot Q + h \cdot u(C_b - C) + X \frac{\partial h}{\partial t}(C_a - C) \tag{5.14}$$

where
  Q is the time average emission flux of pollutants for unit area
  t is time
  C is the concentration at time t within the box
  X is the length of the box along the wind direction
  h is the depth of the box
  u is the time averaged wind speed through the box
  C_a is concentration at the box top boundary
  C_b is concentration at the upwind boundary

In Equation 5.14, the last term is considered only if $\partial h/\partial t > 0$ (box height increasing); the box (Y) width does not appear because it multiplies all the terms of the equation. In Equation 5.14, the first member expresses the velocity at which the pollutant accumulates in the box.

Limited to stationary conditions ($\partial C/\partial t = \partial h/\partial t = 0$), and considering null the upwind concentration ($C_b = 0$), Equation 5.14 can be simplified as

$$C = \frac{Q \cdot X}{h \cdot u} \tag{5.15}$$

"Multibox" models also exist, in which the area of study comprises several contiguous communicating boxes. In each box, the concentration of the pollutant is uniformly distributed and the horizontal flow of exit from a box is the entry one of the contiguous box (Stern, 1976; Zannetti, 1990).

5.3.6 GAUSSIAN MODELS

The Gaussian approach is widely used in air pollution studies to model the statistical properties of the concentration of contaminants emitted in the PBL.

The conditions under which the mean concentration of a pollutant species emitted from a point source can be assumed to have a Gaussian distribution are highly idealized, since they require stationary and homogeneous turbulence. In the PBL, the flow may be assumed to be quasi-stationary for suitably short periods of time (ca. 10 min to 1 h). However, due to the presence of the surface, there are variations with height of both the mean wind and turbulence that cannot always be disregarded.

Much effort has been devoted to the development of non-Gaussian models for handling the nonhomogeneous structures of PBL turbulence. However, they still result in excessively large computer runs, either for emergency response applications or for calculating concentration time series over a long time (e.g., a year). The latter is especially important in the evaluation of violations of air pollution standards, which are often expressed in high percentiles.

Conversely, Gaussian models are fast, simple, do not require complex meteorological input, and describe the diffusive transport in an Eulerian framework, making the use of the Eulerian nature of measurements easy.

For these reasons, they are still widely employed for regulatory applications by environmental agencies all over the world. Nonetheless, because of their well-known intrinsic limits, the reliability of a Gaussian model strongly depends on the way the dispersion parameters are determined on the basis of the turbulence structure of the PBL and the model's ability to reproduce experimental diffusion data. A great variety of formulations exist (Hanna et al., 1977; Briggs, 1985; Berkowicz et al., 1986; Hanna, 1986; Bowen, 1994; Erbrink, 1995; Mohan and Siddiqui, 1997).

The Gaussian solution in a system of coordinates where x is along the direction of the wind, y is transversal to wind, z is the height, and source of intensity Q is located at $(0, 0, H)$, can be written as

$$C(x,y,z) = Q/(2\pi u \sigma_y \sigma_z) \exp(-y^2/(2\sigma_y^2))[\exp((z-H)^2/(2\sigma_z^2)) + \exp((z+H)^2/(2\sigma_z^2))]$$

(5.16)

where σ_y and σ_z are functions of the distance from the source and turbulent intensity, and are determined experimentally.

The physical phenomenon of diffusion of emitted material is therefore mathematically described by such models through the faster or slower "broadening" (expressed by an increase in the numerical value of the "sigmas") of a Gaussian curve (see Figure 5.1).

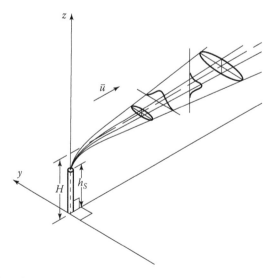

FIGURE 5.1 Gaussian distribution of a plume in a system of reference oriented in the direction of mean wind.

Gaussian models can simulate complex, area or volume sources, both by the spatial integration of the contribution of emissions (made possible because Gaussian models are linear to emissions), and, to obviate the difficulties of integration, by using special algorithms.

In practice, they can be relatively easy to use, and can be applied in numerous conditions (e.g., isolated sources, cities, road traffic, complex terrain). Moreover, bearing in mind that in practical applications, meteorological data are not generally available at both ground level and aloft at high temporal/spatial resolution, their performances are not poorer than those of other models. For these reasons, most operative models are based on the Gaussian approach.

The Gaussian model can be modified so as to extend its applicability to nonstationary and nonhomogeneous conditions, as well as to more complex orography. In particular, the breaking down of the plume into puffs has permitted the simulation of pollutant dispersion in pseudostationary conditions. Such models, illustrated below, decompose the plume into puffs, whose characteristics evolve in time and space together with the changing meteorological and emission conditions.

The various versions of Gaussian models essentially differ in the techniques utilized to calculate the "sigmas" as a function of atmospheric stability and the downwind distance from the emission source.

5.4   SEMIEMPIRICAL EXPRESSIONS OF THE σ

Several schemas exist for the calculation of σ_y and σ_z as functions of stability classes and of the downwind distance from the source.

Stability classes can in fact be calculated with semiempirical techniques using, for example, the method of Pasquill (Pasquill and Smith, 1983) based on simple meteorological observations (Tables 5.1 and 5.2) such as wind velocity, insolation,

---

TABLE 5.1
Stability Classification

		Wind Velocity at the Ground (m/s)				
Insolation/Cloud Cover		<2	≥2 and <3	≥3 and <5	≥5 and <6	≥6
Day	Strong insolation	A	A–B	B	C	C
	Moderate insolation	A–B	B	B–C	C–D	D
	Weak insolation	B	C	C	D	D
Day or night	Overcast	D	D	D	D	D
Night	Thin overcast or ≥0.5		E	D	D	D
	Thin overcast or ≤0.4		F	E	D	D

Source: Pasquill, F. and Smith, F.B., *Atmospheric Diffusion*, Halsted Press, John Wiley & Sons, New York, 1983.

Note: A, strongly unstable; B, unstable; C, weakly unstable; D, neutral; E, weakly stable; and F, stable.

TABLE 5.2

Classification of Atmospheric Stability

Stability Class	Stability Class of Pasquill	σ_θ (°)	Vertical Temperature Gradient (°C/m 10^{-2})	Richardson Number at 2 m	$\sigma_w/\bar{u}$
Very unstable	A	25.0	<−1.9	−0.9	>0.15
Moderately unstable	B	20.0	−1.9 to −1.7	−0.5	0.1 to 0.15
Slightly unstable	C	15.0	−1.7 to −1.5	−0.15	0.1 to 0.15
Neutral	D	10.0	−1.5 to −0.5	0	0.05 to 0.1
Slightly stable	E	5.0	−0.5 to 1.5	0.4	0 to 0.05
Moderately stable	F	2.5	1.5 to 4.0	0.8	0 to 0.05

Source: Zannetti, P., *Air Pollution Modelling*, Computational Mechanics Publications, Southampton, U.K. and Van Nostrand Reinhold, New York, 1990.

Note: σ_θ is the standard deviation of horizontal wind direction. $\bar{U}$ is the mean wind velocity. σ_w is the standard deviation of mean vertical wind velocity.

and, at night, cloud cover. Other techniques adopt measurements of the standard deviations of vertical wind velocity σ_w, horizontal wind direction σ_θ, the vertical gradient of temperature $\Delta T/\Delta z$, and the Richardson number, as illustrated in Table 5.2. Evaluations of stability classes through the standard deviation of wind velocity must be corrected in the nighttime, following Table 5.3, as proposed by Irwin (1980).

Once the stability classes have been evaluated, the "sigmas" are expressed as a function of distance downwind x, using one of the many formulae available in the literature, retrieved from experimental campaigns.

The Pasquill-Gifford sigmas (Gifford, 1961), presented in analytic form by Green et al. (1980) can be written as

$$\sigma_y(x) = \frac{k_1 x}{\left[1 + \left(x/k_2\right)\right]^{k_3}} \tag{5.17a}$$

$$\sigma_z(x) = \frac{k_4 x}{\left[1 + \left(x/k_2\right)\right]^{k_5}} \tag{5.17b}$$

where k_1, k_2, k_3, k_4, and k_5 are constants that vary according to atmospheric stability (see Zannetti, 1990).

TABLE 5.3

Correction of the Unstable Stability at Nighttime

Category Identified by σ_θ	Wind Velocity at 10 m (m/s)	Correct Nighttime Category
A	<2.9	F
	2.9–3.6	E
	>3.6	D
B	<2.4	F
	2.4–3.0	E
	>3.0	D
C	<2.4	E
	≥2.4	D

Source: Irwin, J.S., Estimating plume dispersion—A recommended generalized scheme, *Fourth AMS Symposium on Turbulence and Diffusion*, Reno, NV, 1980.

Note: Nighttime is considered to be from 1 h before sunset to 1 h after sunrise.

The Brookhaven sigmas (Smith, 1968), for which a power law is assumed (for both σ_y and σ_z), are written as

$$\sigma = ax^b \tag{5.18}$$

where the coefficients a and b vary according to the stability classes.

Briggs's sigmas (Briggs, 1973) distinguish between diffusion in the open country and urban environment (Table 5.4).

More recently, the "sigmas" have been expressed as functions of continuous variables of atmospheric turbulence. Most of them are based on the approach proposed by Pasquill (1971). He retained the essential features of Taylor's statistical theory, but evaluated the dispersion parameters in terms of the turbulence quantities and their related timescale using the following expressions:

$$\sigma_y = \sigma_v t S_y \left(t/T_L \right) \tag{5.19a}$$

$$\sigma_z = \sigma_w t S_z \left(t/T_L \right) \tag{5.19b}$$

where

σ_v and σ_w are the standard deviations of the transversal and vertical components of wind velocity

S_y and S_z are universal functions of the diffusion time t and Lagrangian timescale T_L

TABLE 5.4

Briggs's Sigma (1973) in the Open Country and Urban Environment

Pasquill	σ_y (m)	σ_z (m)
Urban dispersion parameters* (for distances between 100 and 10,000 m)		
A–B	$0.32x(1 + 0.0004x)^{-0.5}$	$0.24x(1 + 0.001x)^{0.5}$
C	$0.22x(1 + 0.0004x)^{-0.5}$	$0.20x$
D	$0.16x(1 + 0.0004x)^{-0.5}$	$0.14x(1 + 0.0003x)^{-0.5}$
E–F	$0.11x(1 + 0.0004x)^{-0.5}$	$0.08x(1 + 0.00015x)^{-0.5}$
Rural dispersion parameters (for distances between 100 and 10,000 m)		
A	$0.22x(1 + 0.0001x)^{-0.5}$	$0.20x$
B	$0.16x(1 + 0.0001x)^{-0.5}$	$0.12x$
C	$0.11x(1 + 0.0001x)^{-0.5}$	$0.08x(1 + 0.0002x)^{-0.5}$
D	$0.08x(1 + 0.0001x)^{-0.5}$	$0.06x(1 + 0.0015x)^{-0.5}$
E	$0.06x(1 + 0.0001x)^{-0.5}$	$0.03x(1 + 0.0003x)^{-1.0}$
F	$0.04x(1 + 0.0001x)^{-0.5}$	$0.016x(1 + 0.0003x)^{-1.0}$

Source: Briggs, G.A., Diffusion estimation for small emissions, in environmental research laboratories, air resources atmospheric turbulence and diffusion laboratory, Annual report, 1973. USAEC Report ATDL-106, National Oceanic and Atmospheric Administration, December 1974.

Reported below is a scheme based on the similarity theory and the micrometeorological variables proposed by Irwin (1983), which is included in several regulatory models:

$$\sigma_y = (\overline{v'^2})^{1/2} TF_y(T, T_{Lv}) \tag{5.20}$$

$$\sigma_z = (\overline{w'^2})^{1/2} TF_z(T, T_{Lw}) \tag{5.21}$$

with

$$F_y = \left[1 + 0.9(T/1000)^{1/2}\right]^{-1} \tag{5.22}$$

$$F_z = \left[1 + 0.9(T/500)^{1/2}\right]^{-1} \quad L < 0 \tag{5.23}$$

$$F_z = \left[1 + 0.945(T/100)^{0.806}\right]^{-1} \quad L > 0 \tag{5.24}$$

Modern expression of the "sigmas," in terms of wind variance and the Lagrangian integral timescale, on the basis of an atmospheric turbulence spectra model, are presented in Mangia et al. (1998).

5.5 GAUSSIAN MODEL EXTENSIONS

As mentioned above, the Gaussian model can be modified to allow the simulation of dispersion in certain cases:

- Linear, area, and volumetric sources of emission
- Complex terrains (valleys, cities, and coastal areas)
- Particular meteorological conditions, such as those leading to the phenomena of fumigation or confinement of pollutants
- Diffusion of heavy or reactive pollutants

There also so-called climatological models, in which each concentration value calculated by an equation similar to Equation 5.10 is attributed a weight, depending on the frequency of occurrence of the meteorological conditions corresponding to the given concentration value (Tirabassi et al., 1989; Zannetti, 1990).

To extend the range of applicability of the Gaussian method also to nonhomogeneous and nonstationary conditions, puff models have been developed. Here, the plume emitted is subdivided into a series of independent elements, which evolve as a function of the variation of meteorological conditions in space and time (Zanetti, 1990).

5.5.1 GAUSSIAN PUFF MODELS

Puff models were introduced to simulate the behavior of pollutants in nonhomogeneous and nonstationary meteorological and emission conditions (Zannetti, 1990). The emission is discretized in a temporal succession of puffs, each of which shifts into the area of calculus thanks to a three-dimensional wind field that is time variable.

Gaussian puff models assume that each emission of pollutants in a time interval Δt releases into the atmosphere a mass of pollutants $\Delta M = Q\Delta t$, where Q is the emission rate, which is variable in time.

Each puff contains the mass ΔM, and its baricenter is transported by the wind, which may vary in space and time.

If at time t the center of a puff is localized at $\mathbf{p}(t) = (x_p, y_p, z_p)$, then the contribution of this puff to the calculated concentration in the receptor place at $\mathbf{r} = (x_r, y_r, z_r)$ is given by the following relation:

$$\Delta c = \frac{\Delta M}{(2\pi)^{3/2}\sigma_h^2\sigma_z}\exp\left[-\frac{1}{2}\left(\frac{x_p - x_r}{\sigma_h}\right)^2\right]\exp\left[-\frac{1}{2}\left(\frac{y_p - y_r}{\sigma_h}\right)^2\right]\exp\left[-\frac{1}{2}\left(\frac{z_p - z_r}{\sigma_z}\right)^2\right]$$

$$(5.25)$$

which is Gauss's distribution for the concentration field of a single puff. Equation 5.25 requires the evaluation of σ_z and σ_h for every single puff. The total concentration of a pollutant retrieved by a receptor is calculated by summing the contributions Δc of all the puffs emitted by all the sources present.

A crucial difference between expressions (5.16) and (5.25) should be noted: in the latter a different diffusion term replaces the advection term, with the resulting disappearance of the mean wind velocity u from the expression. This means that in a puff model the wind velocity influences the calculation of the concentration only in the density of puffs in the region of diffusion (the lower the wind velocity, the closer the puffs emitted by a source). For this reason, puff models, unlike the Gaussian model described in Equation 5.16 (the concentration tends to the infinity for u tending toward zero), can be used to simulate diffusion in calm wind conditions.

Puff models are also suitable for the simulation of diffusion over complex orographies: in this case a three-dimensional wind field must be reconstructed from orographic characteristics and available anemometric measurements.

5.6   SPECIAL ALGORITHMS UTILIZED IN GAUSSIAN AND ANALYTIC MODELS

Gaussian and analytic models generally make use of algorithms based on empirical considerations to describe complex real situations. The various models presented in the literature are differentiated according to the algorithms adopted. Listed below are some phenomena relative to such situations.

5.6.1   Plume Rise

Once emitted into the atmosphere, plume tends to continue to rise due to is initial momentum. If hot, the rise is also caused by Archimede's force. In fact, warm fumes are less dense and therefore lighter than the surrounding air. Different formulae are presented in the literature to describe this phenomenon. The most commonly used are those developed by Briggs (Briggs, 1975).

5.6.2   Downwash

This phenomenon arises from the perturbation of the wide field and turbulence caused by structures present in the area of pollutant emission. There are two types of downwash: stack downwash and building downwash. Stack downwash refers to the lowering of the plume due to the perturbation of the wind field caused by the source itself. It is evaluated through an artificial decrease in the emission source height.

Building downwash is a distortion of the plume caused by buildings situated in proximity of the source. The distortion is also evaluated through an empirical modification of the "sigmas."

5.6.3   Plume Trapping

It consists of the entrapment of pollutants in a layer of the atmosphere close to the ground due to a thermal inversion at a level higher than that of fume emission.

Mathematically, the situation is described by Gaussian models as a multiple reflection of pollutants between the terrain and the base of the inversion.

5.6.4   GROUND DEPOSITION

Ground deposition of the diffuse material is evaluated by means of various algorithms. Listed below are those most commonly utilized.

- The decrease of material in the air is expressed by an exponential decrease on the basis of a value of the average life of the pollutant.
- The decrease of material in the air is expressed by a fictitious decrease of the emission flux as a function of the distance from the source.
- A lowering of the plume axis height is introduced as a function of the distance from the source to describe the fall due to the gravitational force of the diffuse material.
- The ground deposition flux is expressed as the concentration at the ground for an assigned constant deposition rate.
- Ground deposition due to meteoric precipitation is parameterized by an exponential decrease linked to the precipitation intensity.

5.6.5   CHEMICAL REACTIONS

The mathematical description of the chemical transformation of the pollutants dispersed in the atmosphere is a highly complex field of study that remains far from any satisfactory solutions. On the one hand, basic problems exist in the knowledge and quantification of the chemical reactions occurring in the atmosphere, while on the other, there are considerable difficulties in representing reactions higher than the first order (particularly, with constant fast reactions). Generally speaking, semiempirical models empirically describe the impoverishment of a reactive pollutant by means of an exponential decrease on the basis of an assigned value of average lifetime of the pollutant. The K models prove to be theoretically most correct in representing the diffusion of chemically reacting material. However, they are limited by the fact that their characteristic time of transport and dispersion is not shorter than that of chemical reactions, which at times make description of dispersion with chemical reactions problematic, even using numerical models (Seinfeld and Pandis, 1997).

5.7   NEW OPERATIVE MODELS

As mentioned above, most operative models for estimating gas and particle dispersion in the atmospheric boundary layer are based on the Gaussian approach. Such models are founded on the hypothesis that the pollutant is dispersed in a homogeneous turbulence. However, due to the presence of terrain, turbulence is generally not homogeneous along the vertical direction. In addition, the inputs of Gaussian models often refer to simple turbulence schemes.

Over the past 20 years, following the works of Holtslag and van Ulden (1983), Weil and Brower (1984), van Ulden and Holtslag (1985), Trombetti et al. (1986),

TABLE 5.5
Values of *a* and *b* in Equation 5.26

Stability Class	a	b
A	−0.0875	−0.1029
B	−0.03849	−0.1714
C	−0.00807	−0.3049
D	0.0	0.0
E	0.00807	−0.3049
F	0.03849	−0.1714

and Beljaars and Holtslag (1990), it has been possible to evaluate the fundamental parameters for the description of the characteristics of the surface and atmospheric boundary layers using measurements close to the ground. This has allowed the development of models that describe pollutant diffusion using as input ground-based meteorological data (which can be acquired by automated networks), but which are able to evaluate directly atmospheric turbulence, through the value of the Monin-Obukhov length and the attrition velocity, rather than empirical classes, like those of Pasquill-Gifford.

A relation between the old stability classes and the new parameterization has been proposed by Golder (1972), through an experimental relation between L and stability classes (see also Trombetti et al., 1986). Algebraically, it is expressed by

$$1/L = a\, z_0^b \qquad\qquad (5.26)$$

where
  z_0 is the ground roughness
  a and b are expressed as in Table 5.5

Among the new-generation models, of particular note are the Danish model OML (Berkowicz et al., 1986) and British model ADMS (Carruthers et al., 1992); the American model HPDM (Hanna and Paine, 1989), and new model AERMOD, proposed by the U.S. EPA. Among the puff models are CALPUFF (Scire et al., 1999), M4PUFF (Tirabassi and Rizza, 1997), and SPM (Tirabassi and Rizza, 1995).

5.8 EVALUATION OF MODEL PERFORMANCES AND RELIABILITY

A correct use of atmospheric transport and diffusion models must be based on a study of their capacity to represent real situations correctly. When possible, it is advisable to test the reliability of the model adopted using the data of topographic and meteorological scenarios characteristic of the areas where they are deployed. Models are evaluated by

- Assessing the soundness of process descriptions
- Examining the computer code for errors

- Examining the reasonableness of the results
- Calculating the sensitivity of model results to change in inputs
- Comparing evaluated concentrations from different air pollution models
- Comparing simulated and observed concentrations

Usually, an air quality measurement network, whether in an urban or industrial site, is designed according to specific criteria, so that it alone cannot provide all the information necessary for the validation of a model: the number of points and typology of measurements, whether of chemical or meteorological parameters, are generally insufficient to provide an overall coverage of the territory, or to guarantee a complete range of information against which to test the model. This is perfectly understandable, since in designing a network that must function stably over an area the tendency is to minimize the number of measurement points, which are normally fixed, in order to give greater evidence to the time evolution of concentrations rather than their spatial distribution.

In general, a monitoring network can contribute to model validation, when it is suitably integrated with other sensors during intensive measurement campaigns specially organized for this purpose.

It must be borne in mind, when using models, that, while they are rather sophisticated instruments that ultimately reflect the current state of knowledge on turbulent transport in the atmosphere, the results they provide are subject to a considerable margin of error. This is due to various factors, in particular the uncertainty of the intrinsic variability of the atmosphere.

Models, in fact, provide values expressed as an average, that is, a mean value obtained by the repeated performance of many experiments, while the measured concentrations are a single value of the sample to which the ensemble average provided by models refer. This is a general characteristic of the theory of atmospheric turbulence and is a consequence of the statistical approach used in attempting to parameterize the chaotic character of the measured data. At the same time, the uncertainty linked to the stochastic character of the parameterization of the atmosphere depends on turbulence intensity and is a function of the mean sampling time. Atmospheric diffusion models ultimately present errors that can be reduced as an uncertainty inherent in the phenomenon they describe. The reducible errors originate from the use of an incorrect or insufficient set of input data and/or from the intrinsic inadequacies of the particular model. As previously noted, irreducible errors are due to the statistical nature of the parameterization of the turbulent fluxes responsible for the dispersion of the material emitted into the atmosphere. However, studies of model performance validation indicate errors in input data (both of emission and meteorology) to be the factor responsible for the greater contribution of the total uncertainty of models. Irwin et al. (1987), using Monte Carlo techniques to simulate to propagation of errors from those of input data, showed that the interval of error of the concentration maximum and of its distance from the source may be double the interval of error of the input data. A model is generally deemed acceptable if the estimated values are within a factor of two of the observed data.

A considerable amount of work has been done, especially in the United States, on the quantitative assessment of model performances. The adopted approach varies according to whether the validation refers to the scientific validity of the model or to the usefulness of the model in environmental management. In validations for management purposes, less importance is attributed to deterministic processes (processes of cause and effect), while attention is focused on the correspondence between observed and predicted values. For such analyses to be objective, it is necessary to utilize different data from those adopted for the parameterization of the model itself. Particular attention is paid, on the whole, to the comparison of concentration maxima both with and without temporal and/or spatial simultaneity.

Generally, the attempt is made to objectivize the validation of model performances by adopting statistical indices that describe their capacity to represent observed data. Among them, the most widely used are (Hanna, 1988):

$$\text{nmse (normalized mean square)} = \overline{(C_o - C_p)^2} / \overline{C_o C_p}$$

$$\text{cor (correlation)} = \overline{(C_o - \overline{C_o})(C_p - \overline{C_p})} / \sigma_o \sigma_p$$

$$fa2 = \text{percentage of data for which } \mathbf{0.5 \leq C_o/C_p \leq 2}$$

$$\text{fb (fractional bias)} = (\overline{C_o} - \overline{C_p})/(0.5(\overline{C_o} + \overline{C_p}))$$

$$\text{fs (fractional standard deviation)} = (\sigma_o - \sigma_p)/0.5(\sigma_o + \sigma_p)$$

where the suffixes o and p respectively refer to observed and predicted concentrations, and the bar indicates the mathematical mean.

REFERENCES

Beljaars, A.C.M. and Holtslag, A.A.M. (1990), A software library for the calculation of surface fluxes over land and sea, *Environ. Soft.* 5, 60–68.

Berkowicz, R.R., Olesen, H.R., and Torp, U. (1986), The Danish Gaussian air pollution model (OML): Description, test and sensitivity analysis in view of regulatory applications. *NATO-CCMS 16th International Meeting on Air Pollution Modelling and Its Applications,* C. De Wispelaere, F.A. Schiermeier, and N.V. Gillani (Eds.), Plenum Press, New York, pp. 453–481.

Berlyand, M.Y. (1975), Contemporary problems of atmospheric diffusion and pollution of the atmosphere. Translated version by NERC, USEPA, Raleigh, NC.

Bowen, B.M. (1994). Long-term tracer study at Los Alamos, New Mexico. Part II: Evaluation and comparison of several methods to determinate dispersion coefficients, *J. Appl. Meteorol.* 33, 1236–1254.

Briggs, G.A. (1973), Diffusion estimation for small emissions, in environmental research laboratories, air resources atmospheric turbulence and diffusion laboratory, Annual report. USAEC Report ATDL-106, National Oceanic and Atmospheric Administration, December 1974.

Briggs, G.A. (1975), Plume rise predictions. In *Lectures on Air Pollution and Environmental Impact Analyses. Workshop Proceedings,* Boston, MA, Sept. 29–Oct. 3, 59–111. American Meteorological Society, Boston, MA.

Briggs, G.A. (1985), Analytical parameterization of diffusion: the convective boundary layer. *J. Clim. Appl. Meteorol.* 24, 1167–1186.

Brown, M.J., Arya, S.P., and Snyder, W. (1997), Plume descriptors from a non-Gaussian concentration model, *Atmos. Environ.* 31, 183–189.

Carruthers, D.J., Holroyd, R.J., Hunt, J.C.R., Weng, W.S., Robins, A.G., Apsley, D.D., Smith, F.B., Thomson, D.J., and Hudson, B. (1992), U.K. atmospheric dispersion modelling system. In *Air Pollution Modeling and its Application IX*, H. van Dop and G. Kallos (Eds.), pp. 15–28, 19th NATO/CCMS International Technical Meeting on Air Pollution Modeling and Its Application, Creta, Greece, Sept. 29–Oct. 4, 1991. Plenum Press, New York.

Catalano, G.D. (1982), An analytical solution to the turbulent diffusion equation with mean vertical wind, In *Proceedings of the 16th Southeastern Seminar on Thermal Science*, Miami, FL, April 19–21, pp. 143–151.

Deardoff, J.W. and Willis, G.E. (1975), A parameterization of diffusion into the mixed layer. *Journal of Appl. Met.* 14, 1451–1458.

Degrazia, G.A. and Moraes, O.L.L. (1992), A model for eddy diffusivity in a stable boundary layer. *Boundary Layer Meteorology* 58, 205–214.

Degrazia, G.A., Rizza, U., Mangia, C., and Tirabassi, T. (1997), Validation of a new turbulent parameterization for dispersion models in convective conditions, *Bound. Lay. Meteorol.* 85, 243–254.

Degrazia, G.A., Anfossi, D., Carvalho, J.C., Mangia, C., Tirabassi, T., and Campos Velho, H.F. (2000), Turbulence parameterisation for PBL dispersion models in all stability conditions, *Atmos. Environ.* 34, 3575–3583.

Demuth, C. (1978), A contribution to the analytical steady solution of the diffusion equation for line sources, *Atmos. Environ.* 12, 1255–1258.

Erbrink, J.J. (1995), Use of boundary-layer meteorological parameters in the Gaussian model "stacks," *Bound. Lay. Meteorol.* 74, 211–235.

Fisher, B.E.A. (1975), The long range transport of sulphur dioxide, *Atmos. Environ.* 9, 1063–1070.

Gifford, F.A. (1961), Use of routine meteorological observations for estimating the atmospheric dispersion, *Nucl. Safety* 47–57.

Golder, D. (1972), Relations among stability parameters in the surface layer, *Bound. Lay. Meteorol.* 3, 47–58.

Green, A.E., Singhal, R.P., and Venkateswar, R. (1980), Analytic extension of the Gaussian plume model, *JAPCA* 30, 773–776.

Hanna, S.R. (1986), Lateral dispersion from tall stacks, *J. Clim. Appl. Meteorol.* 25, 1426–1433.

Hanna, S.R. (1988), Air quality model evaluation and uncertainty, *JAPCA* 38, 406–412.

Hanna, S.R. and Paine, R.J. (1989), Hybrid plume dispersion model development and evaluation. *Journal of Applied Meteorology* 28, 206–224.

Hanna, S.R., Briggs, G.A., Deardoff, J., Egan, B.A., Gifford, F.A., and Pasquill, F. (1977), AMS workshop on stability classification schemes and sigmas curves—Summary of recommendations, *Bull. Am. Meteorol. Soc.* 58, 1305–1309.

Hay, J.S. and Pasquill, F. (1959), Diffusion from a continuous source in relation to the spectrum and scale of turbulence. In *Atmospheric Diffusion and Air Pollution*, Eds. F.N. Frenkiel and P.A. Sheppard. *Advances in Geophysics* 6, 345–365.

Holtslag, A.A.M. and van Ulden, A.P. (1983), A simple scheme for daytime estimation of surface fluxes from routine weather data, *J. Clim. Appl. Meteorol.* 22, 517–529.

Irwin, J.S. (1980), Estimating plume dispersion—A recommended generalized scheme, *4th AMS Symposium on Turbulence and Diffusion*, Reno, NV.

Irwin, J.S. (1983), Estimating plume dispersion—A comparison of several sigma schemes, *J. Clim. Appl. Meteorol.* 22, 92–114.

Irwin, J.S., Rao, S.T., Petersen, W.B., and Turner, D.B. (1987), Relating error bounds for maximum concentration estimates to diffusion meteorology uncertainty, *Atmos. Environ.* 21, 1927–1937.

Lin, J.S. and Hildemann, L.M. (1997), A generalised mathematical scheme to analytically solve the atmospheric diffusion equation with dry deposition, *Atmos. Environ.* 31, 59–71.

Mangia, C., Rizza, U., Degrazia, G.A., and Tirabassi, T. (1998), A new formulation of σ_y and σ_z for air quality dispersion models, In *Air Pollution VI*, C.A. Brebbia, C.F. Ratto, and H. Power (Eds.), Computational Mechanics Publications, Southampton, U.K., pp. 839–848.

Mohan, M. and Siddiqui, T.A. (1977), An evaluation of dispersion coefficients for use in air quality models, *Bound. Lay. Meteorol.* 84, 177–205.

Moreira, D.M., Carvalho, J.C., and Ttirabassi T. (2005a), Plume dispersion simulation in low wind conditions in the stable and convective boundary layers, *Atmos. Environ.* 39, 3643–3650.

Moreira, D.M., Vilhena, M.T., Tirabassi, T., Buske, D., and Cotta, R. (2005b), Near-source atmospheric pollutant dispersion using the new GILTT method, *Atmos. Environ.* 39, 6290–6295.

Nieuwstadt, F.T.M. (1980), An analytical solution of the time-dependent, one-dimensional diffusion equation in the atmospheric boundary layer, *Atmos. Environ.* 14, 1361–1364.

Nieuwstadt, F.T.M. and de Haan, B.J. (1981), An analytical solution of one-dimensional diffusion equation in a non-stationary boundary layer with an application to inversion rise fumigation, *Atmos. Environ.* 15, 845–851.

Pasquill, F. (1971), Atmospheric dispersion of pollution, *Quart. J. Roy. Meteorol. Soc.* 97, 369–395.

Pasquill, F. and Smith, F.B. (1983), *Atmospheric Diffusion*, Halsted Press, John Wiley & Sons, New York.

Pleim, J.E. and Chang, J.S. (1992), A non-local closure model for vertical mixing in the convective boundary layer, *Atmos. Environ.* 26A, 6, 965–981.

Roberts, O.F.T. (1923), The theoretical scattering of smoke in a turbulent atmosphere, *Proc. Roy. Soc.* 104, 640–648.

Rounds, W. (1955), Solutions of the two-dimensional diffusion equation, *Trans. Am. Geophys. Union* 36, 395–405.

Scire, J.S., Robe, F.R., Strimaitis, M.E., and Yamartino, R.J. (1999), *A Users's Guide for the CALPUFF Dispersion Model*, Earth Tech, Inc., Concord, MA.

Scriven, R.A. and Fisher, B.A. (1975), The long range transport of airborne material and its removal by deposition and washout-II. The effect of turbulent diffusion, *Atmos. Environ.* 9, 59–69.

Seinfeld, J.H. and Pandis, S.N. (1997), *Atmospheric Chemistry and Physics*, John Wiley & Sons, New York.

Smith, F.B. (1957a), The diffusion of smoke from a continuous elevated point source into a turbulent atmosphere, *J. Fluid Mech.* 2, 49–76.

Smith, F.B. (1957b), Convection-diffusion processes below a stable layer, Meteorological Research Committee, N. 1048 and 10739.

Smith, M.E. (1968), *Recommended Guide for the Prediction of the Dispersion of Airborne Effluents*, American Society of Mechanical Engineers, New York.

Stern, A.C. (1976), *Air Pollution*, Academic Press, New York.

Tagliazucca, M., Nanni, T., and Tirabassi, T. (1985), An analytical dispersion model for sources in the surface layer, *Nuovo Cimento* 8C, 771–781.

Taylor, G.I. (1921), Diffusion by continuous movement. *Proc. Lond. Math. Soc.* 2, 196–211.

Tirabassi, T. (1989), Analytical air pollution and diffusion models, *Water Air Soil Pollut.* 47, 19–24.

Tirabassi, T. (2003), Operational advanced air pollution modeling, *PAGEOPH* 160(1–2), 5–16.

Tirabassi, T. and Rizza, U. (1995), A practical model for the dispersion of skewed puffs. *J. Appl. Meteorol.* 34, 989–993.

Tirabassi, T. and Rizza, U. (1997), Boundary layer parameterization for a non-Gaussian puff model, *J. Appl. Meteorol.* 36, 1031–1037.

Tirabassi, T., Tagliazucca, M., and Zannetti, P. (1986), KAPPA-G, a non-Gaussian plume dispersion model: Description and evaluation against tracer measurements, *JAPCA* 36, 592–596.

Tirabassi, T., Tagliazucca, M., and Paggi, P. (1989), A climatological model of dispersion in an inhomogeneous boundary layer, *Atmos. Environ.* 23, 857–862.

Trombetti, F., Tagliazucca, M., Tampieri, F., and Tirabassi, T. (1986), Evaluation of similarity scales in the stratified surface layer using wind speed and temperature gradient, *Atmos. Environ.* 20, 2465–2471.

van Ulden, A.P. (1978), Simple estimates for vertical diffusion from sources near the ground, *Atmos. Environ.* 12, 2125–2129.

van Ulden, A.P. and Holstlag, A.A.M. (1985), Estimation of atmospheric boundary layer parameters for diffusion applications, *J. Clim. Appl. Meteorol.* 24, 1196–1207.

Ulke, A.G. (2000), New turbulent parameterisation for a dispersion model in the atmospheric boundary layer. *Atmospheric Environment* 34, 1029–1042.

Weil, J.C. and Brower, R.P. (1984), An update Gaussian plume model for tall stacks, *JAPCA* 34, 818–827.

Yeh, G.T. and Huang, C.H. (1975), Three-dimensional air pollutant modeling in the lower atmosphere, *Bound. Lay. Meteorol.* 9, 381–390.

Zannetti, P. (1990), *Air Pollution Modelling*, Computational Mechanics Publications, Southampton, U.K. and Van Nostrand Reinhold, New York.

6 Analytical Models for the Dispersion of Pollutants in Low Wind Conditions

Pramod Kumar and Maithili Sharan

6.1 INTRODUCTION

Mathematical models are indispensable tools to understand the ad hoc problems fashioned by the adverse effect of the developments in the atmosphere due to the rapid expansion of industrialization, urbanization, and air pollution episodes with the accidental toxic gases, radioactive release, and the growing risk of climate change associated with long-term undesirable effects of urban air. These models are employed to compute the concentration of a contaminant at a given time and position by utilizing available meteorological observations. An investigation of these models assists in the planning of dispersion surveys, air quality analysis, impact assessment studies, and emergency preparedness, and contributes to the development of strategies for future measurement programs.

In practice, most of the estimates of dispersion are based on the Gaussian plume model. The Gaussian model is known to work reasonably well during most of the meteorological regimes but fails to work in weak ($U < 2\,\mathrm{m\ s^{-1}}$) and variable wind conditions (Arya, 1995; Sharan et al., 1996a,b, 2003). Pollutant dispersion in low wind conditions becomes complicated because of the unstructured spatial and temporal behavior of the pollutants in these circumstances. The diffusion of pollutants released from the various emission sources is irregular and indefinite in weak and variable wind conditions. As a consequence, no single plume centerline is obvious and the observed concentration distribution is multipeaked and non-Gaussian (Sagendrof and Dickson, 1974). In these conditions, the state (turbulence and dispersion characteristics) of lower atmosphere is not properly understood, the pollutant is not able to travel far, and, accordingly, the region surrounding the source may be affected adversely. These conditions have been observed to occur for a considerable period of time during the day as well as during the night in most parts of the world and are sensitive due to their great potential for the occurrence of pollution episodes (Sharan et al., 1995a, 2003; Sharan and Gopalakrishnan, 2003). For example, the infamous Bhopal gas leak had also taken place under low wind stable conditions (Sharan et al., 1995a).

Over the years, analytical models have been developed in the literature. These models are essentially constrained by (1) various approximation and assumptions at the level of formulation and (2) parameterization of dispersion parameters. During the development of these models, a number of mathematical approximations are involved. Some of these models have been validated using the limited available observations. Also, some of these are successfully applied to various applications.

Thus, in this chapter, an attempt has been made to provide an overview of analytical models for the treatment of dispersion of air pollutants in low wind conditions.

6.2 GENERAL MATHEMATICAL FORMULATION

The deterministic models for the dispersion of pollutants in atmosphere, based on the advection–diffusion equation and K-theory, can be written as

$$\frac{\partial C}{\partial t} + u\frac{\partial C}{\partial x} + v\frac{\partial C}{\partial y} + w\frac{\partial C}{\partial z} = \frac{\partial}{\partial x}\left(K_x\frac{\partial C}{\partial x}\right) + \frac{\partial}{\partial y}\left(K_y\frac{\partial C}{\partial y}\right) + \frac{\partial}{\partial z}\left(K_z\frac{\partial C}{\partial z}\right) + S + R \tag{6.1}$$

where
   C is the mean concentration of a pollutant
   u, v, and w are the components of the wind velocity in x, y, and z directions, respectively
   S is the source term

R is the removal term that represents pollutant removal by chemical reaction, gravity fallout (settling), etc.

K_x, K_y, and K_z are the eddy diffusivities along x, y, and z directions, respectively

On the left-hand side of Equation 6.1, the first term is the time-dependent term accounting for nonstationary situations and the remaining three terms describe transport due to advection whereas on the right-hand side, the first three terms represent the turbulent diffusion. Equation 6.1 forms the basis for most air pollution dispersion models and allows for the anisotropic diffusion and for the variation of diffusivities as a function of concentration, time, and spatial coordinates. The justification of Equation 6.1 for mean concentration is described in Csanady (1973), Seinfeld (1986), and Arya (1999).

To completely specify a mathematical problem and in order to solve Equation 6.1, one needs to assign the physically relevant boundary and initial conditions depending on the nature of the physical problem. In general, there are three types of boundary conditions (Crank, 1976): (1) the Dirichlet type in which the concentration of the pollutant is prescribed on the boundary, (2) the Neumann type flux condition in which the concentration flux normal to the boundary is prescribed, and (3) the mixed type in which the concentration flux across the boundary is proportional to the difference in the concentrations between the boundary and outside medium. The constant of proportionality is related to the permeability of the boundary. All these types of boundary conditions along with their physical interpretation are discussed in the literature (Crank, 1976; Carslaw and Jaeger, 1986).

The initial conditions to be prescribed are generally expressed in terms of background concentration. Although precise background concentration is normally not available, one can consider arbitrary functional form in terms of spatial coordinates.

6.3 ANALYTICAL MODELS

Analytical models are not only indispensable tools to predict the concentration of air pollutants with high mathematical accuracy but are also the easiest way to describe the unstructured temporal and spatial behavior of the pollutants in the atmosphere. Although it is not always possible to find an analytical solution for every dispersion model, even for some particular form of wind velocities and eddy diffusivities. The treatment of the dispersion of pollutants can be described by two timescales: (1) the duration (t_s) for which the source is released and (2) travel time (t_r) from the source to the receptor. The release can be treated as a continuous release if $t_s > t_r$, otherwise it is instantaneous. Accordingly, a continuous release (plume) or instantaneous release (puff) model may be used for the dispersion of a pollutant emitted from a source (Hanna et al., 1982; Yadav, 1995). These are also often known as the steady- and unsteady- (time-dependent) state models. Here, we describe the features of the various analytical dispersion models.

6.3.1 GAUSSIAN PLUME MODEL

A simple analytical model leading to the widely used Gaussian model has been derived by taking the following assumptions in Equation 6.1:

a. Steady-state conditions are considered (i.e., $(\partial C/\partial t)=0$).
b. The vertical velocity component (w) is smaller in comparison to horizontal velocity components (u and v) and thus, it is neglected.
c. x-Axis is oriented in the direction of mean wind (i.e., $u = U$, $v = 0$).
d. Since the transport due to mean wind is dominated over the downwind diffusion, that is, $|U(\partial C/\partial x)| \gg |(\partial/\partial x)(K_x(\partial C/\partial x))|$ neglecting the downwind diffusion transport in comparison to advection.
e. Removal of pollutants is neglected (i.e., $R = 0$).
f. Wind speed and eddy diffusivity coefficients are assumed constant.
g. A point source located at $(0, 0, H_s)$ with emission rate Q (gm/s) is represented as

$$S = Q\, \delta(x)\delta(y)\delta(z - H_s)$$

where
    H_s is the effective stack height from the ground
    $\delta(\cdot)$ is the Dirac-delta function

Then, these assumptions in Equation 6.1 lead to the following partial differential equation:

$$U\frac{\partial C}{\partial x} = K_y\frac{\partial^2 C}{\partial y^2} + K_z\frac{\partial^2 C}{\partial z^2} + Q\delta(x)\,\delta(y)\,\delta(z - H_s) \tag{6.2}$$

It is a parabolic partial differential equation and subject to the following boundary conditions:

$$C = 0, \quad |y|, z \to \infty \tag{6.3a}$$

$$-K_z\frac{\partial C}{\partial z} = 0, \quad z = 0 \tag{6.3b}$$

$$C(0, y, z) = 0 \quad \text{(deleted neighborhood)} \tag{6.3c}$$

Here deleted neighborhood means the region excluding small neighborhood of the point where the source is located. A closed form solution of the resulting Equation 6.2 in the domain: $0 < x < \infty$, $-\infty < y < +\infty$, $0 < z < \infty$, with boundary conditions (Equation 6.3), is obtained by the method of integral transforms (Seinfeld, 1986) and given as

$$C(x, y, z) = \frac{Q}{4\pi\left(K_y K_z\right)^{1/2} x}\exp\left(-\frac{Uy^2}{4K_y x}\right)\left[\exp\left(-\frac{U(z - H_S)^2}{4xK_z}\right) + \exp\left(-\frac{U(z + H_S)^2}{4xK_z}\right)\right] \tag{6.4}$$

An important aspect of the dispersion problem is the representation of a source in the model formulation. Here notice that the source term is accounted for in the material balance equation. Alternatively, this can be accounted for through the boundary of the domain. Sharan et al. (1999) proposed the equivalent mathematical formulations for accounting the source either through advection–diffusion equation or through one of the boundaries. It has been demonstrated that all such formulations lead to the same solution. Accordingly, by modifying the boundary condition (Equation 6.3c) to account for the source, the solution of Equation 6.2 is found to be the same as given by Equation 6.4 (Yadav et al., 1996; Sharan et al., 2003).

For facilitating the practical application of the analytical solution (Equation 6.4), the eddy diffusivities K_x, K_y, and K_z are expressed in terms dispersion parameters as

$$K_i = \frac{1}{2}\frac{d\sigma_i^2}{dt}, \quad i = x, y, z \tag{6.5}$$

in which parameters σ_x, σ_y, and σ_z are the standard deviations of the concentration distribution in downwind, crosswind, and vertical directions, respectively. These dispersion parameters are the functions of stability and downwind distance and are based on the combination of experimental results and theory (Hanna et al., 1982). For constant K_i's ($i = x, y, z$), Equation 6.5 takes the form:

$$K_i = \frac{\sigma_i^2}{2t} = \sigma_i^2 \frac{U}{2x}, \quad i = x, y, z \tag{6.6}$$

in which t is the travel time in terms of mean wind speed and downwind distance. The analytical solution (Equation 6.4) can be written in terms of the dispersion parameters as

$$C(x,y,z) = \frac{Q}{2\pi U \sigma_y \sigma_z}\exp\left(-\frac{y^2}{2\sigma_y^2}\right)\left[\exp\left(-\frac{(z-H_s)^2}{2\sigma_z^2}\right) + \exp\left(-\frac{(z+H_s)^2}{2\sigma_z^2}\right)\right] \tag{6.7}$$

This is the widely used standard Gaussian plume solution, describing the concentration distribution of a nonreactive pollutant emitted from a continuous point source located at height H_s in an infinite domain. For a ground-level source, the concentration of pollutant can be estimated by taking the limit $H_s \rightarrow 0$ in Equation 6.7. The properties of this solution are discussed in the literature (Seinfeld, 1986; Arya, 1999).

6.3.2 Gaussian Puff Model

Relaxing the assumptions (a) and (d) in standard Gaussian plume model (Section 6.3.1), the differential Equation 6.2 includes the time-dependent and horizontal diffusion terms and can be written as

$$\frac{\partial C}{\partial t} + U\frac{\partial C}{\partial x} = K_x\frac{\partial^2 C}{\partial x^2} + K_y\frac{\partial^2 C}{\partial y^2} + K_z\frac{\partial^2 C}{\partial z^2} + q\delta(x)\delta(y)\delta(z-H_s)\delta(t) \tag{6.8}$$

For the emission from an instantaneous point source of strength q (in grams) located at $(0, 0, H_s)$, the analytical solution of Equation 6.8 with the boundary and initial conditions:

$$\text{(i) } C \rightarrow 0 \quad \text{as } |x|, |y|, |z| \rightarrow \infty, \forall t > 0 \qquad (6.9a)$$

$$\text{(ii) } C(x,y,z) \rightarrow 0 \quad \text{as } t \rightarrow 0 \text{ for all } x, y, \text{ and } z \qquad (6.9b)$$

can be written as (Seinfeld, 1986)

$$C(x, y, z, t) = \frac{q}{8(\pi t)^{3/2} (K_x K_y K_z)^{1/2}} \exp\left[-\frac{(x-Ut)^2}{4K_x t} - \frac{y^2}{4K_y t} - \frac{(z-H_s)^2}{4K_z t} \right] \qquad (6.10)$$

Expressing K_i's in terms of σ_i's, the solution (Equation 6.10) becomes

$$C(x, y, z, t) = \frac{q}{(2\pi)^{3/2} \sigma_x \sigma_y \sigma_z} \exp\left[-\frac{(x-Ut)^2}{2\sigma_x^2} - \frac{y^2}{2\sigma_y^2} - \frac{(z-H_s)^2}{2\sigma_z^2} \right] \qquad (6.11)$$

Equation 6.11 is known as Gaussian puff equation for an elevated release in an infinite medium.

In the presence of boundary located at the ground $z = 0$ because of the complete reflection of the pollutants at the plane, the boundary conditions (Equation 6.9a) can be reformulated by changing the domain in z from $(-\infty, \infty)$ to $(0, \infty)$ and written as

$$C \rightarrow 0, \quad \text{as } |x|, |y|, z \rightarrow \infty \qquad (6.12a)$$

$$-K_z \frac{\partial C}{\partial z} = 0, \quad \text{at } z = 0 \qquad (6.12b)$$

The closed form analytical solution of Equation 6.8 with the initial condition (6.9b) and the boundary conditions (6.12) is given as (Sharan et al., 1996c)

$$C(x,y,z,t) = \frac{q}{8(\pi t)^{3/2} (K_x K_y K_z)^{1/2}} \exp\left[-\left(\frac{(x-Ut)^2}{4K_x t} + \frac{y^2}{4K_y t} \right) \right]$$

$$\times \left\{ \exp\left[-\frac{(z-H_s)^2}{4K_z t} \right] + \exp\left[-\frac{(z+H_s)^2}{4K_z t} \right] \right\} \qquad (6.13)$$

The above solution can be expressed in terms of the dispersion parameters (Equation 6.6):

$$C(x,y,z,t) = \frac{q}{(2\pi)^{3/2}\sigma_x\sigma_y\sigma_z} \exp\left[-\left(\frac{(x-Ut)^2}{2\sigma_x^2} + \frac{y^2}{2\sigma_y^2}\right)\right]$$

$$\times \left\{\exp\left[-\frac{(z-H_s)^2}{2\sigma_z^2}\right] + \exp\left[-\frac{(z+H_s)^2}{2\sigma_z^2}\right]\right\} \qquad (6.14)$$

This is called as the Gaussian puff formula for an elevated source at $(0, 0, H_s)$ with total reflection at the ground level. It provides the distribution of concentration inside a puff. It can be obtained in a Lagrangian frame of reference assuming a stationary, homogeneous Gaussian flow field as illustrated in Seinfeld (1986). The solution (Equation 6.14) of Equation 6.8 gets modified if the ground reflects/absorbs partially the pollutant.

6.3.3 NON-GAUSSIAN MODELS

The Gaussian models are derived by assuming the wind speed and eddy diffusivity constants. This assumption has been relaxed over the years by taking (1) K as a function of time, (2) U as a function of height above the ground, and (3) K as a function of either downwind distance from the source or height above the ground. For particular forms of U and K, the resulting analytical solutions with physically relevant boundary conditions have been derived (Sharan et al., 2003) over the years by various researchers. These solutions are no longer Gaussian. For example, Lin and Hildemann (1996) described an analytical solution for Equation 6.1 with assumptions (a)–(e) in Section 6.3.1, by considering the following power-law profiles of the wind speed and eddy diffusivities:

$$U(z) = az^\alpha \qquad (6.15a)$$

$$K_z(z) = bz^\beta \qquad (6.15b)$$

$$K_y(x,z) = f(x)z^\gamma \qquad (6.15c)$$

where
  $f(x)$ is an integrable function of downwind distance x
  α, β, γ are constants depending upon the atmospheric stability and surface roughness

For n point sources located at (x_s^i, y_s^i, z_s^i) $(i = 1, 2, \ldots, n)$ in a Cartesian coordinate system with total reflection from ground at $z = 0$ and from top of the unbounded inversion layer at $z \to \infty$, and power-law profiles (Equation 6.15) of wind and eddy diffusivities with assumption $\alpha = \gamma$, the analytical solution is derived (Lin and Hildemann, 1996) in the form of Green's function and can be written as

$$C(x,y,z) = \sum_{i=1}^{n} Q^i G_z^i(x,z; x_S^i, z_S^i)\, G_y^i(x,y; x_S^i, y_S^i) \qquad (6.16)$$

where

$$G_z^i\left(x,z;x_S^i,z_S^i\right)=\frac{(zz_S^i)^{(1-\beta)/2}}{b(\alpha-\beta+2)(x-x_S^i)}\,I_{-\mu}\left[\frac{2a\,(zz_S^i)^{(\alpha-\beta+2)/2}}{b\left(\alpha-\beta+2\right)^2\left(x-x_S^i\right)}\right]$$

$$\times\exp\left[-\frac{a\left((z)^{\alpha-\beta+2}+(z_S^i)^{\alpha-\beta+2}\right)}{b(\alpha-\beta+2)^2(x-x_S^i)}\right] \tag{6.17}$$

with $\mu=(1-\beta)/(\alpha-\beta+2)$, in which $I_{-\mu}$ is the modified Bessel function of the first kind of order $-\mu$, and

$$G_y^i\left(x,y;x_S^i,y_S^i\right)=\frac{\sqrt{a}}{\sqrt{4\pi\left(X-X_S^i\right)}}\exp\left(-\frac{a\left(y-y_S^i\right)^2}{4\left(X-X_S^i\right)}\right) \tag{6.18}$$

with

$$X=\int_{x_c}^{x}f(\tau)\,d\tau,\quad X_S^i=\int_{x_c}^{x_S^i}f(\tau)\,d\tau \tag{6.19}$$

in which x_c is some reference location in Cartesian coordinates. Thus, the solution (Equation 6.16) represents a non-Gaussian distribution.

6.3.4 Limitations of Gaussian Models in Low Wind Conditions

The diffusion of pollutant released from the various emission sources is irregular and indefinite in weak and variable wind conditions. No single plume centerline is obvious, and the observed concentration distribution is multipeaked and non-Gaussian (Sagendrof and Dickson, 1974), especially in stable conditions. In these conditions, the turbulence is not generated solely by the mechanical components created by frictional shear at the surface and a thermal component arising from vertical heat flux. Light wind stable conditions were excluded by Pasquill (1961) from the original stability classification because the diffusing plume is unlikely to have any definable travel.

The Gaussian plume models may not be applicable to the limiting conditions of low wind conditions, because (1) the downwind diffusion is neglected in comparison with the advection; (2) the concentration is inversely proportional to U, and, therefore, the concentration approaches to infinity as the wind speed tends to zero; (3) the average conditions are stationary (Anfossi et al., 1990); and (4) the nonavailability of dispersion parameters in low winds. These models have limitations at two levels: (1) at the level of model formulation that involves the approximations and assumptions and (2) the nonavailability of dispersion parameters.

From both these aspects, attempts have been made (Yadav et al., 1996; Sharan and Gopalakrishnan, 2003; Sharan et al., 2003) to deal with the problem of atmospheric dispersion in low wind conditions. Here, we briefly describe the analytical models for the dispersion of air pollutants in low wind conditions.

6.4   ANALYTICAL MODELS IN LOW WIND CONDITIONS

6.4.1   Constant *K*-Models

First, we briefly describe the models derived by considering the downwind diffusion and assuming U and K_i's as constant.

6.4.1.1   Unsteady-State Models

6.4.1.1.1   Puff Model

The Gaussian puff model (Equation 6.14) accounts for the advection and the downwind diffusion term. In this model, the concentration of the material inside puff is assumed to be distributed according to the Gaussian distribution. One of the advantages of the puff approaches is that it can handle very low or even zero wind speeds because of the disappearances of U from the denominator. In other words, in puff models, the wind speed affects the concentration computation only by controlling the density of puffs in the region (i.e., the lower the wind speed, the closer a puff is to the next one generated by the same source). Therefore, the puff approach allows, at least theoretically, the treatment of calm or low wind conditions (Zannetti, 1990). Inclusion of downwind diffusion is another of its advantage over the Gaussian plume equation (Cirillo and Poli, 1992). However, as pointed out in Section 6.3.4, the problem may be encountered with the availability of dispersion parameters in low wind conditions.

6.4.1.1.2   Segment–Puff Model

During weak and variable wind conditions, the dispersion process is generally nonhomogeneous and nonstationary because of the complex meteorological and/ or terrain situations. Sharan et al. (1996c) described a time-dependent mathematical model to treat the nonstationary and nonhomogeneous dispersion from a point source release. This approach involves coupling of plume segment and Gaussian puff methods (Equation 6.14). In this approach, a continuous plume is approximated by a finite number of segments, and each segment is represented by a series of contiguous puffs. The Gaussian puff model allows for the diffusion of pollutants in all three directions (including along wind) and described the pollutant dynamics by the temporal evolution of plume segments, while the plume segment approach helps in tracking the trajectory of the plume. The characteristics of plume segments are updated at each dispersion time interval. Zannetti (1986) also presented a mixed segment approach, which considered the plume to be divided into a series of elements that are either segments or puffs.

The segment–puff model used to simulate the tracer tests (Sagendorf and Dickson, 1974) over a short range (up to 400 m) in the flat area during low wind stable conditions. The overall agreement between the observed and the simulated curves found to be encouraging. This model can handle even nonhomogeneous meteorological input as nonhomogeneity is accounted for indirectly by following the trajectory of the plume.

6.4.1.2 Steady-State Models

a. Steady-state concentration is obtained by integrating the concentration obtained from the unsteady models with respect to time t from 0 to ∞

$$C_S(x,y,z) = \int_0^\infty C(x,y,z,t)\,dt \qquad (6.20)$$

By integrating the unsteady solution (Equation 6.13) with respect to time, the steady-state concentration of a pollutant can be given as

$$C(x,y,z) = \frac{Q}{4\pi\left(K_x K_y K_z\right)^{1/2}}\exp\left(\frac{Ux}{2K_x}\right)\left[\frac{1}{f_1}\exp\left(-\frac{U}{2K_x^{1/2}}f_1\right) + \frac{1}{f_2}\exp\left(-\frac{U}{2K_x^{1/2}}f_2\right)\right]$$

$$(6.21)$$

where

$$f_1 = \left[\frac{x^2}{K_x} + \frac{y^2}{K_y} + \frac{(z+H_s)^2}{K_z}\right]^{1/2}$$

and

$$f_2 = \left[\frac{x^2}{K_x} + \frac{y^2}{K_y} + \frac{(z-H_s)^2}{K_z}\right]^{1/2}$$

This solution was essentially derived for the first time by Roberts (1923). The subscript "s" is suppressed in the relation (Equation 6.21).

b. In case (a), it was mentioned that the steady-state solution can be obtained by integrating the unsteady solution with respect to time from 0 to ∞. This solution can also be obtained (Sharan et al., 1995b) by taking the steady-state advection diffusion equation:

$$U\frac{\partial C}{\partial x} = K_x\frac{\partial^2 C}{\partial x^2} + K_y\frac{\partial^2 C}{\partial y^2} + K_z\frac{\partial^2 C}{\partial z^2} + Q\delta(x)\delta(y)\delta(z-H_s) \qquad (6.22)$$

with the boundary conditions:

$$C \to 0, \quad \text{as } |x|,|y|, z \to \infty \qquad (6.23a)$$

$$-K_z\frac{\partial C}{\partial z} = 0 \quad \text{at } z = 0 \qquad (6.23b)$$

For ground-level source, the concentration distribution is deduced by taking the limit $H_s \to 0$ in Equation 6.21. In low wind conditions, the formula yielding the Gaussian plume solution (Equation 6.4) is approximate whereas the solution (Equation 6.21) seems more accurate as it accounts for the downwind diffusion. The solution (Equation 6.4) can also be deduced by taking the limit $K_x \to 0$ (ignoring the downwind diffusion) in Equation 6.21.

We define the ratio R of the solution (Equation 6.4) to the solution (Equation 6.21) for a ground-level source:

$$R = \sqrt{(1+p)} \exp\left[\frac{\beta}{2}\left\{(1+p)^{1/2} - \left(1+\frac{p}{2}\right)\right\}\right] \tag{6.24}$$

where

$$p = \frac{K_x}{x^2}\left(\frac{y^2}{K_y} + \frac{z^2}{K_z}\right) \quad \text{and} \quad \beta = \frac{Ux}{K_x} \tag{6.25}$$

are dimensionless parameters, which can take only positive values. This ratio (R) essentially represents the overprediction/underprediction by the Gaussian model in treating the dispersion in low wind conditions (Sharan et al., 1996b).

Notice that β resembles the well-known Peclet number P_e, and it essentially represents the ratio of advective transport to diffusive transport. Physically, p represents the region of interest relative to plume centerline and its small values, close to zero, indicates the region in the proximity of plume centerline, whereas the magnitude of β indicates the atmospheric conditions in terms of the strength of winds. Small magnitudes of β are indicating the weakening of winds when downwind diffusion becomes important. Based on p and β, it was suggested (Sharan et al., 1996b) that Gaussian models lead to an overestimation up to 25%.

For facilitating the practical applications of these constant K models, K_i's are expressed in terms of dispersion parameters (σ's) (Equation 6.5). A number of parameterizations of σ's are proposed in the literature (Hanna et al., 1982; Yadav, 1995; Yadav and Sharan, 1996; Arya, 1999; Sharan et al., 2003). For near source dispersion, σ's are proportional to time t, whereas far away from the source, these are proportional to $\sqrt{t}$ (Sharan et al., 2003) and correspondingly from Taylor's hypothesis, the spread of plume in the crosswind and vertical directions starts off with a linear form of downwind distance (proportional to x) and ultimately tends to a parabolic form. Various parameterizations available in the literature for the horizontal and vertical plume dispersion can be broadly classified into (1) methods based on the power-law functions (Briggs, 1973; Green et al., 1980), (2) methods based on the statistical parameters such as σ_θ (standard deviation of horizontal wind direction) and σ_ϕ (standard deviation of vertical wind direction), (Draxler, 1976; Irwin, 1979; Gryning et al., 1987), and (3) methods based on the similarity theory (Hanna et al., 1982; Seinfeld, 1986).

6.4.2 Variable K-Models

In Section 6.4.1, the concentration in low wind conditions is derived by taking the eddy diffusivities and mean wind as constants. However, this assumption is no longer physically relevant for dispersion of pollutants in the nonhomogeneous and nonstationary atmosphere, where the varying turbulent structure of atmosphere plays a pivotal role. In the constant K models discussed in Section 6.4.1, the solution was deduced by assuming K's constant and later on, their dependency on downwind distance was incorporated through dispersion parameters at the level of application. Thus, assuming K's constant in derivation of the solution of advection–diffusion equation and later on introducing their dependency on x is mathematically inconsistent (Liewelyn, 1983). However, this approach is accepted in the applications of dispersion models. Thus, attempts have been made to formulate variable K-models by relaxing the assumption of constant eddy diffusivities (or dispersion parameters) in advection–diffusion equation.

6.4.2.1 Integrated Puff Model with Dispersion Parameters as the Linear Functions of Time

While integrating the concentration (Equation 6.14), σ's are assumed to be independent of time. Cirillo and Poli (1992) obtained a steady-state solution by integrating the Gaussian puff formula (Equation 6.14) with respect to time from 0 to ∞, assuming dispersion parameters to be linear function of time in the following form:

$$\sigma_x = \alpha t, \quad \sigma_y = \beta t, \quad \text{and} \quad \sigma_z = \gamma t \tag{6.26}$$

where α, β, and γ are constants of proportionality, in which α and β are computed on the basis of the measured deviations of the horizontal wind direction, σ_θ (Green et al., 1980) and γ can be deduced from the expressions of σ_z recommended by Briggs (1973). These expressions for diffusion parameters are usually valid for a diffusion time range of up to a few hours (Okamoto and Shiozawa, 1987). After integrating Equation 6.14 with time, the solution is given as (Cirillo and Poli, 1992)

$$C(x,y,z) = \sum_{i=1,2} \frac{Q}{(2\pi)^{3/2} \alpha\beta\gamma T_i^2} \exp\left(-\frac{U^2}{2\alpha^2}\right)$$

$$\times \left\{ 1 + \sqrt{\frac{\pi}{2}} \frac{Ux}{2\alpha^2 T_i^2} \exp\left(-\frac{U^2 x^2}{2\alpha^4 T_i^2}\right) \text{erf}\left(-\frac{Ux}{\sqrt{2}\alpha^2 T_i}\right) \right\} \tag{6.27}$$

where erf(.) is the error function and

$$T_1^2 = \frac{x^2}{\alpha^2} + \frac{y^2}{\beta^2} + \frac{(z+H_s)^2}{\gamma^2}, \quad T_2^2 = \frac{x^2}{\alpha^2} + \frac{y^2}{\beta^2} + \frac{(z-H_s)^2}{\gamma^2}$$

The Expression 6.27 can be used for computing the concentration distribution released from an elevated point source in the absence of inversion layer.

6.4.2.2 Model with Eddy Diffusivities as Linear Functions of Downwind Distance

In general, and especially for horizontal diffusion from point sources, the gradient transfer theory with constant diffusivity yields erroneous results for dispersion close to the source where the size of the dispersed material is smaller than the most energetic turbulent eddies. In the early stage of plume dispersion from a point source, Taylor's (1921) statistical theory of diffusion and dimensional analysis suggest that the eddy diffusion coefficients (K) may be taken as linear functions of downwind distance. Scale analysis of atmospheric advection–diffusion also supports this for near source dispersion and accordingly, the eddy diffusion coefficients are proportional to Ux (product of mean wind speed and the downwind distance) and can be written as

$$K_x = \alpha Ux, \quad K_y = \beta Ux, \quad K_z = \gamma Ux \tag{6.28}$$

where α, β, and γ are constants of proportionality and represent the turbulence parameters. This parameterization accounts for the stability through these turbulent parameters, which essentially represent turbulent intensities (Arya, 1995). Arya (1995) has also argued that for small travel times (i.e., small distances from the source) diffusivities can be expressed as linear functions of downwind distance.

Sharan et al. (1996a) formulated a variable K-model for the dispersion of air pollutants in low wind conditions by taking the linear functional form (Equation 6.28) of eddy diffusivity coefficients. After introducing an elevated source of height H_s through $x = 0$ plane, the analytical solution of modified Equation 6.1 with assumptions (a) through (c) and (e) in Section 6.3.1, the eddy diffusivity coefficients (Equation 6.28) and boundary conditions:

$$C(x, y, z) = 0, \quad x, |y|, z \to \infty \tag{6.29a}$$

$$-K_z \frac{\partial C}{\partial z} = 0, \quad z = 0 \tag{6.29b}$$

$$UC(0, y, z) = Q\delta(y)\delta(z - H_s) \tag{6.29c}$$

is obtained using the integral transform (Sharan et al., 1996a; Sharan and Yadav, 1998) and given as

$$C(x, y, z) = \frac{Q}{2U\pi\sqrt{\beta\gamma}x^2}\left[F_{z+H_S} + F_{z-H_S}\right] \tag{6.30}$$

where

$$F_{z+H_S} = \left[1 + \frac{\alpha}{x^2}\left(\frac{y^2}{\beta} + \frac{(z + H_s)^2}{\gamma}\right)\right]^{-(\mu+1)}$$

and

$$F_{z-H_S} = \left[1 + \frac{\alpha}{x^2}\left(\frac{y^2}{\beta} + \frac{(z - H_s)^2}{\gamma}\right)\right]^{-(\mu+1)}$$

in which $\mu = \dfrac{1}{2\alpha}$.

To compute the concentration close to the plume centerline, the analytical solution (Equation 6.30) can be easily converted into slender plume approximation by taking $\alpha \to 0$ in Equation 6.30 (Yadav, 1995). During the evaluation, Sharan and Yadav (1998) have found similar results for the models (Equations 6.27 and 6.30), though they are based on different approaches. The reason for the similar results is attributed to the similar type of parameterizations used in both the approaches.

Turbulent intensities in Equation 6.28 can be calculated directly from the measurements. In the absence of such measurements, turbulent intensities may be parameterized in terms of friction velocity (u_*) stable conditions (Sharan and Yadav, 1998) and convective velocity (w_*) in unstable conditions (Sharan et al., 1996a). The use of parameterization of turbulent intensities in terms of w_* provides an under-prediction (Sharan et al., 1996a) of the concentration observed in IIT diffusion experiment (Singh et al., 1991) in unstable conditions. However, the parameterization based on u_* provides a significant improvement (Sharan et al., 2002). The eddy diffusivities in Equation 6.28 tend to vanish as downwind distance x from the source or the wind speed U approaches to zero. Vanishing U will introduce a singularity in the solution (Equation 6.30). In addition, this parameterization does not account for upstream diffusion. To overcome these shortcomings, there is a need to explore other alternative parameterizations of eddy diffusivities in terms of travel time, turbulent kinetic energy in order to develop the realistic variable-K models for describing low wind dispersion.

6.4.3 MODELS FOR STEADY SOURCE OF SHORT DURATION

There are many diffusion models for the dispersion of pollutants in atmosphere for continuous and instantaneous source releases. However, dispersion of short-term releases of pollutants from industrial plants, potential consequences of leakage such as fire and explosion, is a matter of great concern of research because of the lack of reliability of most of the dispersion models, mainly in terms of actual duration time of releases, as well as in terms of exposure time for human beings in environments. Thus, some models to realize the dispersion from steady-state source of short duration are described as follows:

a. Palazzi et al. (1982) presented a useful mathematical technique to fill the gap between the instantaneous and continuous point sources diffusions models. The mean concentration field due to continuous point source in an infinite medium is obtained by superimposing the concentrations in sequentially released, ensemble-averaged puffs and is given by integrating Equation 6.14 with respect to time for the finite duration. It can be written as

$$C(x,y,z,t) = \frac{q}{(2\pi)^{3/2}} \int_0^t \frac{1}{\sigma_x \sigma_y \sigma_z} \exp\left[-\left(\frac{(x - Ut')^2}{2\sigma_x^2} + \frac{y^2}{2\sigma_y^2} \right) \right]$$

$$\times \left\{ \exp\left[-\frac{(z - H_s)^2}{2\sigma_z^2} \right] + \exp\left[-\frac{(z + H_s)^2}{2\sigma_z^2} \right] \right\} dt' \qquad (6.31)$$

They solved the integral Equation 6.31 by assuming that σ_i's ($i = x, y, z$) depends only on the downwind distance x from the source to receptor. This assumption on σ_i's is unimportant when U is not too low (Slade, 1968).

Bianconi and Tamponi (1993) presented an extension of the above model (Palazzi et al. 1982), for the calculation of the time evolution of the concentration field and the duration of the exposure to emissions in order to cover the case of substance that may be subject to chemical–physical decay in finite mixing layer.

b. For short-time releases, Arya (1995) made an attempt for modeling and parameterization of near-source diffusion in weak wind, including the calm conditions. From Taylor's theory, for small diffusion or travel times, the dispersion parameters (σ_i's) vary linearly with time. But at large travel or diffusion times, these dispersion parameters are predicted to grow more slowly and are proportional to square root of time. Using the mixed-layer similarity theory, Arya (1995) interpolated the dispersion parameters from the above two parameterizations for different timescales, as continuous function of time t. These new interpolated parameters cover the whole wide range of small, intermediate, and large diffusion times in convective mixed layer and can be expressed as

$$\sigma_x = \frac{\sigma_u t}{\left[1+0.5\left(t/T_{Lu}\right)\right]^{1/2}}, \quad \sigma_y = \frac{\sigma_v t}{\left[1+0.5\left(t/T_{Lv}\right)\right]^{1/2}},$$

$$\text{and} \quad \sigma_z = \frac{\sigma_w t}{\left[1+0.5\left(t/T_{Lw}\right)\right]^{1/2}} \tag{6.32}$$

where T_{Lu}, T_{Lv}, and T_{Lw} are Lagrangian timescales.

In the integration of the right-hand side of Equation 6.31, he used these parameters (Equation 6.32) for near-source dispersion in convective boundary layer. This model was used in conjunction with the mixed-layer similarity scaling to calculate the maximum ground-level concentration and its distance from the source as functions of the effective source height and the dimensionless mean velocity (U/w_*). In low wind conditions, the results are found to be very sensitive to (U/w_*) and also strongly depends upon the downwind distance from the source. This model is valid only for the convective boundary layer over a flat and homogeneous terrain and still be limited to near-neutral and stably stratified conditions (Arya, 1995). In addition, the model accounts for the upstream diffusion.

6.4.4 MODELS IN FINITE LAYER

In the study of pollutant dispersion in atmospheric boundary layer (ABL), it is recognized that ABL is often capped by inversion that reflects back the material reaching the inversion base (Beyrich, 1997; Arya, 1999). Many models have been developed (Robson, 1983; Lin and Hildemann, 1996) for the dispersion in the finite layer. However, these may not be appropriate to deal with the low wind dispersion. The earlier models (Arya, 1995; Sharan et al., 1996a,b; Sharan and Yadav, 1998) described above for the dispersion in low winds allow unrestricted diffusion of plume in the vertical direction. This does not normally happen in the real atmosphere where a finite inversion layer/mixing layer of vanishing turbulence at the top of ABL restricts vertical diffusion. Thus, attempts have been made in the literature to formulate the mathematical models for the dispersion of pollutants in a finite inversion layer in low wind conditions. Here, we describe some of the features of these models.

6.4.4.1 Constant *K*-Models

For dispersion of pollutants in low wind conditions in finite layer, Sharan and Modani (2005) updated the model described in Section 6.4.1.2(b) by modifying the boundary condition on upper boundary as

$$-K_z \frac{\partial C}{\partial z} = 0, \quad z = h \tag{6.33}$$

in which h is the thickness of the inversion/mixed layer. The analytical solution is given by method of integral transform and can be written as (Sharan and Modani, 2005)

$$
C(x,y,z) = \frac{Q}{2\pi h \sqrt{K_x K_y}} \exp\left(\frac{Ux}{2K_x}\right) \left[K_0 \left\{ \frac{U}{2\sqrt{K_x}} \left(\frac{x^2}{K_x} + \frac{y^2}{K_y} \right)^{1/2} \right\} \right.
$$

$$
+ \sum_{n=1}^{\infty} K_0 \left\{ \left(\frac{U^2}{4K_x} + \frac{n^2\pi^2}{h^2} K_z \right)^{1/2} \left(\frac{x^2}{K_x} + \frac{y^2}{K_y} \right)^{1/2} \right\}
$$

$$
\left. \times \left\{ \cos\left(\frac{n\pi}{h}(z - H_s) \right) + \cos\left(\frac{n\pi}{h}(z + H_s) \right) \right\} \right] \tag{6.34}
$$

where K_0 is the modified Bessel function of second kind of order zero. They have pointed out that a solution of Equation 6.22 for $U = 0$ (i.e., in the absence of wind speed) in finite layer does not exist.

A number of particular cases corresponding to specific physical situations can be deduced. Solution (Equation 6.34) can be simplified to (1) slender plume approximation by taking $K_x \to 0$ in Equation 6.34, (2) when inversion is absent by taking $h \to \infty$ (Sharan et al., 1995b), and (3) for a ground-level source by taking $H_s \to 0$.

6.4.4.2 Variable *K*-Models

a. Sharan and Modani (2006) updated the model described in Section 6.4.2.2 by modifying the boundary condition on upper boundary by Equation 6.33. The analytical solution of the updated model is given by eigenfunction expansion (Sharan and Modani, 2006) and can be expressed as

$$C(x,y,z) = \frac{Q}{Uh} \frac{1}{\Gamma(\mu)r\sqrt{\beta\pi}} \left(\frac{x^2}{\alpha r^2}\right)^{\mu} \left[\Gamma\left(\mu+\frac{1}{2}\right) + 2\left(r\sqrt{\gamma}\right)^{\left(\mu+\frac{1}{2}\right)} \right.$$

$$\left. \times \sum_{n=1}^{\infty} \left(\frac{\rho_n}{2}\right)^{\mu+\frac{1}{2}} K_{\mu+\frac{1}{2}}\left\{\rho_n r\sqrt{\gamma}\right\} \left\{\cos\left(\rho_n(z+H_s)\right) + \cos\left(\rho_n(z-H_s)\right)\right\} \right]$$

(6.35)

in which $\mu = 1/(2\alpha)$, K_μ is the modified Bessel function of second kind of order μ, $\rho_n = (n\pi/h)$ and $r^2 = (x^2/\alpha) + (y^2/\beta)$.

The solution (Equation 6.35) can be converted into (Modani, 2006; Sharan and Modani, 2006) (1) for a ground-level source by taking $H_s \to 0$ and (2) when there is no inversion (i.e., $h \to \infty$), the solution (Equation 6.35) is transformed to the same solution (Equation 6.30).

b. A steady-state model similar to Cirillo and Poli (1992) is proposed by Modani (2006) by integrating the Gaussian puff (IGP) formula in finite vertical domain with respect to time between 0 to ∞, by considering the parameterization (Equation 6.26). The expression is given as

$$C(x,y,z) = \frac{Q}{2\pi h\alpha\beta} \exp\left(-\frac{U^2}{2\alpha^2}\right) \left[\sqrt{\frac{\pi}{2}} \frac{1}{\sqrt{r}} \exp\left(\frac{U^2 x^2}{2\alpha^4 r}\right) \mathrm{erf}\left(-\frac{Ux}{\sqrt{2\alpha^2}\sqrt{r}}\right) \right.$$

$$+ \sum_{n=1}^{\infty} \left\{\cos\left(\rho_n(z+H_s)\right) + \cos\left(\rho_n(z-H_s)\right)\right\}$$

$$\left. \times \int_0^{\infty} \frac{1}{t'^2} \exp\left(-\frac{1}{2t'^2} r + \frac{Ux}{\alpha^2 t'} - \rho_n \frac{\gamma^2 t'^2}{2}\right) dt' \right]$$ (6.36)

where $r^2 = (x^2/\alpha) + (y^2/\beta)$.

This IGP model and the variable *K* model (Equation 6.35) yield almost similar qualitative and quantitative results (Modani, 2006). It is expected as both the models have the similar basic structure and use the same type of parameterizations (Sharan and Yadav, 1998).

6.4.5 MODELS BASED ON CROSSWIND INTEGRATED CONCENTRATIONS

In another approach, the steady-state concentration in three-dimensional domain can be described by assuming the Gaussian concentration distribution in crosswind direction and is given as

$$C(x, y, z) = C_y(x, z) \frac{\exp\left(-y^2/2\sigma_y^2\right)}{\sqrt{2\pi}\sigma_y} \tag{6.37}$$

where C_y is the crosswind integrated concentration and can be obtained from the equation:

$$U \frac{\partial C_y}{\partial x} = \frac{\partial}{\partial x}\left(K_x \frac{\partial C_y}{\partial x}\right) + \frac{\partial}{\partial z}\left(K_z \frac{\partial C_y}{\partial z}\right) + Q\delta(x)\delta\left(z - H_s\right) \tag{6.38}$$

Equation 6.38 can also be obtained from Equation 6.1 with assumptions (a)–(c) and (e) (Section 6.3.1) and integrating it with respect to y from $-\infty$ to ∞. This fact has been exploited to develop the models for the behavior of dispersion in low wind conditions. Equation 6.38 is a second order two-dimensional partial differential equation and is relatively easier to solve for the realistic parameterization of eddy diffusivities and wind speed.

On the basis of this approach, Moreira et al. (2005) have presented a steady-state mathematical model for dispersion of contaminants in low wind conditions in a finite layer. For accounting the spatial dependences of U and K, they have divided the domain into multi-layers in vertical and downwind directions. In each layer, the eddy diffusivities and wind speed are assumed to be constant by taking their averaged values. Then the closed form analytical solution of the transformed equation is obtained by Laplace transform.

It is noticed through the above discussion that the behavior of dispersion of pollutants in low wind conditions is non-Gaussian. These models can be extended by relaxing the various assumptions. Most of these models have been evaluated with the available diffusion data in both stable conditions such as Hanford (Nichola, 1977) and Idaho National Engineering Laboratories (Sagendorf and Dickson, 1974), and convective conditions such as IIT diffusion experiment (Singh et al., 1991), EPRI plume validation experiment (Hudischewskyj and Reynolds, 1983), and Copenhagen experiment (Gryning, 1981). In addition, uncertainties are associated with the parameterization of dispersion parameters used in the dispersion models in low wind conditions.

Diffusion data are required for evaluation of mathematical models and empirically estimating the dispersion or diffusion parameters. The diffusion data available for the validation of these models in low wind conditions is limited. In the last 45 years, considerable efforts were devoted to conduct the diffusion experiments such as Prairie Grass (Barad, 1958), Idaho National Engineering Laboratories (Sagendorf and Dickson, 1974), low wind diffusion experiment in Japan (Adachi and Ohta, 1978), and IIT diffusion experiment (Singh et al., 1991). These experiments differ from each other on various counts such as on the height of the release (near-surface, elevated), underlying terrain, and the atmospheric stability. The comprehensive reviews on some of these experiments are given in the literature (Pasquill and Smith, 1983; Draxler, 1984; Arya, 1999). Sharan et al. (2003) have briefly described the salient features of some of these diffusion experiments. It is necessary to plan the extensive field program involving both tracer and meteorological measurements in low wind conditions for evaluation and updating these models.

We have discussed the analytical models for the dispersion of pollutants in low wind conditions. In addition, numerical models and Lagrangian particle models have been used in the literature for simulation in the low wind conditions (Anfossi et al., 1990; Brusasca et al., 1992; Arya, 1999; Oettl et al., 2001; Sharan and Gopalakrishnan, 2003). Since the aim of the study is to deal with the analytical models; we have restricted our discussion in reference to the analytical studies for the near source or low wind dispersion.

6.5 SUMMARY

In view of the limitations of the applicability of dispersion models in low wind conditions, an attempt has been made to provide an overview on analytical models for dispersion of air pollutants in these conditions. In the treatment of dispersion in low wind conditions, limitations are associated with (1) parameterization of eddy diffusivities, (2) parameterization of dispersion parameters, (3) incorporation of a realistic wind profile, and (4) availability of tracer and turbulent observations. The commonly used parameterization schemes for sigma are lacking for estimation of dispersion parameters in low wind conditions because (1) the state of the weak wind ABL is not well defined, (2) plume spread in the horizontal direction is increased because of meandering, and (3) nonavailability of adequate observations. There is a need for an extensive field program comprising (1) diffusion measurements and (2) meteorological measurements including surface/upper air and turbulence observations. A general parameterization of eddy diffusivities for the treatment of near-source dispersion is desired for the formulation of a more realistic dispersion model valid for all wind speeds. Since the aim of the study is to deal with the analytical models, we have restricted our discussion in reference to the analytical studies for the near source or low wind dispersion.

REFERENCES

Adachi, T. and Ohta, S.: (1978), Atmospheric diffusion under very low wind speed, very stable conditions, *Fourth US/JAPAN Joint Meeting on Air Pollution Related Meteorology*, December 11–15, U.S. Environmental Protection Agency, Washington, DC.

Anfossi, D., Brusasca, G., and Tinarelli, G.: (1990), Simulation of atmospheric diffusion in low wind speed meandering conditions by a Monte-Carlo dispersion model, *Il Nuovo Cimento* **13**, 277–282.

Arya, S. P.: (1995) Modeling and parameterization of near-source diffusion in weak winds, *J. Appl. Meteorol.* **34**, 1112–1122.

Arya, S. P.: (1999), *Air Pollution Meteorology and Dispersion*, Oxford University Press, New York. 320 pp.

Barad, M. L.: (1958), Project Prairie Grass, A field program in diffusion, *Geophys. Res. Paper* **59**, vols. I and II, AFCRF-TR-58-235.

Beyrich, F.: (1997), Mixing height estimation from Sodar data—A critical discussion, *Atmos. Environ.* **31**, 3941–3953.

Bianconi, R. and Tamponi, M.: (1993), A mathematical model of diffusion from a steady source of short duration in a finite mixing layer, *Atmos. Environ.* **27A**, 781–792.

Briggs, G. A.: (1973), Diffusion estimation for small emissions, U.S. National Oceanic and Atmospheric Administration E.R.L. Report, ATDL-106.

Brusasca, G., Tinarelli, G., and Anfossi, D.: (1992), Particle model simulation of diffusion in low wind speed stable conditions, *Atmos. Environ.* **26A**, 707–723.

Carslaw, H. S. and Jaeger, J. C.: (1986), *Heat Conduction in Solids*, Oxford University Press, New York.

Cirillo, M. C. and Poli, A. A.: (1992), An inter comparison of semi empirical diffusion models under low wind speed, stable conditions, *Atmos. Environ.* **26A**, 765–774.

Crank, J.: (1976), *The Mathematics of Diffusion*, Clarendon Press, Oxford, U.K.

Csanady, G. T.: (1973), *Turbulent Diffusion in the Environment*, D. Seidel Pub., Dordrecht, Holland.

Draxler, R. R.: (1976), Determination of atmospheric diffusion parameters, *Atmos. Environ.* **10**, 99–105.

Draxler, R. R.: (1984), Diffusion and transport experiments. In *Atmospheric Science and Power Production* (D. Randerson, ed.), U.S. Department of Energy, Technical Information Center, Oak Ridge, TN, pp. 367–422.

Green, A. E. S., Singhal, R. P., and Venkateswar, R.: (1980), Analytical extensions of the Gaussian plume model, *JAPCA* **30**, 773–776.

Gryning, S. E.: (1981), Elevated source SF6-tracer dispersion experiments in the Copenhagen area, Riso National Laboratory Report No. R-446, Roskilde, Denmark.

Gryning, S. E., Holtslag, A. A. M., Irwin, J., and Sivertsen, B.: (1987), Applied dispersion modelling based on meteorological scaling parameters, *Atmos. Environ.* **21**, 79–89.

Hanna, S. R., Briggs, G. A., and Hosker, R. P., Jr.: (1982), *Handbook on Atmospheric Diffusion*, U.S. Dept. of Energy report COE/TIC-11223, Washington, DC, 102pp.

Hudischewskyj, A. B. and Reynolds, S. D.: (1983), A catalog of data for the EPRI plume model validation and developmental data base—plains site, EA-3080, Research Project 1616-9, Final report, October, 1983, SAI, San Rafael, CA.

Irwin, J. S.: (1979), Estimating plume dispersion–A recommended generalized scheme, *Proceedings of the Fourth Symposium on Turbulence, Diffusion and Air Pollution*, American Meteorological Society, Reno, NV.

Liewelyn, R. P.: (1983), An analytical model for the transport, dispersion and elimination of air pollutants emitted from a point source, *Atmos. Environ.* **17**, 431–436.

Lin, J. S. and Hildemann, L. M.: (1996), Analytical solutions of the atmospheric diffusion equation with multiple sources and height-dependent wind speed and eddy diffusivities, *Atmos. Environ.* **30**, 239–254.

Modani, M.: (2006), Mathematical models for dispersion of atmospheric pollutants in the surface based inversion, PhD thesis, Indian Institute of Technology, Delhi.

Moreira, D. M., Tirabassi, T., and Carvalho, J. C.: (2005), Plume dispersion simulation in low wind conditions in stable and convective boundary layers, *Atmos. Environ.* **39**, 3643–3650.

Nichola, P. W.: (1977), The Hanford 67-series: A volume of atmospheric fields diffusion measurements, PNL-2433, Pacific Northwest Laboratory, Richland, WA.

Oettl, D., Almbauer, R. A., and Sturm, P. J.: (2001), A new method to estimate diffusion in low wind stable conditions, *J. Appl. Meteorol.* **40**, 259–268.

Okamoto, S. and Shiozawa, K.: (1987), A trajectory plume model for simulating air pollution transients, *Atmos. Environ.* **21**, 2145–2152.

Palazzi, E., De Faveri, M., Fumarola, G., and Ferralodo, G.: (1982), Diffusion from a steady source of short duration, *Atmos. Environ.* **16**, 2785–2790.

Pasquill, F.: (1961), The estimation of the dispersion of windborne material *Meteorol. Mag.* **90**, 33–49.

Pasquill, F. and Smith, F. B.: (1983), *Atmospheric Diffusion*, John Wiley & Sons, New York.

Roberts, O. F. T.: (1923), The theoretical; scattering of smoke in a turbulent atmosphere, *Proc., Roy. Soc. Ser. A*, **104**, 640–654.

Robson, R. E.: (1983), On the theory of plume trapping by an elevated inversion, *Atmos. Environ.* **17**, 1923–1930.

Sagendorf, J. and Dickson, C. R.: (1974), Diffusion under low wind speed inversion conditions, NOAA Technical Memo-ERL-ARL-52, Air Resources Labs, Silver Spring.

Seinfeld, J. H.: (1986), *Atmospheric Chemistry and Physics of Air Pollution*, John Wiley & Sons, New York, 738 pp.

Sharan, M. and Gopalakrishnan, S. G.: (2003), Mathematical modelling of diffusion and transport of pollutants in the atmospheric boundary layer, *Pure. Appl. Geophys.* **160**, 357–394.

Sharan, M. and Modani, M.: (2005), An analytical study for the dispersion of pollutants in a finite layer under low wind conditions, *Pure. Appl. Geophys.* **162**, 1861–1892.

Sharan, M. and Modani, M.: (2006), Variable K-model for the dispersion of air pollutants in low wind conditions in the surface based inversion, *Atmos. Environ.* **40**, 3469–3489.

Sharan, M. and Yadav, A. K.: (1998), Simulation of diffusion experiments under light wind, stable conditions by a variable K-theory model, *Atmos. Environ.* **32**, 3481–3492.

Sharan, M., McNider, R. T., Gopalakrishnan, S. G., and Singh, M. P.: (1995a), Bhopal gas leak: A numerical simulation of episodic dispersion, *Atmos. Environ.* **29**, 2061–2074.

Sharan, M., Yadav, A. K., and Singh, M. P. (1995b), Comparison of various sigma schemes for estimating dispersion of air pollutants in low winds, *Atmos. Environ.* **29**, 2051–2059.

Sharan, M., Yadav, A. K., and Singh, M. P.: (1996c), Plume dispersion simulation in low-wind conditions using coupled plume segment and Gaussian puff approaches, *J. Appl. Meteorol.* **35**, 1625–1631.

Sharan, M., Singh, M. P., and Yadav, A. K.: (1996a), A mathematical model for the atmospheric dispersion in low winds with eddy diffusivities as linear functions of downwind distance, *Atmos. Environ.* **30**, 1137–1145.

Sharan, M., Singh, M. P., Yadav, A. K., Agarwal, P., and Nigam, S.: (1996b), A mathematical model for dispersion of air pollutants in low wind conditions, *Atmos. Environ.* **30**, 1209–1220.

Sharan, M., Yadav, A. K., Singh, M. P., and Gupta, S.: (1999), Accounting for the source strength in the solution of the diffusion equation: Alternative mathematical formulations, *Atmos. Environ.* **33**, 1327–1330.

Sharan, M., Yadav, A. K., and Modani, M.: (2002), Simulation of short-range diffusion experiment in low-wind convective conditions, *Atmos. Environ.* **36**, 1901–1906.

Sharan, M., Modani, M., and Yadav, A. K.: (2003), Atmospheric dispersion: An overview of mathematical modeling framework, *Proc. Indian National. Sci. Acad.* **69A**, 725–744.

Singh, M. P., Agarwal, P., Nigam, S., and Gulati, A.: (1991), Tracer experiments—A report, Tech. Report, CAS, IIT, Delhi.

Slade, D. H.: (1968), *Meteorology and Atomic Energy*, U.S. Atomic Energy Commission, Washington, DC.

Taylor, G. I.: (1921), Diffusion by continuous movements, *Proc. London Math. Soc. Ser 2*, XX, 196–212.

Yadav, A. K.: (1995), Mathematical modelling of dispersion of air pollutants in low wind conditions, PhD thesis, Indian Institute of Technology, Delhi.

Yadav, A. K. and Sharan, M.: (1996), Statistical evaluation of sigma schemes for estimating dispersion in low wind conditions, *Atmos. Environ.* **30**, 2595–2606.

Yadav, A. K., Sharan, M., and Singh, M. P.: (1996), Atmospheric dispersion in low wind conditions. In *Proceedings of First World Congress of Nonlinear Analysts* (V. Lakshmikantham, ed.), Walter de Gruyter, Berlin, pp. 3567–3593.

Zannetti, P.: (1986), A new mixed segmented-puff approach for dispersion modeling, *Atmos. Environ.* **20**, 1121–1130.

Zannetti, P.: (1990), *Air Pollution Modelling*, Computational Mechanics Publications. Southampton, U.K. 444pp.

7 On the GILTT Formulation for Pollutant Dispersion Simulation in the Atmospheric Boundary Layer

Davidson Martins Moreira, Marco Túllio
M. B. de Vilhena, and Daniela Buske

CONTENTS

7.1 INTRODUCTION

In the last years, special attention has been given to the issue of searching analytical solutions for the advection–diffusion equation in order to simulate the pollutant dispersion in the atmospheric boundary layer (ABL). We must recall that the solution of the advection–diffusion equation can be written either in integral or in series formulations, with the main property that both solutions are equivalent (Moreira et al., 2009b). We are aware of the existence of analytical solutions in the literature, but for specific and particular problems. Among them, we mention the works of Rounds (1955), Smith (1957), Scriven and Fisher (1975), Demuth (1978), van Ulden (1978), Nieuwstadt and de Haan (1981), Tagliazucca et al. (1985), Tirabassi (1989), Tirabassi and Rizza (1994), Sharan et al., (1996a), Lin and Hildemann (1997), and Tirabassi (2003). In fact, all these solutions are valid for very specialized practical situations with restrictions on wind and eddy diffusivities vertical profiles. Further, in the last decade the ADMM (advection diffusion multilayer method) approach appeared in the literature (Costa et al., 2006), which solves the multidimensional advection–diffusion equation for more realistic physical scenario. The main idea relies on the discretization of the ABL in a multilayer domain, assuming in each layer that the eddy diffusivity and wind profile take averaged values. The resulting advection–diffusion equation in each layer is then solved by the Laplace transform technique. For more details about this methodology, see the revision work done by Moreira et al. (2006a). In this chapter, we focus our attention to the revision and updating of the series solution of the advection–diffusion equation, known in the literature as the GILTT (generalized integral Laplace transform technique) approach. The main idea of this methodology comprehends the following steps: expansion of the concentration in series of eigenfunctions attained from an auxiliary problem; replacing this equation in the advection–diffusion equation and taking moments, we come out with a matrix ordinary differential equation that is then solved analytically by the Laplace transform technique. This methodology skips the multilayer discretization of the height z appearing in the ADMM approach.

To reach our objective, we begin presenting the solution of the time-dependent, two-dimensional (2D) advection–diffusion equation in Cartesian geometry by the GILTT approach, assuming non-Fickian flows and considering that the eddy diffusivity coefficients depend on the x and z variable meanwhile the vertical wind profile depends on the z variable. Once we construct the solution for this general problem, in the sequel, we show how to obtain the solution for simplified models. We mean that we consider the solution for the following particular problems: time-dependent Fickian flow model, time-dependent Fickian flow without longitudinal diffusion, stationary Fickian flow problems, and approximated three-dimensional (3D) GILTT solution. We also present numerical simulations and future perspectives of this methodology.

7.2 THE ADVECTION–DIFFUSION EQUATION AND THE GILTT METHOD

The advection–diffusion equation of air pollution in the atmosphere is essentially a statement of conservation of the suspended material and it can be written as

$$\frac{\partial \bar{c}}{\partial t} + \bar{u}\frac{\partial \bar{c}}{\partial x} + \bar{v}\frac{\partial \bar{c}}{\partial y} + \bar{w}\frac{\partial \bar{c}}{\partial z} = -\frac{\partial \overline{u'c'}}{\partial x} - \frac{\partial \overline{v'c'}}{\partial y} - \frac{\partial \overline{w'c'}}{\partial z} + S \qquad (7.1)$$

where

$\bar{c}$ denotes the average concentration of a passive contaminant (g/m³)

$\bar{u}$, $\bar{v}$, $\bar{w}$ are the mean wind (m/s) components along the x-, y-, and z-axes, respectively

S is the source term

The terms $\overline{u'c'}$, $\overline{v'c'}$, and $\overline{w'c'}$ represent, respectively, the turbulent fluxes of contaminants (g/s m²) in the longitudinal, crosswind, and vertical directions.

Observe that Equation 7.1 has four unknown variables (the concentration $\bar{c}$ and turbulent fluxes) that lead us to the known turbulence closure problem. One of the most widely used closures for Equation 7.1 is based on the gradient transport hypothesis (or K-theory), which, in analogy with Fick's law of molecular diffusion, assumes that turbulence causes a net movement of material down the gradient of material concentration at a rate that is proportional to the magnitude of the gradient (Seinfeld and Pandis, 1998). So

$$\overline{u'c'} = -K_x \frac{\partial \bar{c}}{\partial x}; \quad \overline{v'c'} = -K_y \frac{\partial \bar{c}}{\partial y}; \quad \overline{w'c'} = -K_z \frac{\partial \bar{c}}{\partial z} \tag{7.2}$$

where K_x, K_y, and K_z are the Cartesian components of eddy diffusivity (m²/s) in the x-, y-, and z-directions, respectively. In the first-order closure, all the information on the turbulence complexity is contained in the eddy diffusivities.

Equation 7.2, combined with the continuity equation of mass, leads to the advection–diffusion equation. For a Cartesian coordinate system, we rewrite the advection–diffusion equation like (Blackadar, 1997):

$$\frac{\partial \bar{c}}{\partial t} + \bar{u}\frac{\partial \bar{c}}{\partial x} + \bar{v}\frac{\partial \bar{c}}{\partial y} + \bar{w}\frac{\partial \bar{c}}{\partial z} = \frac{\partial}{\partial x}\left(K_x \frac{\partial \bar{c}}{\partial x}\right) + \frac{\partial}{\partial y}\left(K_y \frac{\partial \bar{c}}{\partial y}\right) + \frac{\partial}{\partial z}\left(K_z \frac{\partial \bar{c}}{\partial z}\right) + S \tag{7.3}$$

The simplicity of the K-theory of turbulent diffusion has led to the widespread use of this theory as mathematical basis for simulating pollutant dispersion (open country, urban, photochemical pollution, etc.). But K-closure has its own limits. In contrast to molecular diffusion, turbulent diffusion is scale-dependent. This means that the rate of diffusion of a cloud of material generally depends on the cloud dimensions and the intensity of turbulence. As the cloud grows, larger eddies are incorporated in the expansion process, so that a progressively larger fraction of turbulent kinetic energy is available for the cloud expansion.

Another problem is that the down-gradient transport hypothesis is inconsistent with observed features of turbulent diffusion in the upper portion of the mixed layer, at convective cases where counter-gradient material fluxes are known to occur (Deardoff and Willis, 1975). Because counter-gradient fluxes are thought to be indicative of boundary layer scale eddies, as opposed to small-scale ones, such fluxes are often called nonlocal fluxes. Local K-theory is a method for parameterizing the effects of turbulent mixing based on how small eddies will mix quantities along a local gradient of the transported quantity.

Some decades ago, it was noted that in the upper part of convectively driven boundary layers, the flux of scalars are counter to the gradient of the mean scalar profile (Deardoff, 1966). The mean potential temperature gradient and the flux change sign at different levels introduce a certain region in the convective boundary layer (CBL) where they have the same sign. This was in contrast with the common view in first-order turbulent closure that turbulent diffusion is down gradient. In order to describe diffusion also in these regions, Ertel (1942) and Deardoff (1966, 1972) proposed to modify the usual applied flux–gradient relationship in K-theory approach according to

$$\overline{w'c'} = -K_z \left(\frac{\partial \overline{c}}{\partial z} - \gamma \right) \tag{7.4}$$

where γ represents the counter-gradient term.

Many schemes and parameterizations for counter-gradient term have been developed (e.g., Wyngaard and Brost, 1984; Fiedler and Moeng, 1985; Holtslag and Moeng, 1991; Wyngaard and Weil, 1991; Holtslag and Boville, 1993; Hamba, 1993; Robson and Mayocchi, 1994; Zilitinkevich et al., 1999). In this chapter, without losing generality, we use the parameterization proposed by van Dop and Verver (2001), which is based on the work of Wyngaard and Weil (1991):

$$\left[1 + \left(\frac{S_k T_{Lw} \sigma_w}{2} \right) \frac{\partial}{\partial z} + \tau \frac{\partial}{\partial t} \right] \overline{w'c'} = -K_z \frac{\partial \overline{c}}{\partial z} \tag{7.5}$$

where

S_k is the skewness of the vertical turbulent velocity (w'), that is, $S_k = \overline{w'^3}/(\overline{w'^2})^{3/2}$
σ_w is the vertical turbulent velocity standard deviation (m/s)
T_{Lw} is the Lagrangian timescale (s)
τ is the relaxation time (s)

The second term in the operator (in the brackets) represents the nonlocal counter-gradient term.

Using Equations 7.4 and 7.5, the turbulence closure problem is solved without obeying Fick's law, being called non-Fickian closure (also known as nonlocal closure). The non-Fickian closure allows the investigation of more energetic eddies in different heights and the effect of the asymmetric transport in the computation of the pollutant concentration considering in a more complete way the structure of the turbulent dispersion.

Applying Equations 7.2 and 7.5 in Equation 7.1, the crosswind integrated transient advection–diffusion equation, in the Eulerian framework, for a Cartesian coordinate system in which the x-direction coincide with that of the average wind, is written as (Buske et al., 2009)

$$\frac{\partial c(x,z,t)}{\partial t} + \bar{u}\frac{\partial c(x,z,t)}{\partial x} + \bar{w}\frac{\partial c(x,z,t)}{\partial z} = \frac{\partial}{\partial x}\left(K_x\frac{\partial c(x,z,t)}{\partial x}\right) + \frac{\partial}{\partial z}\left(K_z\frac{\partial c(x,z,t)}{\partial z}\right) +$$

$$-\frac{\partial}{\partial z}\left(\beta\frac{\partial c(x,z,t)}{\partial t}\right) - \frac{\partial}{\partial z}\left(\beta\bar{u}\frac{\partial c(x,z,t)}{\partial x}\right) - \frac{\partial}{\partial z}\left(\beta\bar{w}\frac{\partial c(x,z,t)}{\partial z}\right) +$$

$$-\tau\frac{\partial^2 c(x,z,t)}{\partial t^2} - \frac{\partial}{\partial t}\left(\tau\bar{u}\frac{\partial c(x,z,t)}{\partial x}\right) - \frac{\partial}{\partial t}\left(\tau\bar{w}\frac{\partial c(x,z,t)}{\partial z}\right) +$$

$$+\frac{\partial}{\partial z}\left(\beta\frac{\partial}{\partial x}\left(K_x\frac{\partial c(x,z,t)}{\partial x}\right)\right) + \frac{\partial}{\partial t}\left(\tau\frac{\partial}{\partial x}\left(K_x\frac{\partial c(x,z,t)}{\partial x}\right)\right) \qquad (7.6)$$

where $\beta=0.5S_k\sigma_wT_l$, for $0<z<h$, $x>0$ and $t>0$. Equation 7.6 is subjected to the boundary conditions:

$$K_z\frac{\partial c(x,z,t)}{\partial z} = 0 \quad \text{at } z = 0, h \qquad (7.6a)$$

to the initial condition:

$$c(x,z,0) = 0 \quad \text{at } t = 0 \qquad (7.6b)$$

to the source condition:

$$\bar{u}c(0,z,t) = Q\delta(z - H_s) \quad \text{at } x = 0 \qquad (7.6c)$$

and for far away from the source we have

$$\frac{\partial c(L_*,z,t)}{\partial x} = 0 \quad \text{at } x = L_* \qquad (7.6d)$$

where
  c now represents the crosswind integrated concentration (g/m^2)
  h is the boundary layer height (m)
  H_s is the height of the source (m)
  L_* is far away from the source (m)
  Q is the emission rate (g/s)
  K_x and K_z are the longitudinal and vertical eddy diffusivities (m^2/s), respectively
  δ is the Dirac delta function

Here $\bar{u}$, $\bar{w}$ are functions of height z and K_x and K_z are also functions of source distance x.

In order to solve problem (Equation 7.6), taking advantage of the well-known solution of the stationary problem with advection in the x-direction by the GILTT method (Moreira et al., 2005b), we apply the Laplace transform technique in the t variable. This procedure leads to the stationary problem:

$$r\overline{C}(x,z,r) + \overline{u}\frac{\partial\overline{C}(x,z,r)}{\partial x} + \overline{w}\frac{\partial c(x,z,r)}{\partial z} = \frac{\partial}{\partial x}\left(K_x\frac{\partial\overline{C}(x,z,r)}{\partial x}\right) +$$

$$+ \frac{\partial}{\partial z}\left(K_z\frac{\partial\overline{C}(x,z,r)}{\partial z}\right) - \frac{\partial}{\partial z}(\beta r\overline{C}(x,z,r)) - \frac{\partial}{\partial z}\left(\beta\overline{u}\frac{\partial\overline{C}(x,z,r)}{\partial x}\right) +$$

$$- \frac{\partial}{\partial z}\left(\beta\overline{w}\frac{\partial\overline{C}(x,z,r)}{\partial z}\right) - \tau r^2\overline{C}(x,z,r) - \tau\overline{u}r\frac{\partial\overline{C}(x,z,t)}{\partial x} - \tau\overline{w}r\frac{\partial\overline{C}(x,z,r)}{\partial z} +$$

$$+ \frac{\partial}{\partial z}\left(\beta\frac{\partial}{\partial x}\left(K_x\frac{\partial\overline{C}(x,z,r)}{\partial x}\right)\right) + \tau r\frac{\partial}{\partial x}\left(K_x\frac{\partial\overline{C}(x,z,r)}{\partial x}\right) \tag{7.7}$$

where $\overline{C}$, denotes the Laplace transform technique of the concentration in the t variable, that is, $\overline{C}(x, z, r) = L\{c(x, z, t); t \to r\}$. Next, we rewrite the above equation in an appropriate form to apply the GILTT technique, that is

$$\overline{u}\frac{\partial\overline{C}(x,z,r)}{\partial x} + \overline{w}\frac{\partial c(x,z,r)}{\partial z} = K_x\frac{\partial^2\overline{C}(x,z,r)}{\partial x^2} + K_x'\frac{\partial\overline{C}(x,z,r)}{\partial x} +$$

$$+ K_z\frac{\partial^2\overline{C}(x,z,r)}{\partial z^2} + K_z'\frac{\partial\overline{C}(x,z,r)}{\partial z} - \beta r\frac{\partial\overline{C}(x,z,r)}{\partial z} - (\beta r)'\overline{C}(x,z,r) +$$

$$- \beta\overline{u}\frac{\partial^2\overline{C}(x,z,r)}{\partial z\partial x} - (\beta\overline{u})'\frac{\partial\overline{C}(x,z,r)}{\partial x} - \beta\overline{w}\frac{\partial^2\overline{C}(x,z,r)}{\partial z^2} - (\beta\overline{w})'\frac{\partial\overline{C}(x,z,r)}{\partial z} +$$

$$- \tau r^2\overline{C}(x,z,r) - \tau\overline{u}r\frac{\partial\overline{C}(x,z,t)}{\partial x} - \tau\overline{w}r\frac{\partial\overline{C}(x,z,r)}{\partial z} + \beta K_x\frac{\partial^3\overline{C}(x,z,r)}{\partial z\partial x^2} +$$

$$+ (\beta K_x)'\frac{\partial^2\overline{C}(x,z,r)}{\partial x^2} + \beta K_x'\frac{\partial^2\overline{C}(x,z,r)}{\partial z\partial x} + (\beta K_x)'\frac{\partial\overline{C}(x,z,r)}{\partial x} +$$

$$+ \tau r K_x\frac{\partial^2\overline{C}(x,z,r)}{\partial x^2} + \tau r K_x'\frac{\partial\overline{C}(x,z,r)}{\partial x} - r\overline{C}(x,z,r) \tag{7.8}$$

Following the works of Buske et al. (2007a,b), we pose that the solution of problem (7.8) has the form:

$$\overline{C}(x,z,r) = \sum_{n=0}^{N}\overline{c}_n(x,r)\Psi_n(z) \tag{7.9}$$

where $\Psi_n(z)$ are the eigenfunctions of the associated Sturm–Liouville problem, that is, $\Psi_n(z) = \cos(\lambda_n z)$ where $\lambda_n = n\pi/h$ ($n = 0, 1, 2\ldots$) are the respective eigenvalues.

To determine the unknown coefficient $\bar{c}_n(x,r)$, we replace Equation 7.9 with Equation 7.8 and taking moments, that is, applying the operator $\int_0^h(\ )\Psi_m(z)\,dz$, we come out with the result:

$$\sum_{n=0}^N \bar{c}'_n(x,r)\int_0^h \bar{u}\ \Psi_n(z)\Psi_m(z)dz + \sum_{n=0}^N \bar{c}_n(x,r)\int_0^h \bar{w}\ \Psi'_n(z)\Psi_m(z)dz =$$

$$\sum_{n=0}^N \bar{c}''_n(x,r)\int_0^h K_x\Psi_n(z)\Psi_m(z)dz + \sum_{n=0}^N \bar{c}'_n(x,r)\int_0^h K'_x\Psi_n(z)\Psi_m(z)dz +$$

$$-\sum_{n=0}^N \bar{c}_n(x,r)\lambda_n^2\int_0^h K_z\Psi_n(z)\Psi_m(z)dz + \sum_{n=0}^N \bar{c}_n(x,r)\int_0^h K'_z\Psi'_n(z)\Psi_m(z)dz +$$

$$-\sum_{n=0}^N \bar{c}_n(x,r)r\int_0^h \beta\Psi'_n(z)\Psi_m(z)dz - \sum_{n=0}^N \bar{c}_n(x,r)r\int_0^h \beta'\Psi_n(z)\Psi_m(z)dz +$$

$$-\sum_{n=0}^N \bar{c}'_n(x,r)\int_0^h \beta\bar{u}\ \Psi'_n(z)\Psi_m(z)dz - \sum_{n=0}^N \bar{c}'_n(x,r)\int_0^h (\beta\bar{u})'\Psi_n(z)\Psi_m(z)dz +$$

$$+\sum_{n=0}^N \bar{c}_n(x,r)\lambda_n^2\int_0^h \beta\bar{w}\ \Psi_n(z)\Psi_m(z)dz - \sum_{n=0}^N \bar{c}_n(x,r)\int_0^h (\beta\bar{w})'\Psi'_n(z)\Psi_m(z)dz +$$

$$-\sum_{n=0}^N \bar{c}_n(x,r)\tau r^2\int_0^h \Psi_n(z)\Psi_m(z)dz - \sum_{n=0}^N \bar{c}'_n(x,r)\tau r\int_0^h \bar{u}\ \Psi_n(z)\Psi_m(z)dz +$$

$$-\sum_{n=0}^N \bar{c}_n(x,r)\tau r\int_0^h \bar{w}\Psi'_n(z)\Psi_m(z)dz + \sum_{n=0}^N \bar{c}_n(x,r)\int_0^h \beta K_x\Psi'_n(z)\Psi_m(z)dz +$$

$$+\sum_{n=0}^N \bar{c}''_n(x,r)\int_0^h (\beta K_x)'\Psi_n(z)\Psi_m(z)dz + \sum_{n=0}^N \bar{c}'_n(x,r)\int_0^h \beta K'_x\Psi'_n(z)\Psi_m(z)dz +$$

$$+\sum_{n=0}^N \bar{c}'_n(x,r)\int_0^h (\beta K'_x)\Psi_n(z)\Psi_m(z)dz + \sum_{n=0}^N \bar{c}'_n(x,r)\tau r\int_0^h K_x\Psi_n(z)\Psi_m(z)dz +$$

$$+\sum_{n=0}^N \bar{c}'_n(x,r)\tau r\int_0^h K'_x\Psi_n(z)\Psi_m(z)dz - \sum_{n=0}^N \bar{c}_n(x,r)r\int_0^h \Psi_n(z)\Psi_m(z)dz \qquad (7.10)$$

Recasting Equation 7.10 as a matrix ordinary differential equation, we read

$$B_1(x) \cdot Y''(x,r) + B_2(x) \cdot Y'(x,r) + B_3(x) \cdot Y(x,r) = 0 \tag{7.11}$$

where $Y(x,r)$ is the column vector whose components are $\{\bar{c}_n(x,r)\}$ and the entries of matrices B_1, B_2, and B_3 are respectively given by

$$
(b_1)_{n,m} = \int_0^h K_x \Psi_n(z) \Psi_m(z) dz + \int_0^h \beta K_x \Psi_n'(z) \Psi_m(z) dz +
$$

$$
+ \int_0^h (\beta K_x)' \Psi_n(z) \Psi_m(z) dz + \tau r \int_0^h K_x \Psi_n(z) \Psi_m(z) dz
$$

$$
(b_2)_{n,m} = -\int_0^h \bar{u} \Psi_n(z) \Psi_m(z) dz - \int_0^h \beta \bar{u}\, \Psi_n'(z) \Psi_m(z) dz - \int_0^h (\beta \bar{u})'\, \Psi_n(z) \Psi_m(z) dz +
$$

$$
+ \int_0^h K_x' \Psi_n(z) \Psi_m(z) dz + \int_0^h \beta K_x' \Psi_n'(z) \Psi_m(z) dz + \int_0^h (\beta K_x')' \Psi_n(z) \Psi_m(z) dz +
$$

$$
- \tau r \int_0^h \bar{u} \Psi_n(z) \Psi_m(z) dz + \tau r \int_0^h K_x' \Psi_n(z) \Psi_m(z) dz
$$

and

$$
(b_3)_{n,m} = \int_0^h K_z' \Psi_n'(z) \Psi_m(z) dz - \lambda_n^2 \int_0^h K_z \Psi_n(z) \Psi_m(z) dz - \int_0^h \bar{w}\, \Psi_n'(z) \Psi_m(z) dz +
$$

$$
- r \int_0^h \beta \Psi_n'(z) \Psi_m(z) dz - r \int_0^h \beta' \Psi_n(z) \Psi_m(z) dz + \lambda_n^2 \int_0^h \beta \bar{w} \Psi_n(z) \Psi_m(z) dz +
$$

$$
- \int_0^h (\beta \bar{w})'\, \Psi_n'(z) \Psi_m(z) dz - \tau r^2 \int_0^h \Psi_n(z) \Psi_m(z) dz +
$$

$$
- \tau r \int_0^h \bar{w}\, \Psi_n'(z) \Psi_m(z) dz - r \int_0^h \Psi_n(z) \Psi_m(z) dz
$$

To solve Equation 7.11, we proceed likewise the work of Moreira et al. (2006b), performing a stepwise approximation of the entries of the matrices $B_1(x)$, $B_2(x)$, and $B_3(x)$ by taking average values for the eddy diffusivity and its derivative in the x variable. Here it is important to mention that no approximation is made on the derivatives appearing in the advection–diffusion equation. From this procedure, it turns out that problem 7.11 simplifies to a set of ordinary differential equations, in which the B_1, B_2, and B_3 matrices have constant components. Indeed, we have

$$Y''(x,r) + F \cdot Y'(x,r) + G \cdot Y(x,r) = 0 \tag{7.12}$$

where the matrix F is defined like $F = B_1^{-1} B_2$, and the matrix G like $G = B_1^{-1} B_3$. The integrals appearing in B_1, B_2, and B_3 are solved numerically via Gauss Legendre quadrature.

Similar procedure leads to the boundary conditions:

$$Y(0,r) = \overline{c}_n(0,r) = \frac{Q}{r} \Psi_m(H_s)A^{-1} \text{ and } Y'(L_*,r) = \overline{c_n'}(L_*,r) = 0$$

where A^{-1} is the inverse of matrix A having the entry: $a_{n,m} = \int_0^h \overline{u}\, \Psi_n(z)\Psi_m(z)\mathrm{d}z$.

Applying the standard procedure for order reduction to Equation 7.12, we come out with the following result:

$$Z'(x,r) + HZ(x,r) = 0 \tag{7.13}$$

where $Z(x, r)$ is the column vector $Z(x, r) = \mathrm{col}(Z_1 (x, r), Z_2 (x, r))$ and the matrix H has the block form $H = \begin{bmatrix} 0 & -I \\ G & F \end{bmatrix}$, subjected to the respective boundary conditions for the vector components:

$$Z_1(0,r) = \frac{Q}{r} \Psi_m(H_s)A^{-1} \text{ and } Z_2(L_*,r) = 0 \tag{7.13a}$$

Now, we are in a position to solve problem (Equation 7.13), following the work of Moreira et al. (2009a), by the combined Laplace transform technique and diagonalization of the matrix H ($H = XDX^{-1}$). By this procedure, we come out with the following result:

$$\overline{Z(s,r)} = X \left(sI + D \right)^{-1} X^{-1}Z(0,r) \tag{7.14}$$

where $\overline{Z(s,r)}$ denotes the Laplace transform of the vector $Z(x, r)$. Here X is the matrix of the eigenvectors of the matrix H and X^{-1} it is the inverse. The matrix D is the diagonal matrix of the eigenvalues of the matrix H and the entry of the matrix $(sI + D)$ has the form $\{s + d_n\}$. Performing the Laplace transform inversion of Equation 7.14, we come out with the following result:

$$Z(x,r) = X \cdot G(x,r) \cdot X^{-1} \cdot Z(0,r) = M(x,r)\xi \tag{7.15}$$

where $M(x, r) = X \cdot G(x, r)$ and $G(x, r)$ is the diagonal matrix with components $e^{-d_n x}$. Further the new unknown arbitrary constant vector ξ is given by $\xi = X^{-1} \cdot Z(0)$.

Applying the boundary conditions (Equation 7.13a), we determine the unknown components of the arbitrary column vector $\xi = col(\xi_1, \xi_2)$, solving the equation:

$$\begin{pmatrix} M_{11}(0) & M_{12}(0) \\ M_{21}(L_*) & M_{22}(L_*) \end{pmatrix} \begin{pmatrix} \xi_1 \\ \xi_2 \end{pmatrix} = \begin{pmatrix} Z_1(0) \\ Z_2(L_*) \end{pmatrix} \tag{7.16}$$

Once these unknown coefficients are evaluated, we can construct the analytical solution to problem (Equation 7.6) applying the inverse Laplace transform definition. This procedure yields to the following analytical result:

$$c(x, z, t) = \frac{1}{2\pi i} \sum_{n=0}^{N} \int_{\gamma - i\infty}^{\gamma + i\infty} \overline{c}_n(x, r) \Psi_n(z) e^{rt} ds \tag{7.17}$$

By analytical, we mean that no approximation is made along the derivation of solution (Equation 7.17). To overcome the drawback of evaluating the line integral appearing in Equation 7.17, in the sequel, we report a closed-form solution for this integral, using the Gaussian quadrature scheme. By this procedure, we get:

$$c(x, z, t) = \sum_{k=1}^{M} \frac{P_k}{t} A_k \sum_{n=0}^{N} \overline{c}_n\left(x, \frac{P_k}{t}\right) \Psi_n(z) \tag{7.18}$$

where A_k and P_k are the weights and roots of the Gaussian quadrature scheme tabulated in the book by Stroud and Secrest (1966). Regarding the issue of the adopted Laplace numerical inversion scheme, it is important to mention that this approach is exact if the integrand is a polynomial of degree $2M - 1$ in the $1/r$ variable. We are aware of the existence in the literature of methods to invert numerically the Laplace-transformed functions (Valkó and Abate, 2004; Abate and Valkó, 2004), but we restrict our attention in the problem considered to the Gaussian quadrature scheme. The motivation for this choice comes besides the simplicity the good results achieved.

7.3   PARTICULAR SOLUTIONS

In the sequel, taking advantage of the generality of the discussed solution, we report simplified solutions for specific physical scenarios, readily obtained from Equation 7.17, by just taking limits.

7.3.1   TIME-DEPENDENT FICKIAN FLOW MODEL

From Equation 7.6, we promptly realize that the advection–diffusion equation governed by non-Fickian flow is readily obtained by making the parameters β and τ, responsible for nonlocal transport, to vanish (β and τ goes to zero). This problem is then modeled as

$$\frac{\partial c(x,z,t)}{\partial t} + \bar{u}\frac{\partial c(x,z,t)}{\partial x} + \bar{w}\frac{\partial c(x,z,t)}{\partial z} = \frac{\partial}{\partial x}\left(K_x\frac{\partial c(x,z,t)}{\partial x}\right) + \frac{\partial}{\partial z}\left(K_z\frac{\partial c(x,z,t)}{\partial z}\right)$$

(7.19)

and the solution is also easily attained, by taking the limit when β and τ goes to zero in the solution of non-Fickian flow expressed by Equation 7.17. For more details, see the work of Buske et al. (2006, 2007c).

7.3.2 Time-Dependent Fickian Flow without Longitudinal Diffusion

We may determine the solution of the time-dependent, 2D advection–diffusion equation for Fickian flow regime without longitudinal diffusion for problems where the advection transport term in the x-direction is dominant over the diffusive term, that is, $\left|\bar{u}\frac{\partial c}{\partial x}\right| \gg \left|\frac{\partial}{\partial x}\left(K_x\frac{\partial c}{\partial x}\right)\right|$. Therefore, we can construct the solution for this sort of problem taking the limit of the solution for Fickian flows (Equation 7.19), by just making K_x identically null (Moreira et al., 2006b).

7.3.3 Stationary Fickian Flow Problems

To establish the solution of the stationary, 2D, advection–diffusion equation under Fickian flow regime, with and without longitudinal diffusion, we just need to take respectively the limit of the solutions of items (1) and (2) when t goes to infinity, which is equivalent to make r goes to zero (Buske et al., 2007a). By similar procedure, we come out with the results for the stationary non-Fickian problems (Buske et al., 2007b). Further simplifications we disregard in this chapter because they can be determined in straightforward manner following the works of Wortmann et al. (2005); Moreira et al. (2005b, 2009b); Buske et al. (2008), and Tirabassi et al. (2008, 2009).

7.3.4 Approximated Three-Dimensional GILTT Solution

For physical scenarios in which the turbulence is reasonable approximated by a Gaussian model in the y-direction, we can construct an approximated 3D GILTT solution by assuming that this solution reads like the product of the 2D solution by a Gaussian solution in the y-direction. This assumption leads to

$$\bar{c}(x,y,z,t) = c(x,z,t)\frac{e^{\left(-y^2/2\sigma_y^2\right)}}{\sqrt{2\pi}\sigma_y},$$

(7.20)

where $c(x, z, t)$ is expressed by the discussed solutions. For a better comprehension of the validity of the approximation considered, see the work of Moreira et al. (2009a).

7.4 TURBULENT PARAMETERIZATIONS AND DISPERSION DIFFUSION EXPERIMENTS

In the sequel, we report the parameterizations adopted in the simulations reported in this chapter. In fact, we need to recall that the choice of the turbulent parameterization represents a fundamental aspect for pollutant dispersion modeling (Moreira et al., 2005c). Indeed, to evaluate the 3D concentration in the ground-level centerline (Equation 7.20), we need to know the lateral dispersion parameter σ_y. For a CBL, we used the lateral dispersion parameter σ_y derived by Degrazia et al. (1998a), which has the form:

$$\frac{\sigma_y^2}{h^2} = \frac{0.21}{\pi} \int_0^\infty \sin^2(2.26\psi^{1/3}Xn') \frac{dn'}{(1+n')^{5/3}n'^2} \tag{7.21}$$

Here X is an adimensional distance ($X = xw_*/uh$), w_* is the convective velocity scale, n' is adimensional frequency, and h is the top of the CBL. Equation 7.21 contains the unknown function ψ; the molecular dissipation of turbulent velocity is a leading destruction term in equations for the budget of second-order moments, and according to Højstrup (1982), reads like

$$\psi^{1/3} = \left[\left(1 - \frac{z}{h}\right)^2 \left(\frac{z}{-L}\right)^{-2/3} + 0.75 \right]^{1/2} \tag{7.22}$$

where L is the length of Monin–Obukhov defined in the surface boundary layer.

In terms of the convective scaling parameters, the vertical eddy diffusivity can be formulated as (Degrazia et al., 1997)

$$\frac{K_z}{w_*h} = 0.22\left(\frac{z}{h}\right)^{1/3}\left(1 - \frac{z}{h}\right)^{1/3}\left[1 - \exp\left(-\frac{4z}{h}\right) - 0.0003\exp\left(\frac{8z}{h}\right)\right] \tag{7.23}$$

To represent the near-source diffusion under weak wind conditions, the eddy diffusivities should be considered as functions not only of turbulence (e.g., large eddy length and velocity scales), but also of distance from the source (Arya, 1995). Following this idea, Degrazia et al. (2002) proposed for the unstable boundary layer the ensuing algebraic formulation for the eddy diffusivities:

$$K_\alpha = \frac{0.583w_*hc_i\psi^{2/3}(z/h)^{4/3}X^*\left[0.55(z/h)^{2/3} + 1.03c_i^{1/2}\psi^{1/3}\left(f_m^*\right)_i^{2/3}X^*\right]}{\left[0.55(z/h)^{2/3}\left(f_m^*\right)_i^{1/3} + 2.06c_i^{1/2}\psi^{1/3}\left(f_m^*\right)_i X^*\right]^2} \tag{7.24}$$

where $c_{v,w} = 0.36$, $c_u = 0.3$, $(f_m^*)_i$ is the normalized frequency of the spectral peak, namely,

$$(f_m^*)_w = \frac{z}{(\lambda_m)_w} = 0.55 \left(\frac{z}{h}\right) \left[1 - \exp\left(-\frac{4z}{h}\right) - 0.0003 \exp\left(\frac{8z}{h}\right)\right]^{-1} \quad (7.25)$$

for the vertical component. Further, $(\lambda_m)_w = 1.8h[1 - \exp(-4z/h) - 0.0003 \exp(8z/h)]$ is the value of the spectral peak of vertical wavelength. The longitudinal component according to Olesen (1995) is $(f_m^*)_u = 0.67$.

On the other hand, in our simulations, we use the wind speed profiles described either by the similarity or the power law. According to Panofsky and Dutton (1984), the similarity law has the form:

$$u = \frac{u_*}{k}\left[\ln\frac{z}{z_0} - \psi_m\left(\frac{z}{L}\right)\right] \quad (7.26)$$

where
  u_* is the scale velocity relative to mechanical turbulence
  k the von Karman constant
  z_0 is the roughness length
  ψ_m is the stability function expressed in Businger relations

$$\psi_m\left(\frac{z}{L}\right) = -4.7\frac{z}{L} \quad \text{for } 1/L \geq 0$$

$$\psi_m\left(\frac{z}{L}\right) = \ln\left(\frac{1+x^2}{2}\right) + \ln\left(\frac{1+x}{2}\right)^2 - 2\arctan x + \frac{\pi}{2} \quad \text{for } 1/L < 0$$

with $x = (1 - 15z/L)^{1/4}$, meanwhile the power law is written as

$$\frac{\bar{u}_z}{\bar{u}_1} = \left(\frac{z}{z_1}\right)^n \quad (7.27)$$

where $\bar{u}_z$ and $\bar{u}_1$ are the mean wind velocity respectively at heights z and z_1, while n is an exponent that is related to the intensity of turbulence (Irwin, 1979).

In order to illustrate the aptness of the discussed formulation to simulate contaminant dispersion in the ABL, we evaluate the performance of the discussed solutions against experimental ground-level concentration using different dispersion experiments available in the literature. Below we briefly discuss the Copenhagen, Prairie-Grass, and IIT dispersion experiments, which allow us to validate the results encountered by the mentioned solutions.

The first experiment is carried out in the northern part of Copenhagen, described by Gryning and Lyck (1984). It consisted of a tracer released without buoyancy from a tower at a height of 115 m, and collection of tracer sampling units at the ground-level positions at the maximum of three crosswind arcs. The sampling units were positioned at 2–6 km from the point of release. The site was mainly residential with a roughness length of 0.6 m.

In the Prairie–Grass experiment, according to Barad (1958), the tracer SO_2 was released without buoyancy at a height of 0.46 m, and collected at a height of 1.5 m at five downwind distances (50, 100, 200, 400, and 800 m) at O'Neill, Nebraska in 1956. The Prairie-Grass site was quite flat and much smooth with a roughness length of 0.6 cm. In this chapter, we consider the experimental data appearing in the paper of Nieuwstadt (1980).

Experiments of pollutant dispersion under low wind conditions were done at Indian Institute of Technology and described by Sharan et al. (1996a,b; 2002). The contaminant was released without buoyancy at a height of 1 m and the concentration measured near the ground (0.5 m). The emission rate of the SF_6 tracer ranged from 30 to 50 mL/min. The sample period for each experiment was 30 min. The wind and temperature were measured at four heights (2, 4, 15, and 30 m) from a meteorological tower of 30 m. The samplers were positioned in arcs with radius of 50 and 100 m with a source distance of 100 m.

7.5   APPLICATIONS

Next we report the numerical results attained by the GILTT formulation for the discussed solutions. Tables 7.1 through 7.5 present some performance evaluations of the model results for the above experiments, using the statistical evaluation procedure described by Hanna (1989) and defined in the following way:

NMSE (normalized mean square error) $= \overline{(C_o - C_p)^2}/\overline{C_p}\,\overline{C_o}$,

FA2 = fraction of data (%, normalized to 1) for $0.5 \leq (C_p/C_o) \leq 2$,

COR (correlation coefficient) $= \overline{(C_o - \overline{C_o})(C_p - \overline{C_p})}/\sigma_o\sigma_p$,

FB (fractional bias) $= \overline{C_o} - \overline{C_p}/0.5(\overline{C_o} + \overline{C_p})$,

FS (fractional standard deviations) $= (\sigma_o - \sigma_p)/0.5(\sigma_o + \sigma_p)$,

TABLE 7.1

Statistical Evaluation of Model Results for the Steady-State, Two-Dimensional Advection–Diffusion for Fickian Flows, Copenhagen Experiment (Crosswind Integrated Concentrations), Eddy Diffusivity Equation 7.23, and Similarity Wind Profile

Model	NMSE	COR	FA2	FB	FS
GILTT	0.06	0.92	1.0	−0.14	−0.02
Gaussian	0.08	0.87	1.0	0.10	0.31
ADMM	0.06	0.89	1.0	0.02	0.09

where the subscripts o and p refer to observed and predicted quantities, respectively, and the overbar indicates an averaged value. The statistical index FB says if the predicted quantities underestimate or overestimate the observed ones. The statistical index NMSE represents the model values dispersion in respect to data dispersion. The best results are expected to have values near to zero for the indices NMSE, FB, and FS, and near to one in the indices COR and FA2.

7.5.1 COPENHAGEN EXPERIMENT RESULTS

Table 7.1 shows the performance of the solution of the steady-state, 2D advection–diffusion for Fickian flows, compared with other models considering similarity wind profile and using crosswind integrated ground-level concentration (2D-dataset). The results obtained are presented and compared with other models (Degrazia, 1998b; Mangia et al., 2002). The eddy diffusivity Equation 7.23 was used. The statistical indices of the three tables point out that a good agreement is obtained between experimental data and the GILTT method. Analyzing the statistical indices (Hanna, 1989), we notice that these models simulate satisfactorily the observed concentrations, regarding the NMSE, FB, and FS values relatively near to zero and COR relatively near to 1.

Table 7.2 shows the results of the solution time-dependent, 2D advection–diffusion for Fickian flows. In the simulations of the crosswind integrated concentrations, the Copenhagen experiment with a greater time resolution (Tirabassi and Rizza, 1997) and similarity wind profile were used. Generally, the distributed dataset contains hourly mean values of concentrations and meteorological data. However, in this work, as a test for the time-dependent solution, we also used data with a greater time resolution. In particular, we used 20 min averaged measured concentrations and 10 min averaged values for meteorological data. The results obtained with the GILTT method are compared with the ADMM method (Moreira et al., 2005a) and the M4PUFF model (Tirabassi and Rizza, 1997), which is based on a general technique for solving the K-equation using the truncated Gram-Charlier expansion (type A) of the concentration field and a finite set equation for the corresponding moments.

TABLE 7.2

Statistical Evaluation of Model Results for the Time-Dependent, Two-Dimensional Advection–Diffusion for Fickian Flows, Copenhagen Experiment (Crosswind Integrated Concentrations), Eddy Diffusivity Equation 7.23, and Similarity Wind Profile

Model	NMSE	COR	FA2	FB	FS
GILTT	0.09	0.85	1.00	0.11	0.13
M4PUFF	0.21	0.74	0.90	0.10	0.45
ADMM	0.15	0.81	0.95	0.18	0.38

TABLE 7.3

Statistical Evaluation of Model Results for the Approximated Steady-State, Three-Dimensional Solution for Fickian Flows, Copenhagen Experiment (Centerline Concentrations), Eddy Diffusivity Equation 7.23, and Power Wind Profile

Model	NMSE	COR	FA2	FB	FS
GILTT	0.33	0.80	0.87	0.28	0.09
Gaussian	0.08	0.88	1.00	0.06	0.07
GIADMT	0.15	0.87	0.96	0.01	−0.09

A more detailed inspection of Table 7.2 permits to stress that the GILTT simulates very well the observed concentrations presenting the best values for NMSE, COR, and FA2.

Table 7.3 shows the statistical evaluation of the solution of the approximated steady-state, 3D solution for Fickian flows, power wind profile, and the unstable lateral dispersion parameter described in Section 7.3. The Copenhagen experiment (3D-dataset) is used to obtain the centerline concentrations. The results obtained with the GILTT method are compared with the traditional Gaussian model (Degrazia, 1998b) and the GIADMT method (Costa et al., 2006) (3D-solution of the ADMM model), which consists in the solution of the GITT transformed problem by the ADMM method.

7.5.2 PRAIRIE-GRASS EXPERIMENT RESULTS

Table 7.4 shows the performance of the solution of the steady-state, 2D advection–diffusion for Fickian flow compared with other models considering similarity wind profile. The results obtained are presented and compared with ADMM model (Mangia et al., 2002). The eddy diffusivity Equation 7.23 was used. The statistical indices of the three tables point out that a good agreement is obtained between

TABLE 7.4

Statistical Evaluation of Model Results for the Steady-State, Two-Dimensional Advection–Diffusion for Fickian Flow Using the Prairie-Grass Experiment, Eddy Diffusivity Equation 7.23, and Similarity Wind Profile

Model	NMSE	COR	FA2	FB	FS
GILTT	0.32	0.90	0.72	0.16	0.33
ADMM	0.25	0.92	0.68	0.03	0.20

experimental data and the GILTT method. Analyzing the statistical indices (Hanna, 1989), we notice that these models simulate satisfactorily the observed concentrations, regarding the NMSE, FB, and FS values relatively near to zero and COR relatively near to 1.

7.5.3 IIT EXPERIMENT RESULTS

In Table 7.5 are presented the performance of the approximated 3D GILTT solution (Equation 7.20) applied to low wind conditions in four different situations:

Case 1: Steady-state, 3D advection–diffusion for Fickian flows ($S_k = 0.0$);
Case 2: Steady-state, 3D advection–diffusion for non-Fickian flows ($S_k = 1.0$);
Case 3: Time-dependent, 3D advection–diffusion for Fickian flows ($S_k = 0.0$) with average time of 1 h;
Case 4: Time-dependent, 3D advection–diffusion for Fickian flows ($S_k = 1.0$) with average time of 1 h.

To analyze the influence of the counter-gradient term in the simulations, the value of $S_k = 1.0$ was used as suggested by van Dop and Verver (2001). The eddy diffusivity Equation 7.24 and the power wind profile were used. Here we present the results using the numerical inversion called Fixed-Talbot algorithm (Buske et al., 2009). Giving a closer look to the results appearing in Table 7.5, we promptly note the good agreement between the results attained with the ones of experimental data.

7.5.4 CASE STUDY: SIMULATION OF RADIOACTIVE POLLUTANT USING LARGE EDDY SIMULATIONS (LES)

For a case study, we report numerical results for the time-dependent, approximated 3D advection–diffusion equation for Fickian flows (Equation 7.20). Vertical eddy diffusivity (Equation 7.23) and the lateral dispersion parameter (Equation 7.21) are used

TABLE 7.5

Results of Statistical Indices Used to Evaluate the Model Performance of the Approximated Three-Dimensional GILTT Solution (Equation 7.20) Using the IIT Delhi Experiment, Eddy Diffusivity Equation 7.24, and Power Wind Profile

Model	NMSE	COR	FA2	FB	FS
Case 1	0.32	0.71	0.81	0.08	−0.11
Case 2	0.27	0.71	0.81	−0.07	−0.08
Case 3	0.33	0.71	0.94	−0.02	−0.21
Case 4	0.27	0.71	0.94	−0.01	−0.18

to calculate the ground-level concentration of emissions released from an elevated continuous source point in an unstable/neutral ABL, considering radioactive material released during Angra dos Reis experiment (Biaggio et al., 1985).

Bearing in mind that in this work our aim is to show the feasibility of the proposed models to simulate pollutant dispersion in atmosphere for more realistic problem, we are now in a position to specialize the application of this methodology for a problem with the wind speeds evaluated by the MesoNH research model (Lafore et al., 1998; Cuxart et al., 2000) and the micrometeorological parameters by the LES model.

The MesoNH has different parameterizations and can be run in different modes, from mesoscale to LES. The model uses an inelastic system of equations written with a Gal-Chen and Sommerville vertical system of coordinates. The turbulence closures available are the eddy diffusivity based on the TKE budget equation of Cuxart et al. (2000) and the EDMF (eddy diffusivity/mass-flux) scheme developed by Soares et al. (2004). The convection scheme is based on a bulk mass-flux convection parameterization for deep and shallow convection (Bechtold et al., 2001). MesoNH has a statistical subgrid condensation scheme, based on the distributions of the grid scale values of θ_l and q_t, and their variances, which are supplied by the general turbulence scheme (Cuijpers and Bechtold, 1995). The radiative scheme implemented in MesoNH is the one of the European Center for Medium range Weather Forecasting (ECMWF) model.

This study is based on a simulation with four nested grids, the coarser two run in the regional mode and the inner two grids in LES mode, with two-way interaction between them. The outer grid is forced by re-analysis of the ECMWF model. The main properties of the four grids are the following: Grid 1 (mesoscale, horizontal resolution: 10 km, $(nx, ny, nz) = 60 \times 60 \times 120$ and $\Delta t = 8$ s); Grid 2 (mesoscale, horizontal resolution: 2 km, $(nx, ny, nz) = 60 \times 60 \times 120$ and $\Delta t = 4$ s); Grid 3 (LES, horizontal resolution: 400 m, $(nx, ny, nz) = 120 \times 120 \times 120$ and $\Delta t = 1$ s); Grid 4 (LES, horizontal resolution: 100 m, $(nx, ny, nz) = 96 \times 96 \times 120$ and $\Delta t = 0.5$ s).

The experiment consisted in the controlled releases of radioactive tritiated water vapor from the meteorological tower, 100 m height, close to the power plant in Itaorna Beach, from November 28 to December 4, 1984 (Biagio et al., 1985). The nuclear power plant is located at a latitude −23.0079 and a longitude −44.4612. The total time of emission was 90 min for each day, in all cases around midday LST. The collection of water vapor over cooled aluminum plates in the numbered location took place in three subsequent periods (1, 2, and 3) of 20 min each, 30 min after the beginning of the release, to allow the source and the plume transport to reach a supposed stationary condition on the measurement area. All relevant details, as well as the synoptic meteorological conditions during the dispersion campaign are also described in Biagio et al. (1985). In this work, the simulations were accomplished on the first day (28 November). The roughness length utilized was $z_o = 1$ m and the emission rate $Q = 20.5$ MBq/s.

The micrometeorological dataset, obtained from LES model, used to obtain the numerical results were the friction velocity $u_* = 0.4$ m/s, convective velocity $w_* = 1.6$ m/s, and $h = 1200$ m is the ABL height. Table 7.6 presents the statistical performances obtained in the simulations with wind field and micrometeorological parameters from MesoNH (LES) and semiempirical equations. Here also appears

TABLE 7.6

Statistical Evaluation of the Time-Dependent Approximated Three-Dimensional Advection–Diffusion Equation for Fickian Flows, Using Data from LES Simulations

Model	NMSE	COR	FA2	FA5	FB	FS
GILTT	0.33	0.44	0.75	1.00	0.03	1.37
GILTT*	2.05	0.36	0.25	0.37	0.88	0.08

Note: GILTT* represents result simulations using semiempirical equations to determine micrometeorological parameters and wind field.

the statistical index FA5 (fraction of data [%, normalized to 1] for $0.2 \leq (C_p/C_o) \leq 5$), usually for simulations and complex terrain. Best results are obtained when FA5 is near. Promptly, we observed from Table 7.6 that the model satisfactorily reproduced the concentrations. GILTT* represents result simulations using semiempirical equations to determine micrometeorological parameters and wind field. The best results were obtained with the use of the micrometeorological parameters (LES) and wind field from MesoNH model. The analysis of the results shows a reasonably good agreement between the computed values against the experimental ones using data from MesoNH model.

Observing these results is important to mention that the differences among the experimental data do not depend on the solution of the diffusion equation, but on the equation itself, which is only a model of reality. It must be borne in mind, when using models, that, while they are rather sophisticated instruments that ultimately reflect the current state of knowledge on turbulent transport in the atmosphere, the results they provide are subject to a considerable margin of error. This is due to various factors, including in particular the uncertainty of the intrinsic variability of the atmosphere. Models, in fact, provide values expressed as an average, that is, a mean value obtained by the repeated performance of many experiments, while the measured concentrations are a single value of the sample to which the ensemble average provided by models refer. This is a general characteristic of the theory of atmospheric turbulence and is a consequence of the statistical approach used in attempting to parameterize the chaotic character of the measured data.

7.6 FUTURE GILTT PERSPECTIVES

Keeping us in the track of searching analytical solutions, we conclude this chapter reporting the perspectives of recent advances beginning with the subject of solving, in analytical manner, the advection–diffusion equation for more realistic scenarios, that is, to solve the time-dependent, 3D advection–diffusion equation assuming, generally speaking, non-Fickian flow. We shall also consider the solution of this sort of problem assuming that the velocity field depends on time and z

variables. For such, we circumvent the disability of the Laplace transform to handle this sort of problems by the application of decomposition method (Adomian, 1988). By doing so, we reduce the advection–diffusion equation with the velocity field depending on time variable into a set of recursive diffusion equations with source carrying the time information, which has a well-known solution. The motivation for this procedure comes from the fact that the resulting recursive problems can be straightly solved by the methodology discussed in this chapter. Bearing in mind the aptness of the decomposition method to solve, analytically, nonlinear problems, we are also confident to affirm that this methodology is a promising approach to handle nonlinear problems of pollutant dispersion in atmosphere appearing in problems with pollutants subject to chemical reaction. We hope that with the formulation reported we paved the road to update the GILTT formulation to simulate pollutant dispersion in the described scenarios. We shall focus our future attention in these directions.

REFERENCES

Abate, J. and Valkó, P.P., 2004. Multi-precision Laplace transform inversion. *Int. J. Num. Methods Eng.* 60, 979–993.

Adomian, G., 1988. A review of the decomposition method in applied mathematics, *J. Math. Anal. Appl.* 1(135), 501–544.

Arya, P., 1995. Modeling and parameterization of near-source diffusion in weak winds. *J. Appl. Meteor.* 34, 1112–1122.

Barad, M.L., 1958. *Project Prairie Grass'*. Geophysical Research Paper, No. 59, vols. I and II, GRD, Bedford, MA.

Bechtold, P., Bazile, E., Guichard, F., Mascart, P., and Richard, E., 2001. A mass flux convection scheme for regional and global models. *Quart. J. Roy. Meteor. Soc.* 127, 869–886.

Biagio, R., Godoy, G., Nicoli, I., Nicoli, D., and Thomas, P., 1985. *First Atmospheric Diffusion Experiment Campaign at the Angra Site—KfK 3936*. Karlsruhe and CNEN 1201, Rio de Janeiro.

Blackadar, A.K., 1997. *Turbulence and Diffusion in the Atmosphere: Lectures in Environmental Sciences*, Springer-Verlag, Berlin, Germany, p. 185.

Buske, D., Vilhena, M.T., Moreira, D.M., and Tirabassi, T., 2006. Analytical solution for the transient two-dimensional advection-diffusion equation considering nonlocal closure of the turbulent diffusion. In: K. Hanjalic, Y. Nagano, S. Jakirlic (eds.), *Turbulence, Heat, and Mass Transfer 5*. Begel House Inc., New York, pp. 705–708.

Buske, D., Vilhena, M.T., Moreira, D.M., and Tirabassi, T., 2007a. An analytical solution of the advection-diffusion equation considering non-local turbulence closure. *Environ. Fluid Mech.* 7, 43–54.

Buske, D., Vilhena, M.T., Moreira, D.M., and Tirabassi, T., 2007b. Simulation of pollutant dispersion for low wind conditions in stable and convective planetary boundary layer. *Atmos. Environ.* 41, 5496–5501.

Buske, D., Costa, C., Tirabassi, T., Moreira, D.M., and Vilhena, M.T., 2007c. An unsteady analytical solution of advection-diffusion equation for low wind conditions, *Proceedings do XXVIII CILAMCE*, Porto, Portugal.

Buske, D., Vilhena, M.T., Moreira, D.M., and Bodmann, B., 2008. An analytical solution for the steady-state two-dimensional diffusion-advection-deposition model by the GILTT approach. In: C. Constanda, S. Potapenko (eds.), *Integral Methods in Science and Engineering: Techniques and Applications*, XVI ed. Birkhauser, Boston, MA, pp. 27–36.

Buske, D., Vilhena, M.T., Moreira, D.M., and Tirabassi, T., 2009. An analytical solution for the transient two-dimensional advection-diffusion equation with non-Fickian closure in Cartesian geometry by integral transform technique. In: C. Constanda, S. Potapenko (eds.), *Integral Methods in Science and Engineering*, Vol. 2, Chapter 4.

Costa, C.P., Vilhena, M.T., Moreira, D.M., and Tirabassi, T., 2006. Semi-analytical solution of the steady three-dimensional advection-diffusion equation in the planetary boundary layer. *Atmos. Environ.* 40(29), 5659–5669.

Cuijpers, J.W.M. and P. Bechtold, 1995. A simple parameterization of cloud water related variables for use in boundary layer models. *J. Atmos. Sci.* 52(13), 2486–2490.

Cuxart, J., Bougeault, Ph., and Redelsperger, J.L., 2000. A turbulence scheme allowing for mesoscale and large-eddy simulations. *Q. J. R. Meteorol. Soc.* 126, 1–30.

Deardoff, J.W., 1966. The countergradient heat flux in the lower atmosphere and in the laboratory. *J. Atmos. Sci.* 23, 503–506.

Deardoff, J.W., 1972. Numerical investigation of neutral and unstable planetary boundary layers. *J. Atmos. Sci.* 29, 91–115.

Deardoff, J.W. and Willis, G.E., 1975. A parameterization of diffusion into the mixed layer, *J. Appl. Meteor.* 14, 1451–1458.

Degrazia, G.A., Campos Velho, H.F., and Carvalho, J.C., 1997. Nonlocal exchange coefficients for the convective boundary layer derived from spectral properties. *Contr. Atmos. Phys.*, 57–64.

Degrazia, G.A., Mangia, C., and Rizza, U., 1998a. A comparison between different methods to estimate the lateral dispersion parameter under convective conditions. *J. Appl. Meteor.* 37, 227–231.

Degrazia, G.A., 1998b. Modelling dispersion from elevated sources in a planetary boundary layer dominated by moderate convection. *Il Nuovo Cimento* 21(3), 345–353.

Degrazia, G.A., Moreira, D.M., Campos, C.R.J., Carvalho, J.C., and Vilhena, M.T., 2002. Comparison between an integral and algebraic formulation for the eddy diffusivity using the Copenhagen experimental dataset. *Il Nuovo Cimento* 25C, 207–218.

Demuth, C., 1978. A contribution to the analytical steady solution of the diffusion equation for line sources. *Atmos. Environ.* 12, 1255–1258.

Ertel, H., 1942. Der vertikale turbulenz-wärmestrom in der atmosphäre. *Meteor. Z.* 59, 250–253.

Fiedler, B.H. and Moeng, C.H., 1985. A practical integral closure model for mean vertical transport of a scalar in a convective boundary layer. *J. Atmos. Sci.* 42(4), 359–363.

Gryning, S.E. and Lyck, E. 1984. Atmospheric dispersion from elevated source in an urban area: Comparison between tracer experiments and model calculations. *J. Appl. Meteor.* 23, 651–654.

Højstrup, J., 1982. Velocity spectra in the unstable boundary layer. *J. Atmos. Sci.* 39, 2239–2248.

Hamba, F., 1993. A modified *K* model for chemically reactive species in the planetary boundary layer. *J. Geophys. Res.* 98(3), 5173–5182.

Hanna, S.R., 1989. Confidence limit for air quality models as estimated by bootstrap and jacknife resampling methods. *Atmos. Environ.* 23, 1385–1395.

Holtslag, A. and Boville, B.A., 1993. Local versus nonlocal boundary-layer diffusion in a global climate model. *J. Climate* 6, 1825–1842.

Holtslag, A. and Moeng, C.H., 1991. Eddy diffusivity and countergradient transport in the convective atmospheric boundary layer. *J. Atmos. Sci.* 48, 1690–1698.

Irwin, J.S., 1979. A theoretical variation of the wind profile power-low exponent as a function of surface roughness and stability. *Atmos. Environ.* 13, 191–194.

Lafore, J.P., Stein, J., Asencio, N., Bougeault, P., Ducrocq, V., Duron, J., Fischer, C. et al., 1998. The meso-NH atmospheric simulation system. Part I: Adiabatic formulation and control simulations. *Ann. Geophys.* 16, 90–109.

Lin, J.S. and Hildemann, L.M., 1997. A generalised mathematical scheme to analytically solve the atmospheric diffusion equation with dry deposition. *Atmos. Environ.* 31, 59–71.

Mangia, C., Moreira, D.M., Schipa, I., Degrazia, G.A., Tirabassi, T., and Rizza, U., 2002. Evaluation of a new eddy diffusivity parameterization from turbulent Eulerian spectra in different stability conditions. *Atmos. Environ.* 36, 67–76.

Moreira, D.M., Rizza, U., Vilhena, M.T., and Goulart, A.G., 2005a. Semi-analytical model for pollution dispersion in the planetary boundary layer. *Atmos. Environ.* 39(14), 2689–2697.

Moreira, D.M., Vilhena, M.T., Tirabassi, T., Buske, D., and Cotta, R.M., 2005b. Near source atmospheric pollutant dispersion using the new GILTT method. *Atmos. Environ.* 39(34), 6290–6295.

Moreira, D.M, Carvalho, J.C., Goulart, A.G., and Tirabassi, T., 2005c. Simulation of the dispersion of pollutants using two approaches for the case of a low source in the SBL: evaluation of turbulence parameterizations. *Water Air Soil Pollut.* 161, 285–297.

Moreira, D.M, Vilhena, M.T., Tirabassi, T., Costa, C., and Bodmann, B., 2006a. Simulation of pollutant dispersion in atmosphere by the Laplace transform: The ADMM approach. *Water Air Soil Pollut.* 177, 411–439.

Moreira, D.M., Vilhena, M.T., Buske, D., and Tirabassi, T., 2006b. The GILTT solution of the advection-diffusion equation for an inhomogeneous and nonstationary PBL. *Atmos. Environ.* 40, 3186–3194.

Moreira, D.M., Vilhena, M.T., Buske, D., and Tirabassi, T., 2009a. The state-of-art of the GILTT method to simulate pollutant dispersion in the atmosphere. *Atmos. Res.* 92, 1–17.

Moreira, D.M., Vilhena, M.T., Tirabassi, T., Buske, D., and Costa, C.P., 2009b. Comparison between analytical models to simulate pollutant dispersion in the atmosphere. *Int. J. Environ.* Waste Management, in press.

Nieuwstadt, F.T.M., 1980. An analytical solution of the time-dependent, one-dimensional diffusion equation in the atmospheric boundary layer. *Atmos. Environ.* 14, 1361–1364.

Nieuwstadt, F.T.M. and de Haan, B.J., 1981. An analytical solution of one-dimensional diffusion equation in a nonstationary boundary layer with an application to inversion rise fumigation. *Atmos. Environ.* 15, 845–851.

Olesen, H.R., 1995. Datasets and protocol for model validation. *Int. J. Environ. Pollut.* 5, 693–701.

Panofsky, H.A. and Dutton, J.A., 1984. *Atmospheric Turbulence.* John Wiley & Sons, New York.

Robson, R.E. and Mayocchi, C.L., 1994. A simple model of countergradient flow. *Phys. Fluids* 6(6), 1952–1954.

Rounds, W., 1955. Solutions of the two-dimensional diffusion equation. *Trans. Am. Geophys. Union* 36, 395–405.

Scriven, R.A. and Fisher, B.A., 1975. The long range transport of airborne material and its removal by deposition and washout-II. The effect of turbulent diffusion. *Atmos. Environ.* 9, 59–69.

Seinfeld, J.H. and Pandis, S.N., 1998. *Atmospheric Chemistry and Physics.* John Wiley & Sons, New York, pp. 13–26.

Sharan, M., Singh, M.P., and Yadav, A.K., 1996a. A mathematical model for the atmospheric dispersion in low winds with eddy diffusivities as linear function of downwind distance. *Atmos. Environ.* 30, 1137–1145.

Sharan, M., Singh, M.P., Yadav, A.K., Agarwal, P., and Nigam, S., 1996b. A mathematical model for dispersion of air pollutants in low winds conditions. *Atmos. Environ.* 30, 1209–1220.

Sharan, M., Yadav, A.K., and Modani, M., 2002. Simulation of short-range diffusion experiment in low wind convective conditions. *Atmos. Environ.* 36, 1901–1906.

Smith, F.B., 1957. The diffusion of smoke from a continuous elevated point source into a turbulent atmosphere. *J. Fluid Mech.* 2, 49–76.

Soares, P.M.M., Miranda, P.M.A., Sibesma, A.P., and Teixeira, J., 2004. An eddy-diffusivity/mass-flux for dry and shallow cumulus convection. *Q.J.R. Meteorol. Soc.* 130, 3365–3383.

Stroud, A.H. and Secrest, D., 1966. *Gaussian Quadrature Formulas.* Prentice Hall Inc. Englewood Cliffs, NJ.

Tagliazucca, M., Nanni, T., and Tirabassi, T., 1985. An analytical dispersion model for sources in the surface layer. *Nuovo Cimento* 8C, 771–781.

Tirabassi, T., 1989. Analytical air pollution and diffusion models. *Water Air Soil Pollut.* 47, 19–24.

Tirabassi, T. and Rizza, U. 1994. Applied dispersion modelling for ground-level concentrations from elevated sources. *Atmos. Environ.* 28, 611–615.

Tirabassi, T. and Rizza, U., 1997. Boundary layer parameterization for a non-Gaussian puff model. *J. Appl. Meteor.* 36, 1031–1037.

Tirabassi, T., 2003. Operational advanced air pollution modeling. *PAGEOPH* 160(1–2), 5–16.

Tirabassi, T., Buske, D., Moreira, D.M., and Vilhena, M.T., 2008. A two-dimensional solution of the advection-diffusion equation with dry deposition to the ground. *J. Appl. Meteor. Climatol.* 47, 2096–2104.

Tirabassi, T., Tiesi, A., Buske, D., Moreira, D.M., and Vilhena, M.T., 2009. Some characteristics of a plume from a point source based on analytical solution of the two-dimensional advection-diffusion equation. *Atmos. Environ.* 43(13), 2221–2227.

Valkó, P.P. and Abate, J., 2004. Comparison of sequence accelerators for the Gaver method of numerical Laplace transform inversion. *Comput. Math. Appl.* 48, 629–636.

van Dop, H. and Verver, G., 2001. Countergradient transport revisited. *J. Atmos. Sci.* 58, 2240–2247.

van Ulden, A.P., 1978. Simple estimates for vertical diffusion from sources near the ground. *Atmos. Environ.* 12, 2125–2129.

Wortmann, S., Vilhena, M.T., Moreira, D.M., and Buske, D., 2005. A new analytical approach to simulate the pollutant dispersion in the PBL. *Atmos. Environ.* 39, 2171–2178.

Wyngaard, J.C. and Brost, R.A., 1984. Top-down bottom-up diffusion of a scalar in the convective boundary layer. *J. Atmos. Sci.* 41, 102–112.

Wyngaard, J.C. and Weil, J.C., 1991. Transport asymmetry in skewed turbulence. *Phys. Fluids* A 3, 155–162.

Zilitinkevich, S., Gryanik, V.M., Lykossov, V.N., and Mironov, D.V., 1999. Third-order transport and nonlocal turbulence closures for convective boundary layers. *J. Atmos. Sci.* 56, 3463–3477.

8 An Outline of Lagrangian Stochastic Dispersion Models

Domenico Anfossi and Silvia Trini Castelli

CONTENTS

8.1 INTRODUCTION

The dispersion of airborne pollutant in the planetary boundary layer (PBL) is governed, schematically, by the transport, operated by the mean wind, and by the atmospheric turbulence due to the presence of eddies having different spatial and temporal scales.

Lagrangian stochastic dispersion models (LSDM), also named Lagrangian particle models (LPM), are numerical models aimed at the simulation of these processes, being able to account for flow and turbulence space-time variations. Emissions in the atmosphere are simulated using a number of fictitious particles named "computer particles." Each particle represents a specified pollutant mass. We call particle a fluid portion containing the emitted substance, having dimensions appropriate to follow the motion of the smallest turbulence eddies present in the atmosphere (of the order of the Kolmogorov scale), but containing a number of molecules large enough to allow disregarding the effect of the single molecule. Under the hypothesis, accurately demonstrated, that dispersion due to molecular motion is negligible compared to turbulent dispersion, it can be thought that these particles possess a concentration of their own that is preserved during the motion. It is assumed that the particles passively follow the turbulent motion of air masses in which they are. Particles are moved following the turbulent eddies, thus describing random trajectories. To prescribe this behavior, particle velocities are subject to a random forcing, thus these are stochastic type models, whereas Eulerian models are deterministic. As a consequence, the emitted mass concentration can be calculated from the space distribution of particles at a particular time.

Particle mean motion in the computation domain, which simulates the airborne pollutant motion in the real domain (atmosphere), is prescribed by the local mean wind. Particle dispersion (operated by turbulent eddies) is obtained from random speeds. These last are the solutions of stochastic differential equations, reproducing the statistical characteristics of the local atmospheric turbulence. In such a way, different parts of the plume can be liable to different atmospheric conditions. This approach allows producing more realistic simulations in complex conditions, which can be reproduced by traditional models with difficulty.

The models considered here are single-particle type; this means that the trajectory of each particle represents an individual statistical realization in a turbulent flow characterized by certain initial conditions and physical constraints. Thus, the motion of any particle is independent of the other particles, and consequently the concentration field must be interpreted as an ensemble average. The basic relationship, for an instantaneous source located in x_0 is (Csanady, 1973)

$$C(x, t) = Q\, P(x, t \mid x_0, t_0) \tag{8.1}$$

where
  Q is the total emitted mass
  $C(x, t)$ is the ensemble mean concentration (the mass of particles within a small volume surrounding x at time t)
  $P(x, t \mid x_0, t_0)$ is the probability that a particle that was at x_0 at time t_0 arrives at x at time t

In LSDM, to compute $P(x, t \mid x_0, t_0)$ it is necessary to release a sufficiently large number of particles, to follow their trajectories, and to calculate how many of them arrive in a small control volume surrounding x at time t. It is worth noting that particles move in the computational domain, without any grid, using as input the values of the first two or three, sometimes four, moments of the probability density function (PDF) of the three wind velocity components pertaining to the point at which each particle is. This input information is given on a regular grid and comes either from measurements or from parameterizations appropriate to the actual stability conditions (unstable, neutral, stable), the type of site (flat or complex terrain, coast, etc.), the time, and space scales considered.

Thanks to the present computer power, the use of LSDM is growing even as regulatory models (see, for instance, the official German Federal Environmental Agency air dispersion model—AUSTAL2000: Luft TA 2002).

8.2 GENERALITIES ON LAGRANGIAN STOCHASTIC MODELS

Thomson (1987) showed that the criterion for selecting the correct model for the diffusion of scalars in a turbulent flow is the "well-mixed condition": particles that are initially uniformly distributed must remain so. The LSDM that are based on the generalized Langevin equation satisfy the well-mixed condition. It is worth noting that these models have a unique solution in one dimension only, while they do not have a unique solution in two- or three-dimensional flows (Sawford and Guest, 1988).

The position of each particle, at each time step, is obtained by numerically integrating the following 3-D equations (Thomson, 1987; Anfossi and Physick, 2005; Ferrero, 2005):

$$\mathrm{d}u_i = a_i(x,u,t) + b_{ij}(x,u,t)\mathrm{d}W_j(t) \tag{8.2}$$

$$\mathrm{d}x_i(t) = u_i(t)\cdot\mathrm{d}t \tag{8.3}$$

where
  x_i is the position vector of each particle
  u_i its corresponding Lagrangian velocity vector
  $\mathrm{d}W_j$ is the incremental Wiener process that is Gaussian with zero mean and a variance of $\mathrm{d}t$ (random velocity fluctuation)

The term a_i of Equation 8.2 is a deterministic term, representing the friction force exerted by the flow on the particle, and the term $b_{i,j}$ is a stochastic term, representing the random accelerations caused by pressure fluctuations. $b_{i,j}$ is obtained from the Kolmogorov theory of local isotropy in the inertial subrange (Monin and Yaglom, 1965; Thomson, 1987) and has the following expression:

$$b_{ij} = \delta_{ij}\sqrt{C_0\cdot\varepsilon} \tag{8.4}$$

where
  ε is the dissipation rate of turbulent kinetic energy
  C_0 is a numerical constant

As an alternative choice (Hinze, 1975; Tennekes, 1982), when ε is not known, it is possible to determine

$$\frac{C_0\varepsilon}{2} = \frac{\sigma_i^2}{T_{L_i}} \qquad (8.5)$$

where
  T_L is the Lagrangian decorrelation timescale
  σ_i^2 is the velocity fluctuation variance

a_i depend on the input Eulerian PDF of the velocities $P(x,u)$ and is determined from the corresponding Fokker–Planck equation for stationary conditions:

$$\frac{\partial}{\partial x_i}\big[u_i(x,t)P(x,u)\big] = -\frac{\partial}{\partial u_i}\big[a_i(x,u)P(x,u)\big] + \frac{1}{2}\cdot\frac{\partial^2}{\partial u_i u_j}\big[b_{ij}^2(x)P(x,u)\big] \qquad (8.6)$$

Equation 8.6 can be split into the following two equations:

$$a = \frac{1}{P(x,u)}\cdot\left[\frac{C_0\varepsilon}{2}\cdot\frac{\partial P(x,u)}{\partial u} + \phi(x,u)\right] \qquad (8.7)$$

and

$$\phi(x,u) = -\frac{\partial}{\partial x}\cdot\int_{-\infty}^{u} u\cdot P(x,u)\cdot du \qquad (8.8)$$

where $\phi \to 0$ for $|u| \to \infty$.
In general (Thomson, 1987), u_i refers to the total velocity:

$$u_i(t) = \left(u_i'(t) + \overline{u(t)}\right) \qquad (8.9)$$

where
  $\overline{u(t)}$ is the mean wind velocity (representing the transport)
  $u_i'(t)$ is the velocity fluctuation

Considering a joint nonhomogeneous Gaussian PDF, Thomson (1987) proposed the following solution of Equation 8.7, called the "simplest solution":

$$a_i = -\left(\frac{C_0\varepsilon}{2}\right)\Gamma_{ik}\,(u_k - \overline{u_k}) + \frac{\Phi_i}{g_a} \qquad (8.10a)$$

where, referring to formulation (8.9):

$$\frac{\Phi_i}{g_a} = \frac{\partial \overline{u_i}}{\partial t} + \overline{u_l} \frac{\partial \overline{u_i}}{\partial x_l} + \frac{\partial \overline{u_i}}{\partial x_j} (u_j - \overline{u_j}) + \frac{1}{2} \frac{\partial \tau_{il}}{\partial x_l} + \frac{\Gamma_{lj}}{2} \frac{\partial \tau_{il}}{\partial t} (u_j - \overline{u_j})$$

$$+ \frac{\Gamma_{lj}}{2} \left(\overline{u_m} \frac{\partial \tau_{il}}{\partial x_m} \right) (u_j - \overline{u_j}) + \frac{\Gamma_{lj}}{2} \frac{\partial \tau_{il}}{\partial x_k} (u_j - \overline{u_j})(u_k - \overline{u_k}) \qquad (8.10b)$$

and Γ_{il} is the inverse of the Reynolds stress tensor τ_{il}. By setting in these general solutions (Equations 8.9, 8.10, 8.5, and 8.6):

$$\frac{\partial}{\partial x_1} \neq \frac{\partial}{\partial x_2} \neq \frac{\partial}{\partial x_3} \neq 0 \qquad (8.11)$$

$$\overline{u_1} \neq 0, \quad \overline{u_2} \neq 0, \quad \overline{u_3} \neq 0 \qquad (8.12)$$

$$\Gamma_{lj} = \frac{1}{\sigma_{lj}^2}, \tau_{il} = \sigma_{il}^2 \quad \text{and} \quad \sigma_{1,2}^2 = \sigma_{1,3}^2 = \sigma_{2,1}^2 = \sigma_{2,3}^2 = \sigma_{3,1}^2 = \sigma_{3,2}^2 = 0 \qquad (8.13)$$

that is, considering the nonhomogeneous 3-D case (Equation 8.11), retaining the 3-D mean winds (Equation 8.12), and disregarding the cross-correlation terms (Equation 8.13), Anfossi et al. (2009a) obtained the following 3-D solution:

$$du = \left\{ -\frac{(u - \overline{u})}{T_{Lu}} + \frac{\partial \overline{u}}{\partial x} u + \frac{\partial \overline{u}}{\partial y} v + \frac{\partial \overline{u}}{\partial z} w + \sigma_u \frac{\partial \sigma_u}{\partial x} \right.$$

$$\left. + \frac{(u - \overline{u})}{\sigma_u} \left[\frac{\partial \sigma_u}{\partial x} u + \frac{\partial \sigma_u}{\partial y} v + \frac{\partial \sigma_u}{\partial z} w \right] \right\} dt + \left(\frac{2dt}{T_{Lu}} \right)^{1/2} \sigma_u \xi_u \qquad (8.14a)$$

$$dv = \left\{ -\frac{(v - \overline{v})}{T_{Lv}} + \frac{\partial \overline{v}}{\partial x} u + \frac{\partial \overline{v}}{\partial y} v + \frac{\partial \overline{v}}{\partial z} w + \sigma_v \frac{\partial \sigma_v}{\partial y} \right.$$

$$\left. + \frac{(v - \overline{v})}{\sigma_v} \left[\frac{\partial \sigma_v}{\partial x} u + \frac{\partial \sigma_v}{\partial y} v + \frac{\partial \sigma_v}{\partial z} w \right] \right\} dt + \left(\frac{2dt}{T_{Lv}} \right)^{1/2} \sigma_v \xi_v \qquad (8.14b)$$

$$dw = \left\{ -\frac{(w - \overline{w})}{T_{Lw}} + \frac{\partial \overline{w}}{\partial x} u + \frac{\partial \overline{w}}{\partial y} v + \frac{\partial \overline{w}}{\partial z} w + \sigma_w \frac{\partial \sigma_w}{\partial z} \right.$$

$$\left. + \frac{(w - \overline{w})}{\sigma_w} \left[\frac{\partial \sigma_w}{\partial x} u + \frac{\partial \sigma_w}{\partial y} v + \frac{\partial \sigma_w}{\partial z} w \right] \right\} dt + \left(\frac{2dt}{T_{Lv}} \right)^{1/2} \sigma_v \xi_v \qquad (8.14c)$$

It is worth mentioning that, in most application studies (see, for instance, Brusasca et al., 1992; Carvalho et al., 2002a; Weil, 2008), the variable $u(t)$ of Equation 8.3 represents the velocity fluctuation only, while the mean wind velocity $\overline{u(t)}$ is included and made explicit, following Equation 8.9 formulation, in Equation 8.3, which thus becomes

$$dx_i(t) = \left(u_i(t) + \overline{u(t)} \right) \cdot dt \tag{8.15}$$

In many studies (see, for instance, Wilson et al., 1981; De Baas et al., 1986; Luhar and Britter, 1989; Weil, 1990), only the vertical component of the velocity is computed by means of the Langevin equation, thus obtaining a 1-D model. Note that in this case and in flat terrain, Equations 8.3 and 8.15 coincide since the mean vertical speed is zero. Generally, the 1-D model refers to convective conditions. In this case, the vertical turbulence is nonhomogeneous and asymmetric. Consequently, the PDF is asymmetric (non-Gaussian).

8.3 ASYMMETRIC PDFs

Asymmetric PDFs can be parameterized either by the bi-Gaussian PDF (Baerentsen and Berkowicz, 1984; Luhar and Britter, 1989; Weil, 1990) or by the Gram-Charlier PDF (Anfossi et al., 1996; Ferrero and Anfossi, 1998a,b).

8.3.1 BI-GAUSSIAN PDF

The bi-Gaussian PDF is defined as a linear combination of two normal distributions (Pearson, 1894):

$$P(w,z) = A \cdot N_A(w_A, \sigma_A) + B \cdot N_B(w_B, \sigma_B) \tag{8.16}$$

where $A + B = 1$, $A > 0$, $B > 0$ and N_A, N_B are Gaussian PDFs with means w_A, w_B, and standard deviations σ_a, σ_b. Let us recall that the expression for the Gaussian PDF is

$$N_A = \left[(2\pi)^{1/2} \sigma_A \right]^{-1} \exp\left[-(w - w_A)^2 / (2\sigma_A^2) \right] \tag{8.17}$$

(and similarly for N_B). To compute the A, B, w_A, w_B, σ_A, and σ_B parameters, use is made of the definition of the $P(w, z)$ moments, namely,

$$\overline{w^n} = \int w^n P(w,z)\, dw \tag{8.18}$$

(with $n = 0, 1, 2, 3$), where $\overline{w^n}$ are the corresponding Eulerian moments, thus obtaining

$$A + B = 1 \tag{8.19a}$$

$$Aw_A + Bw_B = 0 \tag{8.19b}$$

$$A(w_A^2 + \sigma_A^2) + B(w_B^2 + \sigma_B) = \overline{w^2} \tag{8.19c}$$

$$A(w_A^3 + 3w_A\sigma_A^2) + B(w_B^3 + 3w_B\sigma_B^2) = \overline{w^3} \tag{8.19d}$$

This system has more unknowns than equations and, consequently, there are various possible closures.

Baerentsen and Berkowicz (1984) assumed that

$$\sigma_A = |w_A| \quad \text{and} \quad \sigma_B = |w_B| \tag{8.20}$$

obtaining

$$w_B = \frac{\left(\overline{w^3} - \sqrt{(\overline{w^3})^2 + 8(\overline{w^2})^3} \right)}{4\overline{w^2}} \tag{8.21a}$$

$$w_A = -\frac{\overline{w^2}}{2w_B} \tag{8.21b}$$

$$A = -\frac{w_A}{w_A - w_B} \tag{8.21c}$$

$$B = 1 - A \tag{8.21d}$$

Weil (1990) generalized the Baerentsen and Berkowicz (1984) closure as follows:

$$\sigma_A = R|w_A| \quad \text{and} \quad \sigma_B = R|w_B| \tag{8.22}$$

thus obtaining

$$w_A = \frac{1}{2}\sqrt{\overline{w^2}} \left[\alpha S + \left(\alpha^2 S^2 + \frac{4}{\beta} \right)^{\frac{1}{2}} \right] \tag{8.23a}$$

$$w_B = \frac{1}{2}\sqrt{\overline{w^2}} \left[\alpha S - \left(\alpha^2 S^2 + \frac{4}{\beta} \right)^{\frac{1}{2}} \right] \tag{8.23b}$$

where $S = \overline{w^3}/(\overline{w^2})^{3/2}$ is the skewness, $\alpha = (1 + R^2/1 + 3R^2)$, $\beta = 1 + R^2$ and $R = 3/2$.

Luhar et al. (1996) proposed a further more generalized closure, also based on the skewness, but that has the correct property that their bi-Gaussian PDF collapses to a Gaussian PDF in the zero skewness limit:

$$w_A = m\,\sigma_A, \quad w_B = -\,m\,\sigma_B, \quad m = (2/3)\,S^{1/3} \tag{8.24}$$

thus obtaining

$$\sigma_A = \sqrt{\overline{w^2}}\left[\frac{B}{A(1+m^2)}\right]^{1/2} \tag{8.25a}$$

$$\sigma_B = \sqrt{\overline{w^2}}\left[\frac{B}{B(1+m^2)}\right]^{1/2} \tag{8.25b}$$

$$A = \frac{1}{2}\left[1-\left(\frac{r}{4+r}\right)^{1/2}\right] \tag{8.25c}$$

$$B = 1 - A \tag{8.25d}$$

$$r = \frac{(1+m^2)^3 S^2}{m^2(3+m^2)^2} \tag{8.25e}$$

8.3.2 GRAM–CHARLIER PDF

The Gram–Charlier PDF, truncated to the fourth order (GC4), has the following form (Kendall and Stuart, 1977):

$$P(x,z) = \frac{e^{-x^2/2}}{\sqrt{2\pi}}\left(1+C_3 H_3 + C_4 H_4\right) \tag{8.26}$$

in which H_3 and H_4 are Hermite polynomials and C_3 and C_4 their coefficients, whose expressions are

$$H_3 = x^3 - 3x \qquad H_4 = x^4 - 6x^2 + 3 \tag{8.27a}$$

$$C_3 = \overline{\mu^3}/6, \quad C_4 = \left(\overline{\mu^4}-3\right)/24 \tag{8.27b}$$

where $\overline{\mu^3}$ and $\overline{\mu^4}$ are the standardized moments of w and $x = w/\sigma_w$.

The use of this PDF in the Lagrangian stochastic models was proposed by Anfossi et al. (1996) and Ferrero and Anfossi (1998a,b). Setting $C_4=0$ in Equation 8.26, one obtains the Gram-Charlier PDF truncated to third order (GC3), whereas by setting $C_3=0$ as well, Equation 8.26 reduces to the Gaussian PDF. Gram-Charlier series expansions, though showing good correspondence to experiments (see, for instance, Frenkiel and Klebanoff, 1967; Durst et al., 1992; Anfossi et al., 1996), can exhibit small negative probabilities in the tails of the distribution. However, numerical experiments (Ferrero and Anfossi, 1998b) found that these unrealistic velocities rarely occur and thus showed that discarding these nonphysical probabilities is inconsequential in practical applications. On the other hand, the main advantages of the Gram-Charlier PDFs are their computational efficiency and their ability to include information on the Eulerian moments directly.

8.4   SOLUTIONS OF THE 1-D LANGEVIN EQUATION

As above anticipated, in many applications concerning dispersion estimate in convective conditions (in which the vertical turbulence is nonhomogeneous and asymmetric) and flat terrain, only the vertical velocity is computed by the Langevin equation; thus a 1-D model is considered.

Using the bi-Gaussian PDF (Equation 8.16) and the closure of Equation 8.20, the following solution for Equation 8.10 is obtained (Luhar and Britter, 1989):

$$
\Phi = AN_A \left[\frac{\sigma_A^2}{A} \frac{\partial A}{\partial z} + \frac{\partial \sigma_A}{\partial z} \left(\sigma_A + \frac{w(w-w_A)}{\sigma_A} \right) + w \frac{\partial w_A}{\partial z} \right]
$$

$$
+ \frac{1}{2} \frac{\partial (Aw_A)}{\partial z} \left[1 - erf\left(\frac{w-w_A}{\sqrt{2}\sigma_A} \right) \right] + BN_B \left[\frac{\sigma_B^2}{B} \frac{\partial B}{\partial z} + \frac{\partial \sigma_B}{\partial z} \left(\sigma_B + \frac{w(w-w_B)}{\sigma_B} \right) + w \frac{\partial w_B}{\partial z} \right]
$$

$$
+ \frac{1}{2} \frac{\partial (Aw_B)}{\partial z} \left[1 - erf\left(\frac{w-w_B}{\sqrt{2}\sigma_B} \right) \right] \tag{8.28}
$$

where $erf(z) = (2/\sqrt{\pi})\int_0^z \exp(-s) \cdot ds$ is the error function.

Solving Equations 8.2 and 8.8 using the GC4 Gram-Charlier PDF (Equation 8.26), the following expressions (Ferrero and Anfossi, 1998a,b) are found:

$$
\Phi = \frac{1}{2} \frac{\partial \sigma_w^2}{\partial z} \frac{e^{-\frac{x^2}{2}}}{\sqrt{2\pi}} \left[1 - C_4 + x^2(1+C_4) - 2C_3x^3 - 5C_4x^4 + C_3x^5 + C_4x^6 \right] \tag{8.29}
$$

and

$$
a = \sigma_w \frac{\frac{1}{T_{Lw}}(T_1) + \frac{\partial \sigma_w}{\partial z}(T_2)}{T_3} \tag{8.30}
$$

where

$$T_1 = -3C_3 - x\left(15C_4 + 1\right) + 6C_3x^2 + 10C_4x^3 - C_3x^4 - C_4x^5$$

$$T_2 = 1 - C_4 + x^2\left(1 + C_4\right) - 2C_3x^3 - 5C_4x^4 + C_3x^5 + C_4x^6 \qquad (8.31)$$

$$T_3 = 1 + 3C_4 - 3C_3x - 6C_4x^2 + C_3x^3 + C_4x^4$$

In 1-D inhomogeneous conditions, in those cases in which a Gaussian PDF can be assumed (i.e., in near-neutral conditions or very light convective conditions), Equation 8.2 has the following expression (Wilson et al., 1983; Thomson, 1987; Rodean, 1996):

$$du_i = -\frac{u_i}{T_{Li}}dt + \frac{1}{2}\left[1 + \left(\frac{u_i}{\sigma_{ui}}\right)^2\right]\frac{\partial\sigma_{u_i}^2}{\partial x_i}dt + \sigma_{ui}\left(\frac{2}{T_{Li}}\right)^{\frac{1}{2}}d\mu \qquad (8.32)$$

In the particular case in which a homogeneous turbulence can be assumed, Equation 8.32 becomes

$$du_i = -\frac{u_i}{T_{Li}}dt + \sigma_{ui}\left(\frac{2}{T_{Li}}\right)^{\frac{1}{2}}d\mu \qquad (8.33)$$

Equation 8.33 is the classical Langevin equation with constant coefficients.

8.5   THE LOW WIND CASE

A critical condition for pollutant dispersion is related to low speed or calm wind regimes. Dispersion in low wind speed conditions is mostly governed by meandering (low-frequency horizontal wind oscillations). Even when the stability reduces the vertical dispersion, meandering disperses plumes over rather wide angular sectors, which can affect, in many cases, all the compass ($0°$–$360°$). Thus, the resulting ground-level concentration is generally much lower than that predicted by standard Gaussian plume models (Sagendorf and Dickson, 1974) and, consequently, it is necessary to use different kinds of models. Among these last, LSDM have been proved to be a reliable modeling tool. For instance, Brusasca et al. (1992) proposed an "ad hoc" algorithm, based on the Gifford (1960) fluctuating plume model, to account for the meandering in their model LAMBDA. On the other hand, Oettl et al. (2001), considering that Eulerian autocorrelation function (EAF), computed for the horizontal components of wind velocity, showed a negative loop attributed to the meandering, proposed a Lagrangian stochastic model that uses a time step of random duration chosen from a uniform distribution (Wang and Stock, 1992). This model used a negative intercorrelation parameter for the horizontal wind components. Both models were satisfactorily tested against the Idaho National Engineering Laboratory (INEL) tracer dataset (Sagendorf and Dickson, 1974). However, both these models present a certain amount of empiricism.

Recently, Anfossi et al. (2005) and Oettl et al. (2005), by studying the low wind speed turbulence and dispersion characteristics from sonic anemometer records, confirmed that EAFs of horizontal wind components show an oscillating behavior with the presence of large negative lobes due to the meandering. They found that the observed EAFs, $R(\tau)$, were correctly fitted by the following relationship:

$$R(\tau) = e^{-\frac{\tau}{(m^2+1)T}} \cos\frac{m\tau}{(m^2+1)T} \qquad (8.34)$$

proposed by Frenkiel (1953). This may also be written (Murgatroyd, 1969) as

$$R(\tau) = e^{-p\tau} \cos(q\tau) \qquad (8.35)$$

by setting

$$p = \frac{1}{(m^2+1)T} \quad \text{and} \quad q = \frac{m}{(m^2+1)T} \qquad (8.36)$$

Equations 8.34 and 8.35 were suggested by Frenkiel (1953) and Murgatroyd (1969) in other contexts. These equations contain two parameters, one (T or p) associated to the classical integral turbulence timescale and the second (m or q) to the meandering characteristics.

Oettl et al. (2005) and Goulart et al. (2007) also provided a new physical explanation of the meandering occurrence. According to these works, meandering is explained as an inherent property of atmospheric flows in low wind speed conditions that, generally, does not need any particular trigger mechanism to be initiated. Meandering is shown to arise when the 2-D flow approaches or near approximate geostrophic balance, and it is damped out and vanishes when the Reynolds stresses are larger. In particular, Oettl et al. (2005) proposed the following set of stochastic Langevin equations for simulating horizontal dispersion in low wind speed conditions in the LSDM frame:

$$du = -(pu + qv)dt + \sigma_u \sqrt{2p\Delta t}\,\xi_u \qquad (8.37a)$$

$$dv = -(-qu + pv)\,dt + \sigma_v \sqrt{2p\Delta t}\,\xi_v \qquad (8.37b)$$

where ξ_u and ξ_v are random Gaussian variates (0,1), σ_u and σ_v are the velocity standard deviations, and p and q are defined in Equation 8.36. Its corresponding Fokker-Plank equation (Gardiner, 1990) is

$$0 = p\left\{\left(\frac{\partial(uP)}{\partial u} + \frac{\partial(vP)}{\partial v}\right) + \frac{q}{p}\left(\frac{\partial(vP)}{\partial u} - \frac{\partial(uP)}{\partial v}\right) + \sigma^2\left(\frac{\partial^2 P}{\partial u^2} + \frac{\partial^2 P}{\partial v^2}\right)\right\} \qquad (8.38)$$

It is important to notice that Equations 8.37a and b collapse on the classical Langevin equation (for homogeneous turbulence) when $m = 0$ (see Equation 8.33), that is, when the wind meandering is absent, namely

$$du = -\frac{u\,dt}{T} + \sigma_u \sqrt{2\frac{\Delta t}{T}}\,\xi_u \tag{8.39a}$$

and

$$dv = -\frac{v\,dt}{T} + \sigma_v \sqrt{2\frac{\Delta t}{T}}\,\xi_v \tag{8.39b}$$

Anfossi et al. (2006) applied this new approach (Equation 8.39) in the LSDM for flat terrain LAMBDA (Ferrero et al., 1995) to the tracer experiments carried out in low wind conditions by Idaho National Engineering Laboratory (INEL, Arco, Idaho) in 1974 (above mentioned) and by the Graz University of Technology and ISAC/CNR-Turin near Graz in 2003. They found a good agreement between prescribed and observed ground-level concentrations, and, in particular, found a noticeable improvement by using Equation 8.39 instead of the standard LAMBDA (not taking into account the meandering effect).

Anfossi et al. (2009a) later proposed a new system of coupled Langevin equations for the more general case of inhomogeneous turbulence and for the total wind velocity, namely,

$$du = \left\{ -p(u-\bar{u}) - q(v-\bar{v}) + \frac{\partial \bar{u}}{\partial x}u + \frac{\partial \bar{u}}{\partial y}v + \sigma_u\frac{\partial \sigma_u}{\partial x} + \frac{(u-\bar{u})}{\sigma_u}\left[\frac{\partial \sigma_u}{\partial x}u + \frac{\partial \sigma_u}{\partial y}v\right] \right\}dt$$

$$+ \sqrt{2pdt}\,\sigma_u\xi_u \tag{8.40a}$$

$$dv = \left\{ q(u-\bar{u}) - p(v-\bar{v}) + \frac{\partial \bar{v}}{\partial x}u + \frac{\partial \bar{v}}{\partial y}v + \sigma_v\frac{\partial \sigma_v}{\partial y} + \frac{(v-\bar{v})}{\sigma_v}\left[\frac{\partial \sigma_v}{\partial x}u + \frac{\partial \sigma_v}{\partial y}v\right] \right\}dt$$

$$+ \sqrt{2pdt}\,\sigma_v\xi_v \tag{8.40b}$$

They also verified that these solutions (Equations 8.37 and 8.40) satisfy the "well-mixed condition."

8.6 INVERSE MODELING

Besides being useful for describing the pollutant impact over urban or rural areas, models (LSDM in the present analysis) can be of great concern also for other aims, such as the pollutant source strength and/or location estimation in case of accidents (see, for instance, Seibert and Stohl, 1999; Roberti et al., 2007) or CO_2 diurnal cycle (Uliasz, 2003). These issues are examples of inverse problems in atmospheric pollution modeling (Seibert and Frank, 2004). The inverse problem is formulated as a nonlinear optimization approach, whose objective function is given by the least square difference between the observed and prescribed (by a dispersion model) pollutant concentration, associated to a regularization operator. The forward problem is iteratively solved for successive approximations of the unknown parameters. The associated forward problem is the solution of the dispersion model.

As pointed out by Seibert (2000), the major obstacle to a good inversion result is the accuracy of the dispersion model and the representativeness of the measurements, since these latter are always ineluctably affected by noise. Inverse problems belong to the class of ill-posed problem, where they are unstable in the presence of noise (small variations in the input data, imply a wide variation in the output data). Thus, regularized inverse solution, as detailed afterward, is a strategy to give good answer (Tikhonov and Arsenin, 1977).

In the case of pollutant source strength as a function of time estimation, it is assumed that the concentration obtained with the dispersion model is given by C^{Mod} $(\vec{r}, \mathbf{Q})$, where $\mathbf{Q} = [Q_1(t), ..., Q_n(t)]^T$ is the vector emission rate and $Q_n(t)$ represents the emission rate of the nth source and $C^{Exp}(\vec{r})$ are data from concentration measurements. The solution of the inverse problem is a function $\mathbf{Q}$ that minimizes the following objective function:

$$J(\lambda, \mathbf{Q}) = \left\| C^{Exp}(\vec{r}) - C^{Mod}(\vec{r}, \mathbf{Q}) \right\|_2^2 + \lambda \Omega(\mathbf{Q}) \tag{8.41}$$

where
   $\Omega(\mathbf{Q})$ is a regularization operator
   λ is the regularization parameter

The regularization operator can be expressed by the Tikhonov scheme (Tikhonov and Arsenin, 1977):

$$\Omega(\mathbf{Q}) = \sum_{m,j=0}^{p} \kappa_{m,j} \left\| \mathbf{Q}^{(m)} \right\|_2^2 \tag{8.42}$$

here $\mathbf{Q}^{(m)}$ denotes the mth difference. In general, the parameter $\kappa_{m,j}$ is chosen as $\kappa_{m,j} = \delta_{mj}$ (Kronecker delta) and the regularization is named Tikhonov-j regularization operator, where j denotes the order of the regularization.

8.7   COMPLEMENTARY INFORMATION

8.7.1   TURBULENCE AND FLOW FIELDS

In order to correctly simulate the dispersion of pollutants in the atmosphere, reliable estimates or parameterizations of the main processes that are responsible for the transport and diffusion are needed. Thus, the key parameters of the turbulent dispersion modeling are the input flow and turbulence fields, namely, the mean wind field, turbulence parameters, surface-layer parameters, and the height of the atmospheric boundary layer. In particular, in 3-D dispersion studies and in complex terrain, it is of great importance to also take into account the spatial variations of wind velocity moments and therefore the complete 3-D structure of the turbulent flow must be considered.

Flow field is generally obtained by meteorological (prognostic) or mass-consistent (diagnostic) models. The latter is built to include observed values too, and prognostic models also can deal with observations through data assimilation techniques. As far as the problem of prescribing the input turbulence fields is concerned, these quantities can be inferred directly from the meteorological model outputs, if that model solves a turbulent kinetic energy equation, or they can be derived from empirical parameterizations based upon both the available measurements and PBL theory. The parameterization schemes prescribed by Hanna (1982) and by Degrazia et al. (2000) are good examples of these parameterizations. These schemes do not include estimations of the skewness of the vertical wind velocity that, as above mentioned, are essential in the simulation of the convective boundary layer. They can be parameterized according to, for instance, Chiba (1978), De Baas et al. (1986), Weil (1990), and Rotach et al. (1996).

8.7.2 BOUNDARY CONDITIONS

In the LSDMs, the simplest and widely used boundary condition is the so-called perfect reflection: when a particle bumps against a boundary (upper or lower), it leaves the boundary in the opposite direction with the same vertical velocity and reversed sign. However, for skewed (non-Gaussian) turbulence, this may lead to an accumulation or deficit of particles at the boundaries.

The correct method of dealing with this problem was stated by Thomson and Montgomery (1994). They assumed that the distribution of particle velocities crossing any level in a fixed time interval must be preserved. This is expressed by the following equation:

$$\int_{w_r}^{\infty} w P_a \mathrm{d}w = - \int_{-\infty}^{w_i} w P_a \mathrm{d}w \qquad (8.43)$$

where
w_r and w_i are the reflected and incident velocities
P_a is the PDF

Equation 8.43 guarantees that the average vertical velocity through any arbitrary level of the domain is zero. Unfortunately, this equation has no analytical solution. Anfossi et al. (1997) proposed two different approximated analytical solutions for practical applications. The first one consists in solving the second integral of Equation 8.43 and expanding the first integral in Taylor series in powers of $(w_r + w_s)$, where $w_s = w_i \left| \dfrac{w_A}{w_B} \right|$ or $w_s = w_i \left| \dfrac{w_B}{w_A} \right|$ for bottom or top reflection. The second solution is a regression curve. They concluded that both methods satisfy the well-mixed condition and do not appreciably depart from the exact solution. The second solution is preferred for practical use since it is less time consuming.

Thomson et al. (1997) also considered, in a way similar to Equation 8.43, the particles crossing through interfaces characterized by largely different values of turbulence parameters (a typical example is the interface between the mixing layer and the capping inversion). Consider the interface is located at z_d and let σ_+ and σ_- (with $\sigma_+ \ll \sigma_-$) be the values of the standard deviations of vertical velocity (σ) in the above and bottom part of the domain. Then, if the particle approaches from above (the side having the smaller value of σ), it is always transmitted; if the particle approaches from below, the velocity w_2 of the transmitted particle (where the incident velocity is w_1), is computed as

$$w_2 = \sqrt{\sigma_+^2 \left[\frac{w_1^2}{\sigma_+^2} + \log\left(\frac{\sigma_+^2}{\sigma_-^2} \right) \right]} \qquad (8.44)$$

If w_2 is negative, the perfect reflection is applied to the particle. On the contrary, if it is positive, the particle is allowed to cross the interface with its velocity changing at the moment it crosses the interface to that given by Equation 8.44. The treatment of particles entering the interface from above is similar.

8.7.3 PLUME RISE

In most Lagrangian stochastic dispersion simulations, one has to account for the rise of buoyant plumes emitted by the industrial plants. Buoyant plumes have a temperature greater than the ambient temperature. The behavior of a chimney plume in the atmosphere is a rather complex process, which is influenced by emission characteristics, and actual wind, turbulence, and stratification profiles (Briggs, 1975; Anfossi et al., 2004).

Plumes emitted into the atmosphere rise under the action of their initial momentum and buoyancy. A plume, moving through the ambient atmosphere, experiences a shear force at its perimeter, where momentum is transferred from the plume to the surrounding air. This causes an increase of the plume diameter and a decrease of its velocity. This phenomenon is known as entrainment. The buoyancy forces help maintaining the motion of the plume as it transfers momentum to the surrounding air. For this reason, buoyant plumes generally rise higher than jet plumes (i.e., non-buoyant plumes). The entrained ambient air mixes with the plume air, thus diluting the plume components and, in the case of buoyant plumes, decreasing the average temperature difference between air and plume. In a calm or very low wind conditions, plumes rise almost vertically, whereas in windy situations they bend over. In this case, the velocity of any plume parcel is the vector composition of horizontal wind velocity and vertical plume velocity in the first stage and then approaches the horizontal wind velocity.

The straightforward method, even if time consuming, of computing the plume rise in the LSDM is based on the numerical integration, at each time step, of a set of differential equations expressing the conservation of the momentum flux, buoyancy flux, and volume flux. As an example, we will briefly review the approach considered in the TAPM LSDM (Hurley, 2005).

An equivalent approach was also proposed by Webster and Thomson (2002) and is included in the NAME LSDM (Maryon et al., 1999). Among many other NAME applications, the modeling study of a very large oil depot fire in England is of peculiar interest (Webster et al., 2006).

The TAPM approach is based on the model proposed by Glendening et al. (1984), as simplified by Hurley and Manins (1995). The equations of conservation of plume volume G, buoyancy and momentum fluxes, F_b and F_m, and the plume particle trajectory are written as follows:

$$\frac{dG}{dt} = 2R\left(\alpha W^2 + \beta U_a W + \gamma u_p E^{1/2}\right) \tag{8.45a}$$

$$\frac{dF_b}{dt} = -s\frac{F_m}{u_p}\left(\frac{1}{2.25}U_a + W\right) \tag{8.45b}$$

$$\frac{dF_m}{dt} = F_b \tag{8.45c}$$

$$\frac{dx_p}{dt} = u \tag{8.45d}$$

$$\frac{dy_p}{dt} = v \tag{8.45e}$$

$$\frac{dz_p}{dt} = W \tag{8.45f}$$

where

$$G = \frac{T_a}{T_s}u_p R^2 \tag{8.46a}$$

$$F_m = \frac{T_a}{T_s}u_p R^2 W \tag{8.46b}$$

$$W = \frac{F_m}{G} \tag{8.46c}$$

$$F_b = u_p R^2 g\frac{(T_s - T_a)}{T_s} \tag{8.46d}$$

$$R = \sqrt{\frac{G + \dfrac{F_b}{g}}{u_p}} \qquad (8.46e)$$

where

g is the acceleration due to gravity,

x_p, y_p, z_p are the particle displacements

T_a is the ambient temperature

T_s the plume temperature

$s = \dfrac{g}{\vartheta_a}\dfrac{\partial \vartheta_a}{\partial z}$ is the stability parameter

$U_a = \sqrt{u^2 + v^2}$ is the horizontal wind speed

$u_p = \sqrt{U_a^2 + W^2}$ is the plume velocity

E is the turbulent kinetic energy

$a = 0.1$, $\beta = 0.6$, and $\gamma = 0.1$ are the vertical plume, bent-over plume, and ambient turbulence entrainment constants, respectively

The initial conditions are

$$F_b = g r_s^2 v_{s0} \frac{(T_{s0} - T_{a0})}{T_{s0}} \qquad (8.47a)$$

$$F_m = v_{s0}^2 r_s^2 \frac{T_{a0}}{T_{s0}} \qquad (8.47b)$$

$$G_s = F_m / v_{s0} \qquad (8.47c)$$

$$R_s = \sqrt{v_{s0} / \sqrt{u_0^2 + v_{s0}^2}} \qquad (8.47d)$$

where

r_s is the radius of the stack outlet

v_{s0} is the effluent emission speed at stack outlet

u_0 is the mean wind speed at the stack outlet height

Tests on these equations showed that they performed correctly and collapsed to the Briggs form for a bent-over plume, and to the Briggs vertical plume model equations for calm conditions (Hurley and Manins, 1995). It is worth mentioning that this method allows dealing with complex atmospheric conditions.

A second simplified and very fast method, also allowing to deal with complex atmospheric conditions, was proposed by Anfossi et al. (1993). In this method, it is assumed that the plume centerline grows according to the following interpolation plume rise formula:

$$\Delta h(t) = 2.6 \left(F_b t^2 / U_a \right)^{1/3} \left(t^2 s + 4.3 \right)^{-1/3} \qquad (8.48)$$

This equation, derived by Anfossi (1985), which describes both the transitional and final phases and different stability conditions, is simply a generalization of the well-known and validated Briggs's formulas (1975).

Then, to each ith particle a buoyancy flux F_b^i is assigned at the stack exit, from a normal distribution having the mean value equal to the mean buoyancy flux $\overline{F_b}$ and the standard deviation equal to $\overline{F_b}/3$. w_b, the vertical velocity contribution of the plume rise, is then computed as follows:

$$w_b = \frac{\Delta z}{\Delta t} = \frac{\left[\Delta h\left(U_a, s, t + \Delta t\right) - \Delta h\left(U_a, s, t\right)\right]}{\Delta t} \tag{8.49}$$

and, consequently, Equation 8.3, written for the vertical component, becomes

$$dz_i(t) = (w_i'(t) + \overline{w(t)} + w_b(t)) \cdot dt \tag{8.50}$$

Equations 8.49 and 8.50 were used to study emissions from power plants (Anfossi et al., 1993) and from ships stacks (Chosson et al., 2008).

It is important to notice that the methods presented here for computing the plume rise in LSDM (but the same considerations hold true for other methods that can be found in the literature) are hybrid, since in order to correctly estimate the plume entrainment one has to account for the characteristics of the ensemble of particles and this contradicts the requirement that the trajectory of any particle is independent of the behavior of the other particles.

8.7.4 CONCENTRATION CALCULATION

Let the emission rate of the considered pollutant be Q_i (kg s^{-1}) and N_p the total number of particles emitted at each time step Δt(s). Dividing the entire computation domain into cells having dimensions Δx, Δy, Δz and counting the number N_i of particle lying in the ith cell, concentration C_i can be calculated by dividing the mass found in that cell, $(N_i Q_i)$, where

$$Q_i = \frac{Q\Delta t}{N_p} \tag{8.51}$$

by the cell volume (Δx, Δy, Δz), thus obtaining

$$C_i = \frac{N_i}{N_p} \frac{Q\Delta t}{\Delta x \Delta y \Delta z} \tag{8.52}$$

To have a more "representative" simulation (especially if the meteorological conditions are not constant), one can record the particles' position N_r times during each simulation period. In this case, Equation 8.52 becomes

$$C_i = \frac{N_i}{N_p N_r} \frac{Q\Delta t}{\Delta x \Delta y \Delta z} \tag{8.53}$$

It is worth noticing that a sufficient number of particles should be released at each time step in order to obtain meaningful concentrations since too few particles give a patchy and incorrect representation of the concentration distribution. On the other hand, releasing too many particles requires a too long computing time. Thus, the calculation of the number of particles N_p to be emitted at each time step Δt in order to have a prefixed concentration precision C_x (minimum concentration associated to a single particle found in a cell) associated to each particle is as follows:

$$N_p = \frac{1}{N_r \, C_x} \frac{Q \, \Delta t}{\Delta x \, \Delta y \, \Delta z} \qquad (8.54)$$

An alternative way to compute ground-level concentration based on the kernel density estimators (Gingold and Monaghan, 1982) may also be used. However, it will not be discussed here since it might have problems in complex terrain, where a sampler may be situated along a hill or mountain side, making it erroneous to account also for particles moving along the other side of the hill/mountain, where the flow and turbulence condition might be completely different.

8.7.5 DENSE GAS DISPERSION

The accidental release and dispersion of hazardous gases and vapors is another application of great concern of atmospheric dispersion models. Very often, because of high molecular weight and/or low release temperature and/or because of high storage pressure and of chemical reactions, these emissions are denser than the ambient air. Initially, these emissions begin to disperse under the action of their own negative buoyancy and arbitrary oriented momentum, then their density excess reduces as ambient air is entrained and, finally, at some distance downwind, transition to passive dispersion occurs.

An important difference from the neutral gas dispersion is the horizontal gravity spreading, together with the cloud slumping in case of sloping terrain, which the dense cloud experiences when it reaches the ground.

It is important to stress that, as above said for the plume rise computation, to also compute the cloud descent, the gravity spreading, and slumping, LSDM have to be hybrid since the motion of each particle again depends on the position and density of the ensemble of particles.

Although correct dispersion simulations of dense gas may be performed by means of computational fluid dynamics (CFD) models (however demanding large CPU times), LSDM (that proved to be fast and reliable models) can be very useful tools, especially when fast emergency response or scenarios in complex terrain and obstacles are needed.

Examples of LSDM applied to dense gas dispersion are QUIC-PLUME Model (Williams and Brown, 2003; Williams et al., 2004) and MSS (Tinarelli et al., 2008; Anfossi et al., 2009b).

8.8   EXAMPLE OF LSDM APPLICATIONS

In order to highlight the possible LSDM applications in different topography and stability conditions and with reference to different typology of applications (validation studies, impact analysis, scenarios, etc.), in this section we briefly introduce the SPRAY model, developed by the team that the authors belong to, and we present a few examples of its application. All these materials were already published in international journals; thus it is not a newly published work and will be referred to the original journals.

8.8.1   SPRAY MODEL

SPRAY (Tinarelli et al., 1994, 2000; Ferrero et al., 2001, 2003; Trini Castelli et al., 2003) is a 3-D model designed to deal with the simulation of passive airborne pollutant dispersion in complex terrain. Basically, it integrates three Langevin equations, one for each Cartesian component of the velocity fluctuations (see Equations 8.2, 8.3, and 8.15) according to the Thomson (1987) scheme. The algorithm to account for the rise of buoyant emissions is the one presented in Section 7.3 (Equations 8.49 and 8.50). Regarding the Eulerian PDF of the vertical turbulent velocities, that is generally skewed, one can choose between bi-Gaussian PDF (Equation 8.16) and the Gram-Charlier PDF (Equation 8.26). The model makes use of the inhomogeneous Gaussian PDF in the horizontal directions (Equation 8.32).

SPRAY also enters an integrated modeling system: RMS, acronym for RAMS, the atmospheric circulation model (Pielke et al. 1992); MIRS, the parameterization interface code (Trini Castelli and Anfossi, 1997; Trini Castelli, 2000), calculating the PBL parameters and Lagrangian turbulence fields not directly supplied by RAMS and processing its outputs for the input to SPRAY, which is the last module of the system.

8.8.2   VALIDATION VERSUS EXPERIMENTS

Model validation is the scientific basis for the development and improvement of the numerical models to make them usable tools both for theoretical studies and for applications in environmental frameworks. SPRAY model was used and tested in several case studies and experiments since 1986. Here we report as examples two works dealing with different approaches, the first considering a comparison with observations collected from physical modeling in a wind tunnel, the second with a real field experiment.

The RUSVAL tracer experiment (Khurshudyan et al., 1990) permits to evaluate the model performances in controlled condition. These type of experiments allow performing sensitivity analyses on specific aspects and parameters in the physics described by the model. In RUSVAL, the flow over a schematic two-dimensional valley was reproduced in a wind tunnel. RUSVAL data were used by Trini Castelli et al. (2001) and Ferrero et al. (2003). The main aim of their work was to suggest proper methods for predicting turbulence field for dispersion models over complex terrain and, more generally, in horizontally nonhomogeneous conditions. In fact, when atmospheric pollutant dispersion is simulated over complex terrain or in urban heat island, the turbulence input parameters are often prescribed according to standard parameterization based on surface layer quantities (see Section 8.7.1). However, these parameterizations may be inadequate in predicting the turbulence field in such horizontally nonhomogeneous boundary layer, due to the essentially local nature of the prescribed turbulence. SPRAY was used in the modeling system RMS. Different and new turbulence closure schemes implemented in RAMS meteorological model were used, in combination with alternative formulations for the Lagrangian turbulent parameters in SPRAY. Their influence on the dispersion of the tracer could thus be evaluated.

Predicted fields of velocity standard deviations and Lagrangian timescales demonstrated to be able to take into account the inhomogeneities due to the valley. The concentrations calculated by the dispersion model, using the turbulent parameters obtained from the new closure models, showed a satisfactory agreement with the observed data and the performances were better than using the usual local parameterizations developed for flat terrain. This is demonstrated in Figure 8.1, where the cumulative frequency distribution of predicted and observed concentration is plotted. The solid line refers to the observations, the dashed line and dotted line refer, respectively, to RMS used with a standard configuration of the turbulence closure and Lagrangian parameterizations with a new closure (Trini Castelli et al., 2001). It can be noticed that the dispersion simulation performed by using the RAMS standard turbulence closure produced a large underestimation of the higher concentrations.

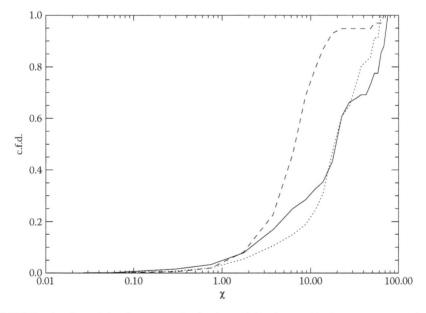

FIGURE 8.1 Cumulative frequency distribution (c.f.d.) of normalized mean concentration χ. Observed data: solid line; RMS with the new closures: dotted line; RMS with standard closure: dashed line.

An application and validation of SPRAY performances, within RMS modeling system, in experiments carried out in real complex terrain was worked out on the TRACT (TRAnsport of air pollutants over complex terrain) field campaign, performed in the Rhine valley, in southern Germany, during September 1992 (Fiedler, 1989; Zimmermann, 1995). Among many other measurements, TRACT included a tracer release and the related concentration measurements (at ground level and aloft). The main objectives of RMS simulations (Carvalho et al. 2002) were to verify its capability to accurately simulate the 3-D transport and diffusion of a passive pollutant in a complex orography. In fact, the TRACT area is a rather complex region characterized by the presence of valleys (such as the Rhine Valley) and mountains (like the Black Forest, the Vosges, and the Swabian Alps).

RAMS simulations were performed using three nested grids, from a 16 km (grid 1) up to a 4 km (grid 2) and 1 km (grid 3) horizontal resolution, and SPRAY was run on all the grids. The tracer emission, near the ground level (source at 8 m) lasted 3 h.

The model system correctly reproduced the general behavior of the plume (that was divided into several tracer puffs), the temporal and spatial distribution of the concentration, and the location of the concentration maxima during the 12 h of observations. Also, the aloft simulated concentration values compared well with data measured by aircraft.

This simulation work allowed demonstrating the feasibility of the complete simulation of a dispersion process (wind field reconstruction, generation of the turbulence field, and reconstruction of the concentration field) in complex terrain. This is of fundamental importance for the air pollution problem and for the assessment of the environmental impact.

As an example of the simulation results, Figure 8.2 shows the computed particle positions (representing the tracer position), plotted over the 10 m wind field, at different hours: 06, 08, 12, and 16 UTC on September 16, from top-left to bottom-right. The more the time passed, the larger was the area involved and, consequently, a different computational grid had to be consider. It can clearly be seen that, at the beginning, the plume is very narrow and follows the wind direction along the Rhine valley, as shown at 06 UTC (grid 3). Approximately 2 h after starting the emission, the plume, though exhibiting a definite principal nucleus, also shows some puffs that travel in different directions. At 08 UTC, the separation from the main plume clearly appears (grid 2). Later the plume appears as a very large cloud and, at 12 and 16 UTC (grid 1), it clearly splits into two parts, one part remains close to the emission source, and the other part moves toward southwest. It is worth mentioning that the clouds' position shown in all these figures correctly reproduced the observed plume trajectory.

8.8.3 SINGLE SOURCES AND LINEAR EMISSIONS: IMPACT ASSESSMENT IN COMPLEX TERRAIN

LSDM are advanced tools that in the last decade are pushing their ways through the impact assessment framework. Lagrangian models offer a much better description of the atmospheric physical processes with respect to simplified models, but still demand

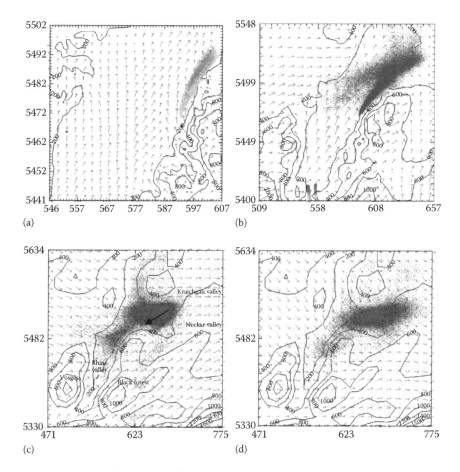

FIGURE 8.2 (See color insert following page 234.) Particle positions and wind field for 10 m terrain-following surface on 16 September 1992 at different hours: (a) grid 3, 06 UTC; (b) grid 2, 08 UTC; (c) grid 1, 12 UTC; (d) grid 1, 16 UTC.

a smaller computational and time effort than CFD models, producing a good quality and accuracy of the dispersion simulation. These aspects make Lagrangian models suitable tools for applicative environmental assessment and emergency response.

The aim of these modeling studies is the assessment of air quality impact, in terms, for instance, of NO_2 or NO_x and PM10, for single source emissions, such as power plants or incinerators, and linear emissions from vehicular traffic in highways and large roads. This is useful for supporting the decision-making process of local authorities for the construction authorization or control of pollutant sources and the protection of the public health. This kind of studies is needed to characterize and quantify the contribution to the environment pollution of emission from stacks, evaluating their impacts on the air quality.

In the following, we summarize some results from real case studies, where SPRAY, within RMS system, was used. As an example, we refer to a recent study

(Trini Castelli et al., 2009a) on a waste incinerator, which is planned to be built in the city of Turin (Northern Italy). This study was aimed at estimating the ground-level concentrations (g.l.c.) distribution during particularly adverse dispersion conditions, possibly causing severe pollution episodes. The rationale of this approach was based on the principle that, if the influence on g.l.c. of the incinerator in the worst dispersion conditions does not bring to exceedances of the law limits imposed on short-term concentrations for the considered pollutants, its construction could be proposed without worsening or strictly causing pollution episodes. The typical wind and stability conditions of Turin very often do not favor the pollutant dispersion because of its geographical position. Turin (220 m a.m.s.l and about one million of inhabitants) is located at the western edge of the Po Valley. It is surrounded by a hill chain on the eastern sector and by the Alps in the other three sectors. This peculiar orographic position typically brings to low wind and/or calm conditions, thermal ground-based inversions during night-time, föhn episodes and fog situations. In such highly complex situations, advanced 3-D modeling systems need to be used, and consequently RMS system was applied. The importance of using advanced models, which can take into account the meteorological variability and the topographical inhomogeneities, is clearly highlighted in Figure 8.3, where a snapshot of the evolution of the plume during an episode of anticyclonic conditions associated to fair weather and local-scale thermal circulation (February 10, 2000) is shown.

We notice that in the beginning of the day, 09 UTC, the plume elongates south-easterly, while after a few hours, at 13 UTC, due to the impact of the plume with the hill chain on the east part of the area, it is split into two main puffs that separated and moved toward north. Simulations were repeated in other severe meteo-dispersive conditions, which are common of the area, allowing to identify the subregions where the pollution was bounded to give the highest impact. It was thus proved that a research-based modeling system can be profitably used for supporting the decision-making process for the control and protection of the environment and health.

Analogous applications were performed to study the atmospheric pollution due to the traffic in mountain valleys, characterized by peculiar meteorological and dispersive characteristics due to the complex topography. Even in these particular conditions, which largely affect the effectiveness of the dispersion of road traffic pollutant, simplified models or parameterizations are not sufficient to properly describe such complexity. Hereafter we report two examples related to studies performed in the Alps, in the Frejus and Brenner alpine transects.

In the frame of ALPNAP Project (Heimann et al., 2008), the pollutant dispersion, related to the emissions from the major traffic routes in Susa (Italy, national roads SS24 and SS25, and highway A32) and Maurienne (France, national road RN6, and highway A43) valleys, was simulated. For a detailed reproduction of the atmospheric circulation in Frejus transect area, a downscaling from the regional to the local scale was performed with RAMS up to 1000 m resolution and with a diagnostic mass-consistent model up to a resolution of 100 m. Three periods, characterized by critical conditions of the dispersive scenarios, were chosen in the reference year 2004. The output data, that is the main meteorological fields, the ground-level pollutant concentration of NO_x and PM10 and the plume dynamics, were transferred to other ALPNAP partners to support the part of the project related to noise study and impact

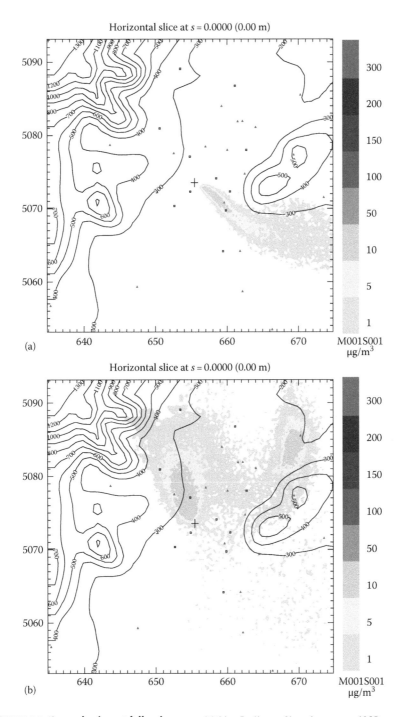

FIGURE 8.3 (See color insert following page 234.)   Isolines of hourly averaged NO_x concentration for the day 10.02.2000 at 09 UTC (a) and 13 UTC (b) (concentration scale in $\mu g/m^3$).

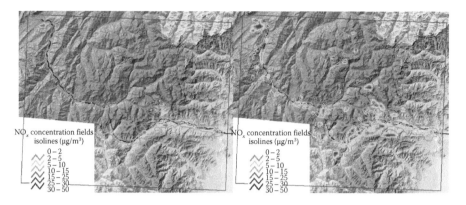

FIGURE 8.4 (See color insert following page 234.) Project ALPNAP. Example of mean (left) and maximum (right) NO_x concentration maps from RMS simulations, year 2004, 3–13 July episode.

assessment. In Figure 8.4, an example of the elaboration of g.l.c data is presented for the mean and maximum values of concentration for a summer period, from July 3 to 13, 2004. Simulations were also repeated considering possible future scenarios for the traffic emissions (Trini Castelli et al., 2009b).

Similar numerical experiments were performed also for the Brenner transect in the frame of the International Project BBT (Oettl et al., 2007), focusing on the area of South Tyrol and considering the A22 highway and the national road S12. In this case, two domains were considered, centered on Vipiteno (Sterzing) and Fortezza (Franzenfeste), respectively, and the final goal was to compute the annual mean of the NO_x and PM10 ground-level concentrations due exclusively to the vehicular traffic. The study was aimed at verifying the atmospheric pollution reduction that would be achieved with the opening of the Railway Brenner Tunnel and the consequent possibility to move part of the vehicular traffic on the railway. Three emission scenarios were simulated, under the proposal of BBT project responsibilities: actual (year 2004), that refers to the present situation at the time of the study, minimum (year 2015), assuming that the tunnel is not built, and consensus (year 2015), in which the new tunnel is considered under operation and some measures aimed at transferring traffic from the road to the railway are established. In Figure 8.5, a comparison of the PM10 g.l.c. annual mean estimated for the actual (left) and consensus (right) scenarios is presented, clearly showing the reduction of the g.l.c. in the future scenario.

This kind of investigation permits to verify whether possible and alternative law measures on traffic management may effectively reduce its pollutant impact and sensibly improve the air quality of the affected regions.

8.8.4 INVERSE MODELING EXAMPLES

As an example of inversion modeling, we refer to a work by Roberti et al. (2007) whose aim was to identify the pollutant emission rate, assumed unknown, knowing

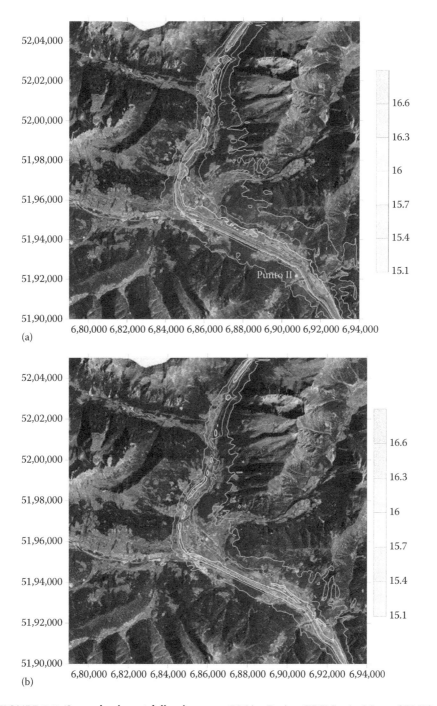

FIGURE 8.5 (See color insert following page 234.)   Project BBT Study. Maps of PM10 g.l.c. annual mean for actual (a) and consensus (b) scenarios in the area including the two domains considered.

TABLE 8.1
Emission Rate Estimation

Time (h:min)	Q^{true} (g s^{-1})	Q^{est} (g s^{-1})
12:05	3.20	3.185
12:15	3.20	3.185
12:25	3.20	3.187
12:35	3.20	3.188
12:45	3.20	3.188

Q^{true} indicates the true emission rate and Q^{est} indicates
the estimated emission rate.

the ground-level concentrations measured at a network of ground-level samplers, following the procedure reported in Equations 8.41and 8.42. The inverse procedure was applied to data from the real field (the Copenhagen tracer experiment, performed on October 19, 1978), where some level of noise in the data is expected. Roberti et al. (2007) made use of the second-order Tikhonov regularization and determined the regularization parameter according to the L-curve scheme (Hansen, 1992).

In the Copenhagen experiment, 39 ground-level samplers were located in three crosswind arcs, located at 2–6 km from the releasing point, and the main meteorological parameters were measured at three heights (10, 120, and 200 m) along a tower and the emission was released from the same tower at 115 m.

The Lagrangian particle model LAMBDA, that is, the SPRAY version for flat terrain, was used to simulate the direct (or forward) problem, while the inverse problem was formulated as an optimization problem.

The time period for the experiment and, consequently, of the emission rate estimation was 50 min. The emission rate was assumed variable with time but it was constant and equal to 3.2 g s^{-1} in the experiment. Thus, in the simulation, the unknown source term could be represented by the vector: $\mathbf{Q} = [Q_1, Q_2, Q_3, Q_4, Q_5]^T$, where $Q_i = Q(t_0 + i\Delta t)$ with $\Delta t = 10$ min. Table 8.1 showing the results suggests that the inverse modeling procedure was quite accurate.

8.9 CONCLUSIONS

In this review, the actual state of the art of LSDMs for the description of airborne dispersion in the PBL is briefly presented. It covers various aspects of their derivation and applications. Their theoretical bases (Langevin equation, Fokker-Plank equation, well-mixed condition, probability density functions, turbulence parameterization) are described and the related technical information (boundary conditions, concentration calculation, plume rise, dense gas dispersion) are presented. Then, the application of this modeling tool to the low wind situations and in the inverse modeling technique are also briefly outlined.

Finally, a few applications of Lagrangian stochastic model simulations performed by the author's team and already published on peer reviewed international journals are briefly presented as examples of LSDM possible applications. These examples illustrate the wide field of application of this kind of modeling technique: going from basic model evaluation to impact assessment analysis and to scenarios' studies, both in complex terrain and in urban environment.

Lagrangian models are proved to be reliable and accurate tools not only for the description and investigation of the physical processes determining the dispersion of pollutants in the atmosphere but also for supporting the assessment and control of the air quality and the protection of the environment and public health.

REFERENCES

Anfossi D. 1985. Analysis of plume rise data from five TVA Steam Plants. *J. Clim. App. Met.* 24:1225–1236.

Anfossi D., E. Ferrero, G. Brusasca, A. Marzorati, and G. Tinarelli. 1993. A simple way of computing buoyant plume rise in a Lagrangian stochastic dispersion model for airborne dispersion. *Atmos. Environ.* 27 A, 1443–1451.

Anfossi D., E. Ferrero, D. Sacchetti, and S. Trini Castelli. 1996. Comparison among empirical probability density functions of the vertical velocity in the surface layer based on higher order correlations. *Bound. Lay. Meteorol.* 82, 193–218.

Anfossi D., E. Ferrero, G. Tinarelli, and S. Alessandrini. 1997. A simplified version of the correct boundary conditions for skewed turbulence in Lagrangian particle models. *Atmos. Environ.* 31:301–308.

Anfossi D., E. Canepa, and H. van Dop. 2004. Plume Rise. Chapter 6 of AIR QUALITY MODELING Theories, Methodologies, Computational Techniques, and Available Databases and Software. Vol. I Fundamentals, P. Zannetti, Ed., The EnviroComp Institute and the Air & Waste Management Association.

Anfossi D. and W. Physick. 2005. Lagrangian Particle Models. Chapter 11 of AIR QUALITY MODELING—Theories, Methodologies, Computational Techniques, and Available Databases and Software. Vol. II—Fundamentals, P. Zannetti, Ed., The EnviroComp Institute and the Air & Waste Management Association.

Anfossi D., D. Oettl, G. Degrazia, and A. Goulart. 2005. An analysis of sonic anemometer observations in low wind speed conditions. *Bound. Lay. Meteorol.* 114:179–203.

Anfossi D., S. Alessandrini, S. Trini Castelli, E. Ferrero, D. Oettl, and G. Degrazia. 2006. Tracer dispersion simulation in low wind speed conditions with a new 2-D Langevin equation system. *Atmos. Environ.* 40:7234.45.

Anfossi D., G. Tinarelli, S. Trini Castelli, E. Ferrero, D. Oettl, G. Degrazia, and L. Mortarini. 2009a. Well mixed condition verification in windy and low wind speed conditions. In press on *Int. J. Environ. Pollut.*

Anfossi D., G. Tinarelli, S. Trini Castelli, and G. Belfiore. 2009b. Proposal of a new Lagrangian particle model for the simulation of dense gas dispersion. In press on *Int. J. Environ. Pollut.*

Baerentsen J.H. and R. Berkowicz. 1984. Monte-Carlo simulation of plume diffusion in the convective boundary layer. *Atmos. Environ.* 18:701–712.

Briggs G.A. 1975. Plume rise predictions. In: Lectures on air pollution and environmental impact analyses. D. Haugen, ed., *Workshop proceedings*, Boston, MA, September 29–October 3, pp. 59–111, American Meteorological Society, Boston, MA.

Brusasca G., G. Tinarelli, and D. Anfossi. 1992. Particle model simulation of diffusion in low wind speed stable conditions. *Atmos. Environ.* 4:707–723.

Carvalho J., D. Anfossi, S. Trini Castelli, and G. A. Degrazia. 2002a. Application of a model system for the study of transport and diffusion in complex terrain to the TRACT experiment. *Atmos. Environt.* 36:1147–1161.

Chiba O. 1978. Stability dependence of the vertical wind velocity skewness in the atmospheric surface layer. *J. Meteorol. Soc. Jpn.* 56:140–142.

Chosson F., R. Paoli, and B. Cuenot. 2008. Ship plume dispersion rates in convective boundary layers for chemistry models. *Atmos. Chem. Phys.* 8:4841–4853.

Csanady G.T. 1973. *Turbulent Diffusion in the Environment.* Reidel, Dordrecht, the Netherlands.

De Baas H.F., H. Van Dop, and F.T.M. Nieuwstadt. 1986. An application of the Langevin equation for inhomogeneous conditions to dispersion in a convective boundary layer. *Quart. J. Roy. Meteor. Soc.* 112:165–180.

Degrazia G.A., D. Anfossi, J.C. Carvalho, C. Mangia, T. Tirabassi, and H.F. Campos Velho. 2000. Turbulence parameterization for PBL dispersion models in all stability conditions. *Atmos. Environ.* 34:3575–3583.

Durst F., J. Jovanovic, and T.G. Johansson. 1992. On the statistical properties of truncated Gram-Charlier series expansions in turbulent wall-bounded flows. *Phys. Fluids. A* 4:118–126.

Ferrero E., D. Anfossi, G. Brusisca, and G. Tinarelli. 1995. Lagrangian particle model LAMBDA: evaluation against tracer data. *Int. J. Environ. Pollut.* 5:360–374.

Ferrero E. and D. Anfossi. 1998a. Sensitivity analysis of Lagrangian Stochastic models for CBL with different PDF's and turbulence parameterizations. *Air Pollution Modelling and its Applications* XII, S.E. Gryning and N. Chaumerliac, eds., Plenum Press, New York, 22:673–680.

Ferrero E. and D. Anfossi. 1998b. Comparison of PDFs, closures schemes and turbulence parameterizations in Lagrangian Stochastic Models. *Int. J. Environ. Poll.* 9:384–410.

Ferrero E., D. Anfossi, and G. Tinarelli. 2001 Simulations of atmospheric dispersion in an urban stable boundary layer. *Int. J. Environ. Poll.* 16:1–6.

Ferrero E., S. Trini Castelli, and D. Anfossi. 2003. Turbulence fields for atmospheric dispersion models in horizontally non-homogeneous conditions. *Atmos. Environ.* 37:2305–2315.

Ferrero E. 2005. *Lagrangian Particle Models (Theory). in* Chapter 11 of *AIR QUALITY MODELING – Theories, Methodologies, Computational Techniques, and Available Databases and Software. Vol. II – Fundamentals,* P. Zannetti, Ed., The EnviroComp Institute and the Air & Waste Management Association.

Fiedler F., 1989. EUREKA Environmental ProjectFEUROTRACT, Proposal of a Subproject, Transport of Air Pollutants over Complex Terrain (TRACT), Karlsruhe.

Frenkiel F.N. 1953. Turbulent diffusion: mean concentration distribution in a flow field of homogeneous turbulence. *Adv. Appl. Mech.* 3:61107.

Frenkiel F.N. and P.S. Klebanoff. 1967. Higher order correlations in a turbulent field. *Phys. Fluids.* 10:507–520.

Gardiner C.W. 1990. *Handbook of Stochastic Methods.* Springer Verlag, Berlin, Germany.

Gifford F.A. 1960. Peak to average concentration ratios according to a fluctuating plume dispersion model. *Int. J. Air Poll.* 3:253–260.

Gingold R.A. and J.J. Monaghan. 1982. Kernel estimates as a basis for general particle methods in hydrodynamics. *J. Comput. Phys.,* 46:429–453.

Glendening J.W., J.A. Businger, and R.J. Farber. 1984. Improving plume rise prediction accuracy for stable atmospheres with complex vertical structure. *J. Air Poll. Control Ass.* 34:1128–1133.

Goulart A., G. Degrazia, O. Acevedo, and D. Anfossi. 2007. Theoretical considerations of meandering wind in simplified conditions. *Bound. Lay. Meteorol.* 125:279–287.

Hanna S.R. 1982. Applications in air pollution modeling. In *Atmospheric Turbulence and Air Pollution Modeling,* F.T.M. Nieuwstadt and H. Van Dop, eds., Reidel, Dordrecht, the Netherlands, Ch. 7.

Hansen P.C. 1992. Analysis of discrete ill-posed problems by means of the L-curve. *SIAM Rev.* 34:561–580.

Heimann D., M. de Franceschi, S. Emeis, P. Lercher, P. Seibert (Eds.), D. Anfossi, G. Antonacci, M. Baulac, G. Belfiore, D. Botteldooren, A. Cemin, M. Clemente, D. Cocarta, J. Defrance, E. Elampe, R. Forkel, E. Grießer, B. Krüger, B. Miège, F. Obleitner, X. Olny, M. Ragazzi, J. Rüdisser, K. Schäfer, I. Schicker, P. Suppan, S. Trini Castelli, U. Uhrner, T. Van Renterghem, J. Vergeiner, and D. Zardi. 2008. Air Pollution, Traffic Noise and Related Health Effects in the Alpine Space: A Guide for Authorities and Consulters, ALPNAP comprehensive report. Università degli Studi di Trento, Dipartimento di Ingegneria Civile e Ambientale, Trento, Italy, 335 pp.

Hinze J.O. 1975. *Turbulence*. Mc Graw Hill, New York, 790 pp.

Hurley P.J. 2005. The Air Pollution Model (TAPM) Version 3. Part1: Technical Description. *CSIRO Atmospheric Research Technical Paper* No. 71.

Hurley P.J. and P.C. Manins. 1995. Plume rise and enhanced dispersion in LADM. ECRU technical note No.4, CSIRO Division of Atmospheric Research, Australia.

Kendall M. and A. Stuart. 1977. *The Advanced Theory of Statistics*, MacMillan, New York.

Khurshudyan L.H., W.H. Snyder, I.V. Nekrasov, R.E. Lawson, R.S. Thompson, and F.A. Schiermeier. 1990. Flow and dispersion of pollutants within two-dimensional valleys, summary report on joint Soviet-American Study, EPA REPORT No. 600/3–90/025.

Luhar A.K. and R.E. Britter. 1989. A random walk model for dispersion in inhomogeneous turbulence in a convective boundary layer. *Atmos. Environ.* 23:1911–1924.

Luhar A., M. Hibberd, and P. Hurley. 1996. Comparison of closure schemes used to specify the velocity PDF in Lagrangian stochastic dispersion models for convective conditions, *Atmos. Environ.* 30:1407–1418.

Maryon R.H., D.B. Ryall, and A.L. Malcolm. 1999. The NAME 4 dispersion model: Science documentation. *Turbulence and Diffusion Note*, 262, Met Office.

Monin A.S. and A.M. Yaglom. 1965. *Statistical Fluid Mechanics: Mechanics of Turbulence*. MIT Press, Cambridge, MA, Vol.1, 225.

Murgatroyd R.J. 1969. Estimations from geostrophic trajectories of horizontal diffusivity in the mid-latitude troposphere and lower stratosphere. *Quart. J. R. Met. Soc.* 95:40–62.

Oettl D., R.A. Almbauer, and P.J. Sturm. 2001. A new method to estimate diffusion in stable, low wind conditions. *J. Appl. Meteorol.* 40:259–268.

Oettl A. Goulart, G. Degrazia, and D. Anfossi. 2005. A new hypothesis on meandering atmospheric flows in low wind speed conditions. *Atmos. Environ.* 39:1739–1748.

Oettl D., P. Sturm, D. Anfossi, S. Trini Castelli, P. Lercher, G. Tinarelli, and T. Pittini. 2007. Lagrangian particle model simulation to assess air quality along the Brenner transit corridor through the Alps. *Developments in Environmental Science 6. Air Pollution Modelling and Its Applications XVIII*, C. Borrego and E. Renner, eds., Elsevier, Amsterdam, the Netherlands, 689–697.

Pearson K. 1894. Contributions to the mathematical theory of evolution. *Philosophical Transactions of the Royal Society of London*. 185—Part I, 71–110.

Pielke R.A., W.R. Cotton, R.L. Walko, C.J. Tremback, W.A. Lyons, L.D. Grasso, M.E. Nicholls, M.D. et al. Copeland. 1992. A comprehensive meteorological modeling system—RAMS. *Meteorol. Atmos. Phys.* 49:69–91.

Roberti D.R., D. Anfossi, H. Fraga de Campos Velho, and G.A. Degrazia. 2007. Estimation of emission rate from experimental data. *Nuovo Cimento* C. 30:177–186.

Rodean H.C. 1996. Stochastic Lagrangian models of turbulent diffusion. *Meteorological Monographs*, 26, American Meteriological Society, Boston, MA.

Rotach M., S. Gryning, and C. Tassone. 1996. A two-dimensional Lagrangian stochastic dispersion model for daytime conditions. *Q. J. R. Meteorol. Soc.* 122:367–389.

Sagendorf J.F. and C.R. Dickson. 1974. Diffusion under low windspeed, inversion conditions. *NOAA Technical Memorandum ERL ARL-52*, 89 pp.

Sawford B.L. and F.M. Guest. 1988. Uniqueness and universality of Lagrangian stochastic models of turbulent dispersion. *Eighth Symposium on Turbulence and Diffusion*, San Diego, CA, American Meteriological Society, 96–99.

Seibert P. 2000. *Inverse Methods in Global Biogeochemical Cycles*, Kasibhatla P. et al., eds., American Geophysical Union, Washington, DC, 147–154.

Seibert P. and A. Frank. 2004 Source-receptor matrix calculation with a Lagrangian particle dispersion model in backward mode. *Atmos. Chem. Phys.* 4:51–63.

Seibert P. and A. Stohl. *Inverse Modelling of the ETEX-1 Release with a Lagrangian Particle Model*. Available at https://www.boku.ac.at/imp/envmet/glor3.html.

Luft TA. 2002. Dispersion calculation according to VDI 3945 part 3 (Environmental meteorology—Atmospheric dispersion models—Particle model) and TA LUFT 2002.

Tennekes H. 1982. Similarity relations, scaling laws and spectral dynamics. In *Atmospheric Turbulence and Air Pollution Modelling*, F.T.M. Nieuwstadt and H. van Dop, eds., Reidel, Dordrecht, the Netherlands, 37–68.

Thomson D.J. 1987. Criteria for the selection of stochastic models of particle trajectories in turbulent flows. *J. Fluid Mech.* 180:529–556.

Thomson D.J. and M.R. Montgomery. 1994. Reflection boundary conditions for random walk models of dispersion in non-Gaussian turbulence. *Atmos. Environ.* 28:1981–1987.

Thomson D.J., W.L. Physick, and R.H. Maryon. 1997. Treatment of interfaces in random walk dispersion models. *J. Appl. Meteor.* 36:1284–1295.

Tikhonov A.N. and V.I. Arsenin. 1977. *Solutions of Ill-posed Problems*, John Wiley & Sons, New York.

Tinarelli G., D. Anfossi, G. Brusasca, E. Ferrero, U. Giostra, M.G. Morselli, J. Moussafir, F. Tampieri, and F. Trombetti. 1994. Lagrangian Particle Simulation of Tracer Dispersion in the Lee of a schematic Two-Dimensional Hill. *J. Appl. Meteorol.* 33:744–775.

Tinarelli G., D. Anfossi, M. Bider, E. Ferrero, and S. Trini Castelli. 2000. A new high performance version of the Lagrangian particle dispersion model SPRAY, some case studies. In *Air Pollution Modelling and its Applications XIII*, S.E. Gryning and E. Batchvarova, eds., Kluwer Academic/Plenum Press, New York, 499–507.

Tinarelli G., D. Anfossi, S. Trini Castelli, and G. Belfiore. 2008. Development of a Lagrangian particle model for dense gas dispersion in urban environment. *Air Pollution Modelling and its Applications XIX*, C. Borrego and A.I. Miranda, eds., Springer, New York, 28–36.

Trini Castelli S. and D. Anfossi. 1997. Intercomparison of 3-D turbulence parameterizations for dispersion models in complex terrain derived from a circulation model. *Il Nuovo Cimento C.* 20:287–313.

Trini Castelli S. 2000. MIRS: A turbulence parameterisation model interfacing RAMS and SPRAY in a transport and diffusion modelling system. *Rap. Int. ICGF/CNR No 412/2000.*

Trini Castelli S., E. Ferrero, and D. Anfossi. 2001. Turbulence closure in neutral boundary layer over complex terrain. *Boun. Lay. Meteorol.* 100:405–419.

Trini Castelli S., D. Anfossi, and E. Ferrero. 2003. Evaluation of the environmental impact of two different heating scenarios in urban area. *Int. J. Env. and Pollut.* 20, 207–217.

Trini Castelli S., D. Anfossi, and S. Finardi. 2009a. Simulations of the dispersion from a waste incinerator in the Turin area in three different meteorological scenarios. *Int. J. Env. and Pollut.*, in press.

Trini Castelli S., G. Belfiore, D. Anfossi, E. Elampe, and M. Clemente. 2009b. Modelling the meteorology and traffic pollutant dispersion in highly complex terrain: The ALPNAP alpine space project. *Int. J. Environ. Pollut.*, in press.

Uliasz M. 2003. A modelling framework to evaluate feasibility of deriving mesoscale surface fluxes of trace gases from concentration data. Available at http://biocycle.atmos.colostate.edu/marek.mesoinversion7c.pdf.

FIGURE 1.1 Palacio del Marqués de Sta. Cruz (Oviedo-Spain), a building with rain streaking and biological and pollutant staining showing the forms that can disfigure architecture. (Photo courtesy of Carlota Grossi.)

FIGURE 1.4 Carved stone at Tower of London exhibiting the warmer tones found today. (Photo courtesy of Carlota Grossi.)

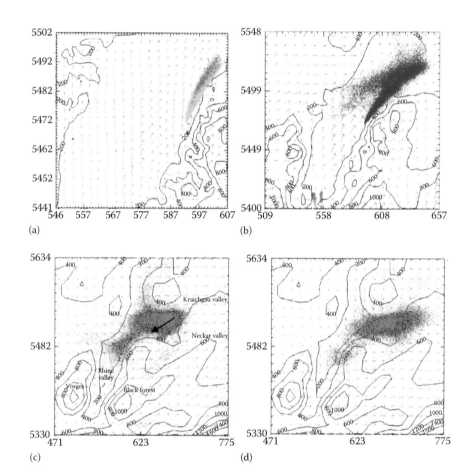

FIGURE 8.2 Particle positions and wind field for 10 m terrain-following surface on 16 September 1992 at different hours: (a) grid 3, 06 UTC; (b) grid 2, 08 UTC; (c) grid 1, 12 UTC; (d) grid 1, 16 UTC.

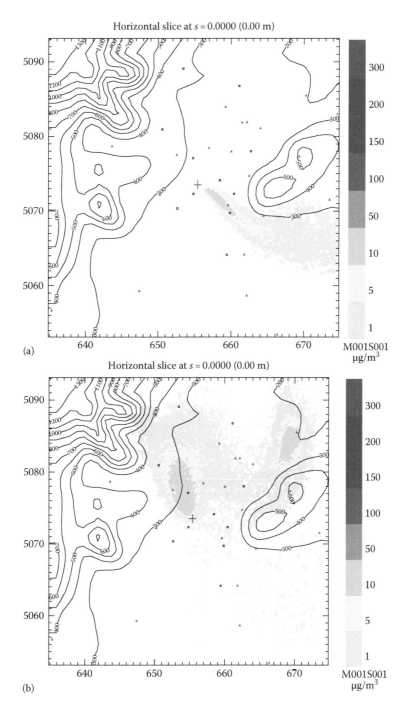

FIGURE 8.3 Isolines of hourly averaged NO$_x$ concentration for the day 10.02.2000 at 09 UTC (a) and 13 UTC (b) (concentration scale in µg/m^3).

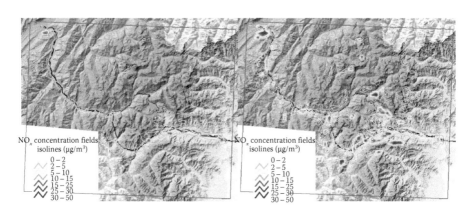

FIGURE 8.4 Project ALPNAP. Example of mean (left) and maximum (right) NO$_x$ concentration maps from RMS simulations, year 2004, 3–13 July episode.

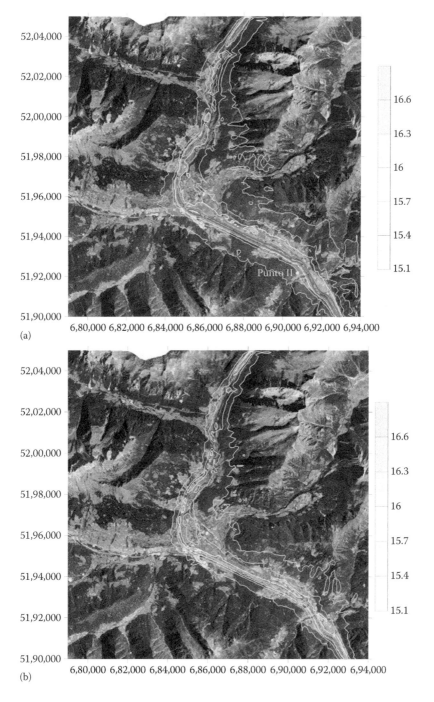

FIGURE 8.5 Project BBT Study. Maps of PM10 g.l.c. annual mean for actual (a) and consensus (b) scenarios in the area including the two domains considered.

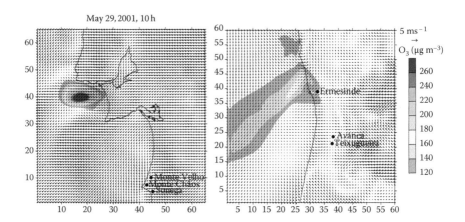

FIGURE 10.6 Hourly averaged wind and ozone surface concentration fields, simulated by CAMx, with 2 × 2 km resolution for the domains Lisbon and Porto.

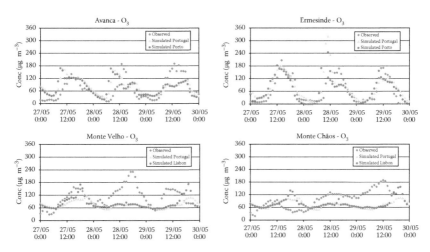

FIGURE 10.7 Temporal evolution of hourly averaged concentrations of ozone ($\mu g\ m^{-3}$), simulated for Portugal, Porto, and Lisbon domains, and comparison with measured data.

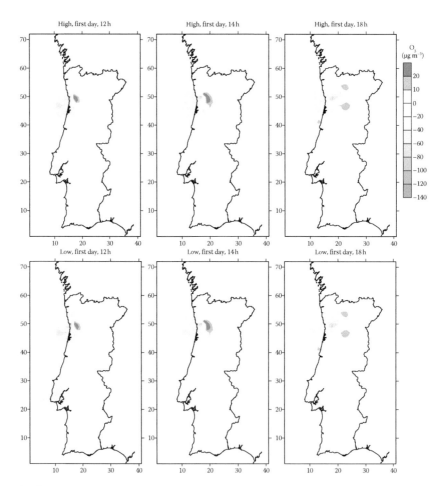

FIGURE 10.8 The differential of ozone concentration between the reference scenario (2001) and both, high and low, scenarios for 2010 given by the National Emission Ceiling Program.

Wang L.-P. and D.E. Stock. 1992. Stochastic trajectory models for turbulent diffusion: Monte-Carlo process versus Markov chains. *Atmos. Environ.* 9:1599–1607.

Webster H.N. and D.J. Thomson. 2002. Validation of a Lagrangian model plume rise scheme using the Kincaid data set. *Atmos. Environ.* 36:5031–5042.

Webster H.N, S.J. Abel, J.P. Taylor, D.J. Thomson, J.M. Haywood, and M.C. Hort. 2006. Dispersion modelling studies of the Buncefield Oil Depot incident. Hadley Centre technical note 69.

Weil J.C. 1990. A diagnosis of the asymmetry in top-down and bottom-up diffusion using a Lagrangian stochastic model. *J. Atmos. Sci.* 47:501–515.

Weil J.C. 2008. Linking a Lagrangian particle dispersion model with three-dimensional Eulerian wind field models. *J. Appl. Meteor. Climatol.* 47:2463–2467.

Williams M.D. and M.J. Brown. 2003. Description of the QUIC-PLUME Model, Los Alamos National Laboratory, LA-UR- 03-1426.

Williams M.D., M.J. Brown, B. Singh, and D. Boswell. 2004. QUIC-PLUME theory guide. Los Alamos National Laboratory. LA-UR-04-0561.

Wilson J., G. Thurtell, and G. Kidd. 1981. Numerical simulation of particle trajectories in inhomogeneous turbulence, II: Systems with variable turbulent velocity scale. *Bound. Lay. Meteorol.* 21:423–441.

Wilson J.D., B.J. Legg, and D.J. Thomson. 1983. Calculation of particle trajectories in the presence of a gradient in turbulent-velocity variance. *Bound. Lay. Meteorol.* 27:163–169.

Zimmermann, H. 1995. Field phase report of the TRACT field measurement campaign EUROTRACT report, Garmisch–Partenkirchen, Germany, 196 pp.

9 Atmospheric Dispersion with a Large-Eddy Simulation: Eulerian and Lagrangian Perspectives

Umberto Rizza, Giulia Gioia, Guglielmo Lacorata, Cristina Mangia, and Gian Paolo Marra

CONTENTS

9.1   INTRODUCTION

The planetary boundary layer (PBL) is the lowest layer of the atmosphere in which the direct effect of the earth's surface on dynamical processes plays a relevant role, and it is also the environment in which almost all human and biological activities (with their consequences) take place. It is therefore clear why the study of the PBL is a fundamental issue for environmental applications, as it can provide strategies for air quality management.

The PBL can be considered as a continuous fluid characterized by a mixture of many subsidiary motions with different scales in space and time. The governing equations of motion of a PBL flow are provided by the conservation principles, like the conservation of mass and momentum. The main difficulties in getting a general solution for the set of equations describing turbulent motions essentially arise from the nonlinearity of the equations of motion, the 3-D character of the velocity field, and the enormous number of scales involved in such motion.

The first point means that it is not possible to find analytical solutions, considering also the insufficient information for boundary and initial conditions. Even the numerical approach is quite complex and expensive: the high computational cost due to the large number of scales involved makes numerical solutions obtainable only for quasi-stable flows (low Reynolds numbers). Hence, historically, attention has shifted to statistical methods for studying a random velocity field. In this context, Reynolds (1895) developed his theory based on an averaging process for eliminating the most random characteristics of the fields. The Reynolds approach on the one hand determined a significant and fruitful turning point, while on the other it brought to the foreground the closure problem of turbulence, which has intrigued researchers for over a hundred years. This remains an outstanding unsolved problem of modern physics. The limited power of computers and the closure problem have induced researchers to use boundary-layer models to provide approximate solutions.

Turbulence models enable effective predictions of turbulent flows. Despite this, they work satisfactorily only in situations that do not differ too much from those used to calibrate them. This means, for example, that models developed for shear flows often do not work properly for convective flows.

It is well known that turbulence and air pollutant dispersion are interdependent phenomena. The development of numerical simulations for turbulence in the PBL during the last 30 years has brought enormous advantages in numerical simulations of pollutant dispersion in the PBL.

Following the notation of Wyngaard and Peltier (1996), the word "modeling" is usually used to represent the turbulence through approximate equations whose solutions have behavioral similarities to turbulence. The term "simulation" generally refers to equations that are derivable from the exact set and, hence, remain faithful to the essential physics.

As a consequence, a direct numerical simulation (DNS), where all turbulent flow scales are solved numerically, is a simulation, while a large-eddy simulation (LES) is a compromise between simulation and modeling. In fact, in an LES, only the energy-containing eddies (ECE) are simulated, while the rest are modeled. The proper distinction between flow scales is accomplished by applying a high-pass filter, with a cut-off length Δ_f, to the Navier–Stokes equations. It is clear that LES will provide satisfactory results when the cut-off length Δ_f is much smaller than the scale of the ECE of the turbulence.

LES was used first in micrometeorology by Deardorff (1970, 1972). The growth in size, speed, and availability of supercomputers has since made LES more and more suitable to PBL applications. The velocity and turbulence fields provided by LES can be used to calculate the transport and dispersion of contaminants. In this way, one can obtain, at the same time, detailed dispersion data and complete information on meteorological and turbulent parameters. Several studies with LES for atmospheric dispersion have been reported in the literature and concern both Eulerian and Lagrangian dispersion simulation approaches (Nieuwstadt and De Valk, 1987; Van Haren and Nieuwstadt, 1989; Kemp and Thompson, 1996; Meeder and Nieuwstadt, 2000; Rizza et al., 2006).

We assume that the theoretical background and the principal differences between the most important instruments in computational fluid dynamics (CFD), that is, DNS, LES, and RANS (Reynolds averaged Navier–Stokes) are known, so we describe very briefly Moeng's LES code used in the present work.

Both approaches for studying turbulent dispersion, the Eulerian and the Lagrangian frameworks, are used by means of LESs. In fact, both of these approaches have peculiar aspects that we must examine. Lagrangian particle models can successfully describe the turbulent dispersion of passive contaminants because they take into account essential aspects of turbulence, but they are limited to a simplified set of reacting species. The Eulerian approach, on the other hand, is based on the conservation equation and can incorporate the numerous second and high-order chemical kinetic equations necessary to describe photochemical smog generation, which currently represents a challenging environmental problem. The critical point is the numerical scheme used to discretize the conservation equation, which can generate nonphysical results.

9.2 LARGE-EDDY SIMULATION: THE MOENG AND SULLIVAN MODEL

9.2.1 DESCRIPTION

The large-eddy model used in this study is the LES model developed by Moeng (1984) and Sullivan et al. (1994). Here, we only give a short outcome.

In the LES technique, the smallest eddies in a large Reynolds number PBL flow are removed by applying a spatial filter function to the Navier–Stokes equations. For each turbulent quantity f, the filtered (or resolved) variable, denoted by an overbar, is defined as

$$\bar{f}(x,t) = \int_D f(y,t)\bar{G}(x-y)\,\mathrm{d}y = \int_D f(x-y,t)\bar{G}(y)\,\mathrm{d}y \qquad (9.1)$$

Integration is done over the flow volume D. The function $\bar{G}$ is a three-dimensional low-pass filter that removes the subgrid scale fluctuations (or small eddies: $f'(x,t)$).

Applying the filtering operator to the incompressible Navier–Stokes equations and making the substitution $f(x,t)=\bar{f}(x,t)+f'(x,t)$, we obtain the governing equations for the filtered variables:

$$\frac{\partial \bar{u}_i}{\partial t} + \frac{\partial \overline{u_i u_j}}{\partial x_j} = -\frac{1}{\rho}\frac{\partial \bar{p}}{\partial x_i} - \frac{\partial \tau_{ij}}{\partial x_j} + \nu \frac{\partial^2 \bar{u}_i}{\partial x_j \partial x_j} + g_i \frac{\theta}{\theta_0} - 2\varepsilon_{ijk}\Omega_j \bar{u}_k \tag{9.2}$$

where

$i \equiv (x,y,z)$

ρ is the density

$\bar{p}$ is the pressure term

$\nu = \mu/\rho$ is the kinematic viscosity

μ is the dynamic viscosity

the gravitational acceleration g_i is nonzero only in the z-direction

θ is the virtual potential temperature

θ_0 is the temperature of some reference state

ε_{ijk} is the permutation tensor

Ω_j is the angular vector of the earth's rotation

The terms

$$\tau_{ij} = \overline{u_i' \bar{u}_j} + \overline{\bar{u}_i u_j'} + \overline{u_i' u_j'} \tag{9.3}$$

are the subgrid scale fluxes that represent the effect of the subgrid scale on the resolved field. The tensor τ_{ij} is modeled following Sullivan et al. (1994):

$$\tau_{ij} = -2\nu_t \gamma S_{ij} - 2\nu_T \langle S_{ij} \rangle \tag{9.4a}$$

$$\tau_{\theta i} = -\nu_\theta \frac{\partial \bar{\theta}}{\partial x_i} \tag{9.4b}$$

where the brackets $\langle\rangle$ denote the average over the (x, y) plane as a surrogate for ensemble average modeling, ν_t and ν_T are respectively the fluctuating and mean-field eddy viscosities, and γ is the isotropy factor.

The eddy viscosity coefficient for the scalar (heat) is $\nu_\theta = [1+(2l_\Delta/\Delta)]\nu_t$, and the eddy viscosity coefficient for the momentum ν_t is expressed as

$$\nu_t = C_k l_\Delta (\bar{e'})^{1/2} \tag{9.5}$$

where C_k is a diffusion coefficient to be determined and l_Δ is an SGS length-scale function of fluid stratification, which for unstable conditions (negative stratification) is $l_\Delta = \Delta = (\Delta x \Delta y \Delta z)^{1/3}$ and for stable conditions (positive stratification) is $l_\Delta = l_s = 0.76(\overline{e'})^{1/2}\left(\dfrac{g}{\theta_0}\dfrac{\partial\overline{\theta}}{\partial z}\right)^{-1/2}$.

The SGS energy evolves following the prognostic equation:

$$\left(\frac{\partial}{\partial t} + \overline{u}_i\frac{\partial}{\partial x_i}\right)\overline{e'} = 2\nu_t\gamma\left(S_{ij} - \langle S_{ij}\rangle\right)\left(S_{ij} - \langle S_{ij}\rangle\right) + \frac{g}{\theta_0}\left(-\nu_\theta\frac{\partial\overline{\theta}}{\partial x_i}\right) - \varepsilon$$

$$+ \frac{\partial}{\partial x_i}\left(2\nu_t\frac{\partial\overline{e'}}{\partial x_i}\right) \tag{9.6}$$

where the different terms on the right-hand side are shear production, buoyancy, dissipation ε, and diffusion. The dissipation rate is given by $\varepsilon = \dfrac{C_\varepsilon(\overline{e'})^{3/2}}{l_\Delta}$.

If the LES grid falls into the inertial subrange, the spectral analysis of Moeng and Wyngaard (1988) shows that $C_k \approx 0.1$ and $C_\varepsilon \approx 0.19 + 0.74 l_\Delta/\Delta s$.

Making use of similarity theory, the expression for the mean-eddy-viscosity ν_T is $\nu_T(z = z_1) = \nu_T^*$ where

$$\nu_T^* = \frac{u_* k z_1}{\phi_m(z_1)} - \langle\nu_t\gamma\rangle - \frac{k z_1}{u_*\phi_m(z_1)}[\langle uw^2\rangle + \langle vw^2\rangle]^{1/2} \tag{9.7}$$

in which z_1 is a reference height, and $\nu_T = \nu_T^* \dfrac{k z_1}{\phi_m(z_1)}\sqrt{2\langle S_{ij}\rangle\langle S_{ij}\rangle}$ at any other height.

The isotropy factor γ is defined as the ratio between the small- and the large-scale strain rates in view of their easy availability in an LES. Here, the horizontally averaged fluctuating resolved strain (small-scale strain from the LES field) is $S' = \sqrt{2\langle\left(S_{ij} - \langle S_{ij}\rangle\right)\left(S_{ij} - \langle S_{ij}\rangle\right)\rangle}$.

The large-scale strain is simply determined from the mean strain as $\langle S\rangle = \sqrt{2\langle S_{ij}\rangle\langle S_{ij}\rangle}$. The isotropy factor is defined as $\gamma = \dfrac{S'}{S' + \langle S\rangle}$. At a fixed S', the asymptotic behavior is: when $\langle S\rangle \to 0 \Rightarrow \gamma \to 1$, and when $\langle S\rangle \to \infty \Rightarrow \gamma \to 0$ (near the wall), so near the wall, the only contribution to the eddy viscosity is from the inhomogeneous ensemble-average field, while far from the boundary, the only contribution comes from the isotropic term computed from the LES field. The isotropy factor varies continuously from the near-wall value close to zero to the unit value far from the wall. This parameter can also facilitate the transition from the SGS to the ensemble average turbulence parameterizations.

9.2.2 Numerical Scheme and Boundary Conditions

The present model uses a pseudospectral representation to calculate the horizontal derivatives and a finite differencing scheme for the vertical ones. From the point of view of homogeneity, atmospheric turbulence may have completely different behaviors in the horizontal and vertical directions. The near-homogeneity in any horizontal plane allows the application of periodic boundary conditions in both the x- and the y-directions. On the other hand, in the vertical direction, sources and sinks of turbulence are not uniformly distributed. A mixed scheme of Fourier expansion in the horizontal plane and finite differencing in the vertical is then appropriate for the numerical algorithm. The pseudospectral technique as developed by Fox and Orzag (1973) was chosen to calculate any horizontal derivative; for example, the x-derivatives of $\bar{u}$ are calculated first by transforming it into Fourier space in the x-direction:

$$\hat{\bar{u}}(k_m, y, z) = \frac{1}{N} \sum_{n=1}^{N} \bar{u}(x_n, y, z) e^{-ik_m x_n}$$

The transform coefficient $\hat{\bar{u}}$ is then multiplied by ik_m as required by the derivative in spectral space, and then $ik_m \hat{\bar{u}}$ is inversely transformed and normalized back to the grid points using

$$\left(\frac{\partial \bar{u}}{\partial x}\right)_n = \sum_{m=-(N/2)+1}^{N/2} ik_m \hat{\bar{u}}(k_m, y, z) e^{ik_m x_n}$$

where
   N is the total number of grid points in the x-direction
   $k_m = 2\pi m/N\Delta x$ is the wave number

This procedure is used to calculate any horizontal derivatives. To advance the solution from one time step to the next, we use an explicit third-order accurate multistage Runge–Kutta scheme (RK3) with a variable time step (Spalart and Moser, 1991).

9.2.3   LATERAL BOUNDARY CONDITIONS

As noted before, the pseudo-spectral method allows the application of periodic boundary conditions in both horizontal directions. Those fields that are provided in the output from one side of the domain are therefore used as input fields in the plane (x, y) for the opposite side of the domain. Even if periodic boundary conditions are convenient from a computational point of view, they are appropriate only for PBLs over homogeneous terrain.

9.2.4   SURFACE BOUNDARY CONDITIONS

For surface boundary conditions, we use the Monin–Obukhov similarity theory to relate surface fluxes to resolved-scale fields at the lowest grid level. The wind gradient at the surface is prescribed by similarity formulas. For the u-component, for example,

$$\frac{\partial U_S}{\partial z} = \frac{u_* \phi_m}{\kappa z}$$

where

$$U_S = \sqrt{\bar{u}^2 + \bar{v}^2}$$

u_* is the friction velocity

ϕ_m is the Monin–Obukhov stability function for the moment

$z = z_1$ is the height of the first grid point

κ is the von Karman constant

The SGS vertical fluxes at the surface are also assigned.

9.2.5 UPPER BOUNDARY CONDITIONS

The top boundary conditions are zero vertical velocity ($\bar{w} = 0$), zero SGS turbulence fields, and

$$\frac{\partial \bar{u}}{\partial z} = 0$$

$$\frac{\partial \bar{v}}{\partial z} = 0$$

$$\frac{\partial \bar{\theta}}{\partial z} = \text{const}$$

across the $\bar{w} = 0$ level. These kinds of conditions do not allow the transmission of gravity waves, which can be generated in the stably stratified layer due to turbulent motions in the PBL. Besides, the upper boundary of a typical LES domain is set to be well above the PBL top where artificial upper boundary conditions on the simulated PBL may arise.

9.3 EULERIAN DISPERSION WITH LARGE-EDDY SIMULATIONS AND EXPERIMENTS

9.3.1 EULERIAN DISPERSION

The Eulerian approach is based on the conservation of the pollutant mass of concentration $c(x,y,z,t)$ in a Cartesian frame:

$$\frac{\partial c}{\partial t} + u \cdot \nabla c + S = 0 \tag{9.8}$$

where the molecular diffusion term is neglected and $S = Q\delta(x)\delta(y)\delta(z - H_s)$ represents a generic source term in which Q is the rate emission and H_S is the source height.

We have already noted that, except for simplified geometries of the release source, the last equation cannot be trivially solved, due to the turbulent nature of variables. In this context, we can use a filter operation to decompose all of those variables into a "resolved" and a "subgrid" part. In particular, using the LES filter, Equation 9.8 can be written as (Andren et al., 1994)

$$\frac{\partial \overline{C}}{\partial t} = -\frac{\partial}{\partial x_i}\overline{u_i C} - \frac{\partial \tau_{ci}}{\partial x_i} + S \tag{9.9}$$

where the overbar denotes the filtered components and τ_{ci} are the SGS turbulent scalar fluxes. To get a solution for Equation 9.9, we consider the following closure model for τ_{ci}:

$$\tau_{ci} = -K_c \frac{\partial \overline{C}}{\partial x_i} \tag{9.10}$$

where K_c is the eddy diffusivity for a scalar quantity. Introducing the SGS Schmidt number $S_c = \dfrac{K_m}{K_c}$, we can express the eddy diffusivity in terms of the eddy viscosity for the momentum:

$$\tau_{ci} = -\frac{K_m}{S_c}\frac{\partial \overline{C}}{\partial x_i} \tag{9.11}$$

Substituting Equation 9.11 into Equation 9.9 leads to

$$\frac{\partial \overline{C}}{\partial t} = -\frac{\partial}{\partial x_i}\overline{u_i C} + \frac{1}{S_c}\frac{\partial}{\partial x_i}\left[K_m \frac{\partial \overline{C}}{\partial x_i}\right] + S \tag{9.12}$$

where $S_c = 0.33$ (Moeng and Sullivan, 1994). In the Cartesian (x, y, z) reference, Equation 9.12 can be rewritten as

$$\frac{\partial \overline{C}}{\partial t} + \frac{\partial \overline{u}\overline{C}}{\partial x} + \frac{\partial \overline{v}\overline{C}}{\partial y} + \frac{\partial \overline{w}\overline{C}}{\partial z} = \frac{1}{S_c}\left[\frac{\partial}{\partial x}K_m \frac{\partial \overline{C}}{\partial x} + \frac{\partial}{\partial y}K_m \frac{\partial \overline{C}}{\partial y} + \frac{\partial}{\partial z}K_m \frac{\partial \overline{C}}{\partial z}\right] + S \tag{9.13}$$

9.3.2 THE NUMERICAL METHOD

This method consists of splitting Equation 9.13 into a set of time-dependent equations, each one locally one-dimensional (LOD). Writing Equation 9.13 as a sum of advective/diffusive differential operators, we get

$$\frac{\partial \overline{C}}{\partial t} = A_x\overline{C} + A_y\overline{C} + A_z\overline{C} + D_x\overline{C} + D_y\overline{C} + D_z\overline{C} \tag{9.14}$$

or equivalently,

$$\frac{\partial \overline{C}}{\partial t} = \Lambda_x\overline{C} + \Lambda_y\overline{C} + \Lambda_z\overline{C} \tag{9.15}$$

where

$$
\begin{cases}
\Lambda_x = A_x + D_x \equiv -\bar{u}\dfrac{\partial}{\partial x} + \dfrac{\partial}{\partial x}\left(K_m\dfrac{\partial}{\partial x}\right) \\[3mm]
\Lambda_y = A_y + D_y \equiv -\bar{v}\dfrac{\partial}{\partial y} + \dfrac{\partial}{\partial y}\left(K_m\dfrac{\partial}{\partial y}\right) \\[3mm]
\Lambda_z = A_z + D_z \equiv -\bar{w}\dfrac{\partial}{\partial z} + \dfrac{\partial}{\partial z}\left(K_m\dfrac{\partial}{\partial z}\right)
\end{cases}
$$

Using Crank–Nicholson time integration, the LOD approximation (McRae et al., 1982) is given by

$$
\bar{C}^{n+1} = \prod_{j=1}^{3}\left[I - \frac{\Delta t}{2}\Lambda_j\right]^{-1}\left[I + \frac{\Delta t}{2}\Lambda_j\right]\bar{C}^n
$$

which can be rewritten as

$$
\bar{C}^{n+1} = \prod_{j=1}^{3} T_j^n \bar{C}^n = T^n \bar{C}^n \tag{9.16}
$$

where I is the unity matrix.

To obtain second-order accuracy, it is necessary to reverse the order of the operators in each alternate step in order to cancel the two noncommuting terms. We have to replace the scheme given by Equation 9.16 with the following double-sequence equations:

$$
\bar{C}^n = \prod_{j=1}^{3} T_j^n \bar{C}^{n-1}
$$

$$
\bar{C}^{n+1} = \prod_{j=3}^{1} T_j^n \bar{C}^n
$$

The effective system used is therefore the following:

$$
\bar{C}^n = [A_x FD_x][A_y FD_y][A_z FD_z]\bar{C}^{n-1} \tag{9.17a}
$$

$$
\bar{C}^{n+1} = [D_z A_z F][D_y A_y F][D_x A_x F]\bar{C}^n \tag{9.17b}
$$

where the operator F represents a filter operation to be applied after each advective step necessary to damp out the small-scale perturbations before they can corrupt the basic solution (Forester, 1979). This scheme is described in detail by Yanenko (1971) and Marcuk (1984).

In the next paragraph, we consider a method based on cubic-spline interpolations for the advective terms (operators A_i), which are usually the most difficult to implement, and a Crank–Nicholson implicit scheme for the diffusive terms (operators D_i).

9.3.3 DETAILS OF THE LOD METHOD

The finite difference algorithm for Equation 9.17a, or its reverse Equation 9.17b, contains three steps, one for each direction. In the following, we only show the numerical scheme for the x direction, as the same scheme is used for the other directions with the appropriate boundary conditions.

We use the notation $j \in [1, N_x]$, $k \in [1, N_y]$, $m \in [1, N_z]$ for increments in the (x, y, z) Cartesian space, so we have

$$x_j = x_0 + j\Delta x$$

$$y_k = y_0 + k\Delta y$$

$$z_m = z_0 + m\Delta z$$

where Δx, Δy, and Δz are the grid sizes and (N_x, N_y, N_z) are the number of grid points in the x-, y-, and z-directions, respectively. For each direction, the scheme (Equations 9.17a and 9.17b) contains three substeps: (a) the advective part, (b) the filtering procedure, and (c) the diffusive part.

(a) The advection is computed using a quasi-Lagrangian cubic-spline method (Long and Pepper, 1981; Pielke, 1984), so for operator A_x, we have

$$\overline{C}_{j,k,m}^{h+\frac{1}{2}} = S^h(x_j - \alpha\Delta x), \quad \text{if } \overline{u}_{j,k,m}^h \geq 0 \tag{9.18}$$

$$\overline{C}_{j,k,m}^{h+\frac{1}{2}} = S^h(x_j + \alpha\Delta x), \quad \text{if } \overline{u}_{j,k,m}^h < 0 \tag{9.19}$$

with $\alpha = \overline{u}_{j,k,m}^h \dfrac{\Delta t}{\Delta x}$, where the superscript h denotes an intermediate fictitious time step between n and $n+1$. This is called the fractional steps (FS) technique.

The interpolation function (cubic spline) S can be expressed in terms of the spline derivatives $P_j^h = \left(\dfrac{\partial \overline{C}}{\partial x}\right)_{j,k,m}^h$ as

$$S^h(x) = P_{j-1}^h \frac{(x_j - x)^2(x - x_{j-1})}{h_j^2} - P_j^h \frac{(x - x_{j-1})^2(x_j - x)}{h_j^2}$$

$$+ \overline{C}_{j-1}^h \frac{(x_j - x)^2\left[2(x - x_{j-1}) + h_j\right]}{h_j^3} + \overline{C}_j^h \frac{(x - x_{j-1})^2\left[2(x_j - x) + h_j\right]}{h_j^3}$$

$$\tag{9.20}$$

where $h_j = x_j - x_{j-1}$.

The spline derivatives are obtained by solving the tri-diagonal algebraic system (Ahlberg et al., 1967; Price and Mac Pherson, 1973):

$$\frac{1}{2}P^h_{j-1} + 2P^h_j + \frac{1}{2}P^h_{j+1} = \frac{3}{2\Delta x}(\bar{C}^h_j - \bar{C}^h_{j-1}) + \frac{3}{2\Delta x}(\bar{C}^h_{j+1} - \bar{C}^h_j) \qquad (9.21)$$

Using Equations 9.20 through 9.21, Equations 9.18 and 9.19 become, respectively

$$\bar{C}^{h+\frac{1}{2}}_{j,k,m} = \bar{C}^h_{j,k,m} - P^h_j h_j \alpha + \left[P^h_{j-1} h_j + 2P^h_j h_j + 3(\bar{C}^h_{j-1,k,m} - \bar{C}^h_{j,k,m}) \right]\alpha^2$$

$$- \left[P^h_{j-1} h_j + P^h_j h_j + 2(\bar{C}^h_{j-1,k,m} - \bar{C}^h_{j,k,m}) \right]\alpha^3 \qquad (9.22a)$$

$$\bar{C}^{h+\frac{1}{2}}_{j,k,m} = \bar{C}^h_{j,k,m} + P^h_j h_{j+1} \alpha - \left[P^h_{j+1} h_{j+1} + 2P^h_j h_{j+1} + 3(\bar{C}^h_{j,k,m} - \bar{C}^h_{j+1,k,m}) \right]\alpha^2$$

$$+ \left[P^h_j h_{j+1} + P^h_{j+1} h_{j+1} + 2(\bar{C}^h_{j,k,m} - \bar{C}^h_{j+1,k,m}) \right]\alpha^3$$

$$(9.22b)$$

(b) After each advective step, a filter operation is applied to the intermediate field to remove any negative concentration that is usually produced by advection:

$$\bar{C}^{h+\frac{1}{2}}_{j,k,m} = F\left(\bar{C}^{h+\frac{1}{2}}_{j,k,m} \right)$$

(c) Finally, the diffusive step (operator D_x) is computed by means of the Crank–Nicholson implicit scheme:

$$\frac{\bar{C}^{h+1}_{j,k,m} - \bar{C}^{h+\frac{1}{2}}_{j,k,m}}{\Delta t} = \frac{1}{2S_c}\left[\frac{K'_M\left(\bar{C}^{h+\frac{1}{2}}_{j+1,k,m} - \bar{C}^{h+\frac{1}{2}}_{j,k,m} \right) - K''_M\left(\bar{C}^{h+\frac{1}{2}}_{j,k,m} - \bar{C}^{h+\frac{1}{2}}_{j-1,k,m} \right)}{(\Delta x)^2} \right]$$

$$+ \frac{1}{2S_c}\left[\frac{K'_M(\bar{C}^{h+1}_{j+1,k,m} - \bar{C}^{h+1}_{j,k,m}) - K''_M(\bar{C}^{h+1}_{j,k,m} - \bar{C}^{h+1}_{j-1,k,m})}{(\Delta x)^2} \right]$$

$$(9.23)$$

where

$$K'_M = K_M(x_{j+1/2})$$

$$K''_M = K_M(x_{j-1/2})$$

9.3.4 BOUNDARY CONDITIONS FOR THE ADVECTIVE TERMS

We assume, for every time step, zero gradient boundary conditions and zero outflow boundary conditions, that is,

$$\text{for } j = 1 \quad \begin{cases} P_0^h = 0 \\ \overline{C}_{0,k,i}^h = 0 \end{cases} \tag{9.24}$$

$$\text{for } j = N_x \quad \begin{cases} P_{Nx}^h = 0 \\ \overline{C}_{Nx+1,k,i}^h = 0 \end{cases} \tag{9.25}$$

Analogous conditions are applied in the y and z-directions.

9.3.5 BOUNDARY CONDITIONS FOR THE DIFFUSION TERMS

We assume, for every time step, zero outflow boundary conditions, that is,

$$\text{for } j = 1 \quad \overline{C}_{0,k,i}^{h+1} = \overline{C}_{0,k,i}^{h+\frac{1}{2}} = 0 \tag{9.26}$$

$$\text{for } j = N_x \quad \overline{C}_{Nx+1,k,i}^{h+1} = \overline{C}_{Nx+1,k,i}^{h+\frac{1}{2}} = 0 \tag{9.27}$$

Analogous conditions are applied in the y-direction. For diffusion in the z-direction, the boundary conditions are zero gradient conditions, that is,

$$\text{for } i = 1 \quad \overline{C}_{j,k,0}^h = \overline{C}_{j,k,1}^h \tag{9.28}$$

$$\text{for } i = N_z \quad \overline{C}_{j,k,Nz+1}^h = \overline{C}_{j,k,Nz}^h \tag{9.29}$$

9.3.6 DISPERSION EXPERIMENTS

In order to point out the differences in dispersion patterns between buoyancy- and shear-driven PBLs, we generated different types of PBLs with the LES model by varying the geostrophic wind speed (U_g, V_g) and the surface heat flux Q_*. Moeng and Sullivan (1994) did a sensitivity analysis over such forcing parameters that generate different turbulent regimes. Following these arguments, we reproduced two types of PBLs: a buoyancy-dominated PBL, hereafter simulation B, and a neutral (pure shear) PBL, hereafter simulation S.

Simulation B is buoyancy-dominated flow with little shear effect confined close to the ground, while simulation S is a shear-dominated flow with no heat flux. Calculations were performed in a rectangular domain arranged in such a way that it could comprise several updrafts at a given time. For simulation B, the dimensions of the box are 10×10 km in the horizontal directions and 2 km in the vertical direction, with a resolution of 128 grid points in the x- and y-directions and 96 grid points in the z-direction. For simulation S, the dimensions of the box are 3×3 km in the

TABLE 9.1

External Simulation Parameters

Simulation	Mesh Grid Points (N_x, N_y, N_z)	Domain Size L_x, L_y, L_z (km)	Geostrophic Wind (U_g, V_g) (m/s)	Surface Heat Flux Q^* (ms^{-1} K)	Initial Inversion Height $(z_i)_0$ (m)
B	(128, 128, 96)	(10, 10, 2)	(10, 0)	0.24	1000
S	(96, 96, 96)	(3, 3, 1)	(15, 0)	0	500

horizontal direction and 1 km in the vertical direction, with a resolution of 96 grid points in each of the three directions.

Simulations started from a laminar flow, with the geostrophic wind constant throughout the numerical domain. In order to have a strong capping inversion above the simulated PBL, the initial mean virtual potential temperature profile was 300 K below an initial PBL height, $(z_i)_0$, that increases by a total of 8 K across six Δz levels and increases with a lapse rate of 3 K/km above that. External parameters like the extent of the domain, the grid size, the geostrophic winds, the surface heat flux, and the initial capping inversion height are summarized in Table 9.1. The extent of the numerical domain is larger (in both the horizontal and the vertical directions) in simulation B than in simulation S in view of the fact that the PBL in highly convective cases is much deeper than the PBL where the shear dominates the buoyancy.

For simulation B, initial $(z_i)_0$, Q_*, and the geostrophic wind were respectively set equal to 1000 m, 0.24 (m/s K), and 10 m/s.

The generation of the neutral PBL (S) was more complex; we first generated a mixed (shear/buoyancy) PBL, then turned off the buoyancy heat flux at the ground and ran the simulation until a steady state was reached. Quasistationary conditions were obtained after the LES model ran for 5000 time steps (more than 2 h of real simulated time), which corresponds to about six turnover times. This time represents the initial time, $t=0$, for dispersion experiments. Table 9.2 provides a summary of the following parameters of the LES runs for the two cases: friction velocity u_*, convective velocity w_*, PBL height h, Monin–Obukhov length L, stability parameter h/L, and large-eddy turnover time.

9.3.6.1 Dispersion from Elevated Sources

At the starting time, we introduced the contaminant from an elevated point source placed at half the height of the PBL inversion into the box domain. In the numerical grid, the point source is approximated by an elementary volume $dV = dxdydz$. The

TABLE 9.2

Internal Parameters for Both Simulations

Simulation	u_* (m/s)	w_* (m/s)	h/L	h (m)
B	0.7	2.1	−18	1100
S	0.54	0	0	480

contaminant is injected at every time step. In this way, our results can be interpreted in terms of plume diffusion from a continuous point source. After the initial time, we continued the integration of the large-eddy model, and the evolution of the given source was calculated simultaneously by solving the conservation equation of the scalar. The time step of our simulation is determined by the LES numerical stability conditions (never greater than 2 s). As mentioned above, we imposed the zero-outflow boundary conditions in both horizontal directions, while at the top and bottom of the simulation domain, we used zero-gradient boundary conditions.

9.3.6.1.1 Convective Dispersion

The main characteristics of passive plume dispersion from an elevated source in the convective boundary layer (CBL) have been demonstrated through numerical predictions (Lamb, 1978, 1979, 1981), field observations (Nieuwstadt, 1981; Moninger et al., 1983), and in a more detailed way through laboratory experiments (Willis and Deardorff, 1976, 1981). In this section, we study the plume dispersion from an elevated continuous point source. Our main goal is to reproduce classical plume behavior in a CBL taking as reference the pioneering works of Deardorff (1972), Lamb (1981), Weil (1988), and Nieuwstadt and De Walk (1987). In the numerical scheme described above, we kept the full set of (advective/diffusive) operators in order to properly describe all of the physical processes involved in the contaminant transport and diffusion.

The contaminant was injected throughout the duration of the simulation (release time = 2000 time steps). The travel time, which is a rough estimate of how long it takes to cross the longitudinal domain (10 km), is about 200 time steps. To get proper plume behavior, both the sampling and the release time must be greater than the travel time. In order to satisfy this constraint, we chose a concentration averaging time of 4000 time steps.

We introduce the crosswind integrated concentration by $C_y = \dfrac{1}{\Delta T} \int_{\Delta T} \int_{L_y} \bar{C}(x, y, z)$ $dy\, dt$ where ΔT is an arbitrary average interval. This concentration is made dimensionless by dividing by Q/Uh, where U is the mean longitudinal wind velocity, Q is the source strength, and the nondimensional distance X_* is defined by $(x/U)/(h/w_*)$. Figure 9.1 shows the evolution of the dispersion process. In Figure 9.1, each graph is obtained by averaging the instantaneous concentration field over 250 time steps. As expected, the contaminant fills the box until it reaches a quasisteady situation (Figure 9.1d) due to the equilibrium between the outflow and inflow of the pollutant mass.

Figure 9.2 shows the crosswind integrated concentration patterns, where U is the mean longitudinal wind velocity. This figure illustrates the main aspect of passive plume dispersion in a CBL.

We can see that the plume centerline from an elevated source descends until it reaches the ground, causing a maximum concentration, remains there for some distance, and then rises. The descent of the plume centerline is due to the size, long life, and organized nature of the downdrafts in a convective mixed layer. Because the downdrafts cover a greater area than the updrafts, the probability of material being released into them is higher.

These characteristics agree qualitatively very well with the laboratory results of Willis and Deardorff (1976, 1981), at least in the range of nondimensional distance

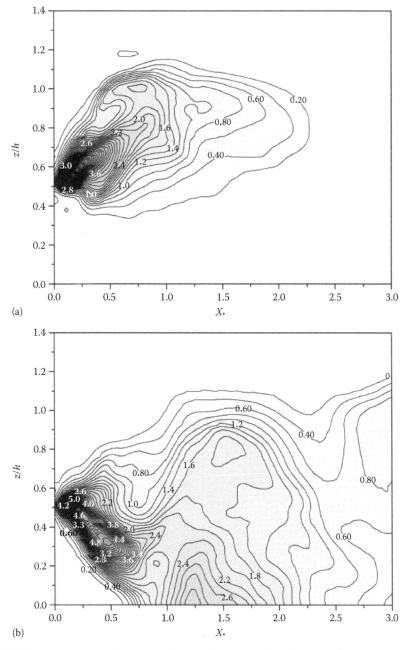

FIGURE 9.1 Isopleths of crosswind integrated concentration for elevated release, (a) average over first 0–1000 time steps, (b) average over 1000–2000 time steps, (c) average over 2000–3000 time steps, and (d) average over 3000–4000 time steps.

(*continued*)

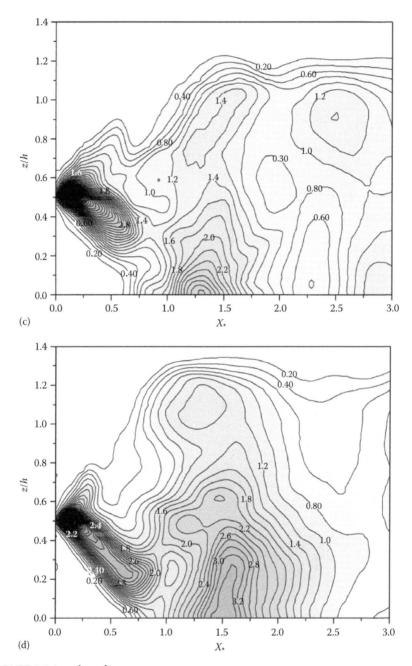

FIGURE 9.1 (continued)

imposed by our simulation domain ($X_* \in [0,2.5]$). Results show that our LES can reproduce the well-established experiments of passive plume dispersion. These results are also comparable with the results obtained with the particle approach of Lamb (1981).

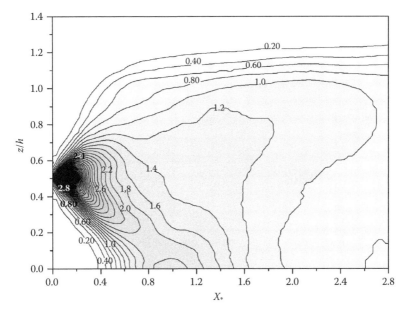

FIGURE 9.2 Isopleths of adimensional crosswind integrated concentration averaged over last 3000 time steps. (From Rizza, U. et al., *Nuovo Cimento Sez. C*, 26, 297, 2003. With permission.)

9.3.6.1.2 Neutral Dispersion

A neutral boundary layer is characterized by a profile of potential temperature being constant with height and by the heat flux Q_* being equal to zero. A neutral PBL is typically observed for short periods (e.g., before sunset and after sunrise), especially when buoyancy forces are weaker than mechanical ones. The neutral boundary layer was generated by LES using the input parameters indicated in Table 9.1. With no heat flux from ground, the mechanical generation of turbulence is the predominant factor influencing the flow structures and flow patterns.

To generate a neutral PBL, we first generated a shear/buoyancy PBL, and then (after 3000 s, approximately equivalent to seven turnover times), we set the surface heat flux to zero. In this shear-driven PBL, we injected the contaminant from an elevated point source ($h/H_s = 0.5$). To build a plume representation, we emitted (emission time) the contaminant throughout the simulation period ($2000\Delta t$) and got concentration statistics (sampling time) just in the second half period ($1000\Delta t$). Since the travel time is much smaller (about $100\Delta t$) than the sampling/emission time, the plume representation of the present dispersion process is appropriate. The mean plume height (red dotted line in Figure 9.3) remains almost constant during the simulation, confirming the absence of strong vertical meandering caused by the largest eddies peculiar of a CBL. The isopleths show a plume distribution very close to a Gaussian shape. This concentration dataset may be very useful to test, for example, Gaussian dispersion parameterizations that are still largely used in air quality.

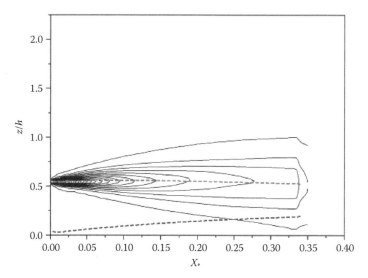

FIGURE 9.3 Isopleths of crosswind integrated concentration averaged between last 1000 time steps, the red dotted line represents the mean plume height and the blue dotted line the vertical dispersion parameter.

9.3.6.2 Dispersion from Near-Surface Sources Convective Dispersion

The CBL is characterized by large-scale flow motions, which consist of strong updrafts of hot air (i.e., thermals) and weaker downdrafts surrounding these thermals. One of the most important features of the CBL is the difference in strength between the updrafts and downdrafts, which causes an asymmetry of the turbulent flow field. This asymmetry makes the dispersion characteristics depend on the location of the source. As a result, the dispersion from a source at the surface, primarily caused by positive velocity fluctuations, is different from the dispersion from an elevated source. Moreover, the analysis of both release heights (near-surface and elevated) for the CBL allows us to compare our results with those from the well-established laboratory experiments of Willis and Deardorff (1976, 1978, 1981). In fact, their classical investigations demonstrated that for an elevated source, the plume descends within a short distance from the source until it reaches the ground. In contrast, the average plume centerline from a near-surface source ascends after a short downwind distance (Sorbjan and Uliasz, 1999).

To investigate convective plume dispersion from a surface source, we injected the contaminant from a point source situated close to the ground ($h/H_S = 0.25$) in the CBL described in the previous section. To get proper plume behavior, the contaminant was emitted throughout the simulation period (i.e., the release time, $2000\Delta t$), while the concentration statistics were determined only in the second half period (i.e., the sampling time, $1000\Delta t$). This allowed us to get an appropriate plume representation, as the travel time (about $100\Delta t$) was much smaller than both the sampling and the emission times. Figure 9.4 shows the evolution of the dispersion process. As expected, the isopleths of crosswind-integrated concentration show that the average plume centerline ascends after a short downwind distance as the material is emitted principally into a region covered almost entirely by updrafts. This is confirmed by Figure 9.5, which shows the isopleths of the adimensional crosswind-integrated concentration averaged over the last 3000 time steps. Figure 9.6 shows the adimensional crosswind-integrated concentration at the surface from an elevated (Figure 9.6a) and

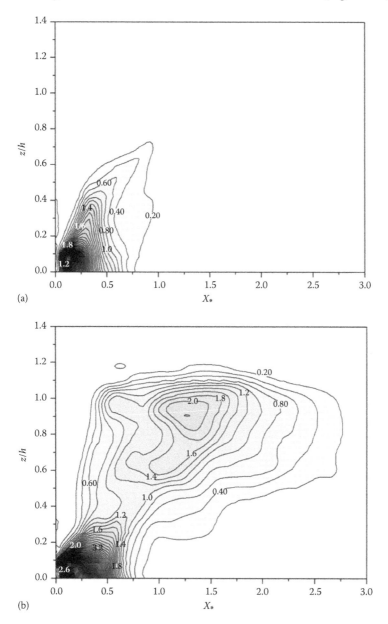

(a)

(b)

FIGURE 9.4   Isopleths of crosswind integrated concentration for low source (a) average over first 0–1000 time steps, (b) average over 1000–2000 time steps, (c) average over 2000–3000 time steps, and (d) average over 3000–4000 time steps.

(continued)

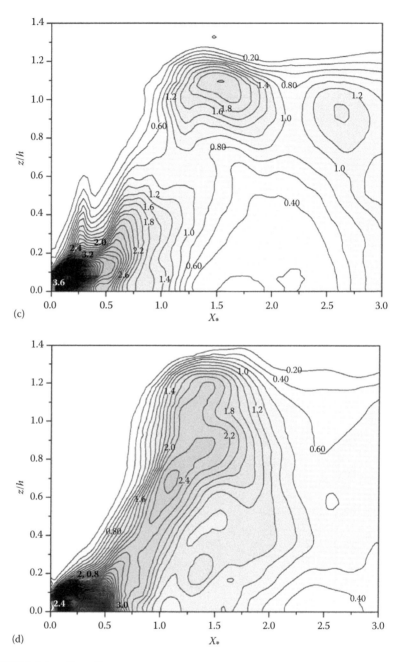

(c)

(d)

FIGURE 9.4 (continued)

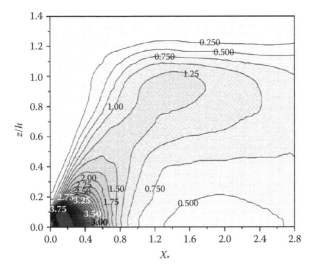

FIGURE 9.5 Isopleths of adimensional crosswind-integrated concentration averaged over last 3000 time steps.

a low source (Figure 9.6b) in the CBL. Points connected with the blue line are data obtained from Willis and Deardorff laboratory dispersion experiments; the red line is obtained from the present Eulerian model.

9.3.7 CONCLUSIONS FOR THE EULERIAN EXPERIMENTS

Results obtained from the simulations indicate that our model can successfully simulate the Eulerian dispersion of passive contaminants in high-resolution turbulent fields provided by LES. The Eulerian approach to studying pollutant dispersion and, particularly, the developed scheme based on a splitting technique to numerically solve the conservation equation, produce good results. In general, the main problem in dealing with this kind of scheme is the numerical discretization of advective terms, which may present many complications. The spline technique used to solve these terms has been preferred, as it can be easily adapted to a domain with irregular grid-spacing (Long and Pepper, 1981). Since our LES code has been accurately tested (Moeng, 1984), we have only tested the numerical scheme and its matching with the LES model. We then simulated pollutant dispersion from an elevated and low continuous point source in a CBL and from an elevated point source in a shear-driven boundary layer. Our dispersion experiment results are in good agreement with the classical ones found in the literature. The nontrivial characteristics of dispersion are adequately captured, and the crosswind-integrated concentration distribution closely resembles the well-established numerical and laboratory experiments found in the literature. Results confirm that the LES constitutes an alternative to field experiments: it can provide databases of dispersion data on which a wide range of dispersion models can be developed and tested, in view of their use in air quality applications.

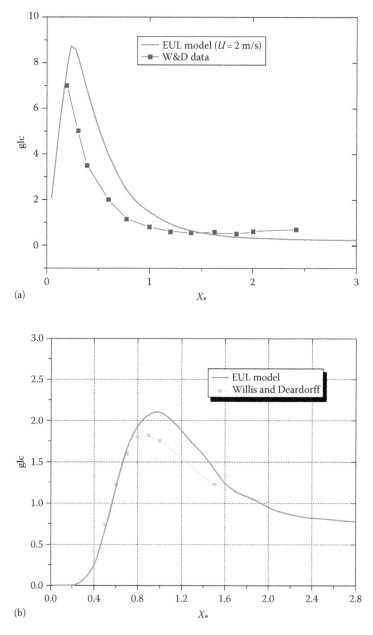

FIGURE 9.6 Nondimensional crosswind-integrated concentration at the surface versus dimensionless downwind distance for passive release from a (a) elevated and (b) low source in the convective boundary layer. Points connected by the blue line are data obtained from Willis and Deardorff laboratory dispersion experiments; the red line is obtained from our Eulerian model.

9.4 LAGRANGIAN DISPERSION WITH LARGE-EDDY SIMULATIONS AND EXPERIMENTS

9.4.1 LAGRANGIAN DISPERSION

This chapter focuses on Lagrangian dispersion with LES. In particular, we investigate the problem of relative dispersion in a neutrally stratified and celebrated Richardson's law. This equation implies that the mean square particle separation grows in time as

$$R^2(t) = C_2 \varepsilon t^3 \qquad (9.30)$$

where
  C_2 is the so-called Richardson constant
  ε is the mean energy dissipation

This study is relevant to describe small-scale motions and to provide important information on the way to parameterize subgrid scales. It is not easy to observe the "t^3" behavior of particle pair separation in a realistic PBL as it is hard to obtain a PBL with a sufficiently extended inertial range of scales in which one can clearly identify the expected law. Moreover, the classical time-dependent approach in isolating Richardson's law with the classical statistical technique has been inconclusive due to its strong dependency on initial conditions (i.e., initial pair separations), which consequently does not permit an accurate estimate of the Richardson constant.

To overcome this problem, we have applied, for the first time in the context of boundary layer physics, a recently established technique coming from the study of dynamical systems theory (Gioia et al., 2004). This exit-time technique, known as the finite scale Lyapunov exponents (FSLE) (Boffetta et al., 2000), has been exploited for treating finite-scale Lagrangian relative dispersion as a finite-error predictability problem (Lacorata et al., 2001). This new adopted strategy has given many important results. First of all, it has permitted us to isolate a clean region of scaling showing the occurrence of Richardson's law. For this reason, a measure of the Richardson constant has become possible.

9.4.2 PAIR DISPERSION IN A LES-GENERATED NEUTRAL PBL

9.4.2.1 The Simulated PBLs

We generated two types of neutral boundary layers, one with a spatial resolution of 128^3 grid points (SN1) and the other with a spatial resolution of 96^3 grid points (SN2). In order to obtain a stationary PBL, we advanced our LES code in time for around six large-eddy turnover times. This time is the starting point for the successive Lagrangian analysis that will be described later. The relevant parameters characterizing our simulated PBLs at $t = 6\tau_*$ are summarized in Table 9.3.

TABLE 9.3

Parameters from LES Simulations

Simulation	Mesh (N_x, N_y, N_z)	Domain (L_x, L_y, L_z) (km)	Geostrophic Wind (U_g, V_g) (m/s)	Surface Heat Flux Q_* (ms^{-1} K)	Initial Inversion Height $(z_i)_0$ (m)	Friction Velocity u_* (m/s)	Turnover Time τ_* (s)
SN1	128^3	(2,2,1)	(15,0)	0	461	0.7	⎵674
SN2	96^3	(2,2,1)	(15,0)	0	440	0.6	⎵734

9.4.2.1.1   Lagrangian Simulations and Pair Dispersion Statistics

To investigate the pair-dispersion statistics from the initial time $t=6\tau_*$ (corresponding to the PBL quasi-stationary regime), we integrated, in parallel with the LES, the equation for the passive tracer trajectories defined by

$$\frac{\mathrm{d}x_i^n(t)}{\mathrm{d}t} = \bar{u}_i(x_i^n(t),t) \tag{9.31}$$

where $x_i^n(t)$ is the vector position of the ith-particle of the nth-couple at time t. The velocity field necessary to integrate this last equation, $\bar{u}_i(x_i^n(t),t)$, was obtained by means of a bilinear interpolation from the eight nearest grid points for which the velocity field is known. For both simulations, we performed a single long run in which the evolution of around 1000 particle pairs was followed starting from two different initial separations: $R(0)=\Delta x$ and $R(0)=2\Delta x$, Δx being the grid mesh spacing whose value is $\Delta x=15.6$ m for SN1 and $\Delta x=20.8$ m for SN2.

At the initial time $(t=6\tau_*)$, the particle pairs are uniformly distributed on a horizontal plane located at $z = h/2$.

Reflection has been assumed at both the top $(z=h)$ and the bottom boundary. For testing purposes, a second run (again started from $t=6\tau_*$) with a greater number of particle pairs was performed. No significant differences in the Lagrangian statistics were observed. In this preliminary investigation, we did not use any subgrid model describing the Lagrangian contribution arising from motions on scales smaller than the grid mesh spacing.

The classical time-dependent approach is based on studying the behavior of the square particle separation $R^2(t)$ with the aim of proving Richardson's "t^3" prediction. Since the same conclusion has been obtained for the SN2 simulation, only the results for the SN1 simulation are presented. Figure 9.7 shows the second moment of relative dispersion $R^2(t)$ for the two initial separations, that is, $R(0)=\Delta x$ and $R(0)=2\Delta x$. The dashed line represents the relative dispersion of pairs of initial separation Δx, while the heavy dashed line represents the expected Richardson's "t^3" law. It is evident that our data are not compatible with this law for an initial separation $2\Delta x$ (see the solid line). This strong dependence on the initial separation is one of the main reasons why

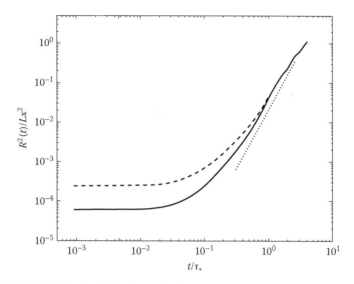

FIGURE 9.7 The behavior of the (dimensionless) mean square relative dispersion versus the (dimensionless) time. Full line: the initial separation is Δx; Dashed line: the initial separation is $2\Delta x$. Dotted line is Richardson's law $C_2 \varepsilon t^3$ with $C_2 = 0.5$ and $\varepsilon = 6 \cdot 10^{-4} \, \text{m}^2 \, \text{s}^{-3}$. (From Gioia, G. et al., *Bound. Layer Meteorol.*, 113, 187, 2004. With permission.)

the time-dependent approach fails. Moreover, the mean square relative dispersion of our LES trajectories seen as a function of time is affected (see Figure 9.7) by overlap effects between different regimes, which implies that those regimes are not well separated and clearly distinguishable.

9.4.2.2 Fixed-Scale Statistics

The aim of this section is to introduce and show the use of an indicator—the FSLE—originally introduced in the context of predictability problems for studying nonasymptotic transport properties in nonideal systems, that is, systems in which the characteristic length-scales are not sharply separated (Boffetta et al., 2000). A dynamical system consists basically of an N-dimensional state vector $\vec{x}$, having a set of N observables as components evolving in so-called phase space, and of an N-dimensional evolution operator $\vec{F}$ related by a first-order ordinary differential equations system:

$$\dot{\vec{x}}(t) = \vec{F}\left[\vec{x}\right] \tag{9.32}$$

If $\vec{F}$ is nonlinear, the system (9.32) can have chaotic solutions, and therefore limited predictability, for which case an infinitesimally small error $\delta\vec{x}$ on a trajectory $\vec{x}$ is exponentially amplified:

$$\delta\vec{x}(t) \sim \delta\vec{x}(0) \exp \lambda t \tag{9.33}$$

with a (mean) growth rate λ known as the maximum Lyapunov exponent (MLE). The Lagrangian description of fluid motion can also be seen from a dynamical systems point of view. In fact, in the Lagrangian framework, the vector $\vec{x}$ is the tracer trajectory, the operator $\vec{F}$ is the velocity field, and the error $\delta\vec{x}$ is the distance between two tracer trajectories. It is therefore straightforward to consider the relative dispersion of Lagrangian trajectories as a problem of finite-error predictability.

The importance of the finite-scale analysis will become clear as this tool permits us to overcome the difficulties noted in the last section and that usually appear when trying to study the relative dispersion in fully developed turbulence (i.e., high Reynolds numbers turbulence) by means of the time-dependent approach. However, in many recent works, the FSLE analysis has been used as diagnostic of transport properties in geophysical systems (e.g., Lacorata et al., 2001). Before introducing the finite size analysis for dispersion problems, we recall what asymptotic regimes hold for N particle pairs advected by a Eulerian velocity field $\vec{u}(\vec{x},t)$ characterized by two typical length-scales: a small-scale l_u and a large-scale L_0:

$$R^2(t) \simeq \begin{cases} r_0 e^{\lambda t} & \text{for } r_0 \ll l_u \\ \\ 2Dt & \text{for } r_0 \gg L_0 \end{cases}$$

where $r_0 = r(0)$ is the initial separation between a pair of particles. Note that

$$R^2(t) = \frac{1}{N} \sum_{i=1}^{N} R_i^2(t)$$

An alternative method to characterize the dispersion properties is to introduce the "doubling time" $\tau(\delta)$ at scale δ, which is a concept that permits us to define the FSLE. Let $R = |\delta\vec{x}|$ be the distance between two trajectories. Considering a given series of thresholds $\delta^{(n)} = r^n \delta^{(0)}$, one can measure the time $T_i(\delta^{(0)})$ it takes for the separation, $R_i(t)$, of the ith couple to grow from $\delta^{(0)}$ to $\delta^{(1)} = r\delta^{(0)}$, and so on for $T_i(\delta^{(2)})$, ..., $T_i(\delta^{(n)})$. The factor r may be any value greater than 1, properly chosen in order to have good separation between scales of motion; that is, r should be not too large. $\tau(\delta)$ is exactly the doubling time only if $r=2$. Once the doubling time experiments have been performed over the N particle pairs, the average doubling time $\tau(\delta)$ at the scale δ can be defined as

$$\tau(\delta) = \langle T(\delta) \rangle = \frac{1}{N} \sum_{i=1}^{N} T_i(\delta) \tag{9.34}$$

At this point, we can define the FSLE in terms of the average doubling time as

$$\lambda(\delta) = \frac{\ln r}{\tau(\delta)} \tag{9.35}$$

which quantifies the average rate of separation between two particles at a distance δ. We stress that $\lambda(\delta)$ is independent of r if r is close to 1, and then that for very small separations (i.e., $\delta \ll l_u$), the FSLE coincides with the MLE λ, since $\lambda = \lim\limits_{\delta \to 0} \dfrac{1}{\tau(\delta)} \ln r$. Note that if the trajectories refer to Lagrangian particles, as in our analysis, λ can also be called the Lagrangian Lyapunov Exponent (LLE).

In general, for finite δ, the FSLE is expected to follow a power law of the type:

$$\lambda(\delta) \sim \delta^{-2/\gamma} \tag{9.36}$$

where the value of γ defines the dispersion regime at the scale δ. In fact,

$$\lambda(\delta) \sim \lambda \quad \text{for } \delta \ll l_u \tag{9.37a}$$

$$\lambda(\delta) \sim \delta^{-2/3} \quad \text{for } l_u \ll \delta \ll L_0 \tag{9.37b}$$

$$\lambda(\delta) \sim \delta^{-2} \quad \text{for } \delta \gg L_0 \tag{9.37c}$$

This means that $\gamma = 3$ refers to the Richardson diffusion within the inertial range and $\gamma = 1$ corresponds to standard diffusion, that is, large-scale uncorrelated spreading of particles. These scaling rules can be explained by means of dimensional arguments. In fact, as the scaling law of the relative dispersion in time is of the form $R^2(t) \sim t^\gamma$, the corresponding scaling law in terms of the FSLE is given considering the inverse of time as a function of space.

Before showing our results, we introduce another quantity that can also provide information about the existence of the inertial range.

This quantity, which is related to the FSLE, is the mean relative Lagrangian velocity at a fixed scale that we indicate with

$$v(\delta) = \left[\left\langle \delta \vec{v}(\delta)^2 \right\rangle \right]^{1/2} \tag{9.38}$$

where

$$\delta \vec{v}(\delta)^2 = \left(\dot{\vec{x}}^{(1)} - \dot{\vec{x}}^{(2)} \right)^2 \tag{9.39}$$

is the square (Lagrangian) velocity difference between two trajectories, $\dot{\vec{x}}^{(1)}$ and $\dot{\vec{x}}^{(2)}$, on the scale δ (i.e., $|\vec{x}^{(1)} - \vec{x}^{(2)}| = \delta$). The quantity $v(\delta)/\delta$ is dimensionally equivalent to $\lambda(\delta)$, so a scaling law of the type:

$$\frac{v(\delta)}{\delta} \sim \delta^{-2/3} \tag{9.40}$$

is compatible with the FSLE inside the inertial range.

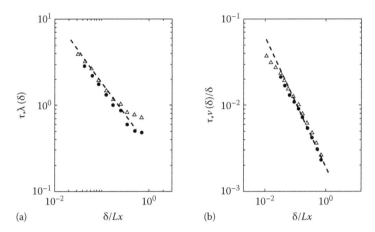

(a) δ/Lx (b) δ/Lx

FIGURE 9.8 (a) FSLE at two different resolutions. Triangles: 128^3 grid points; Circles: 96^3 grid points. The dashed line corresponds to $\alpha\delta^{-2/3}$ with $\alpha=0.1 \text{ m}^2 \text{ s}^{-3}$. (b) The same as in (a) but for the relative velocity. The dashed line has slope $-2/3$. (From Gioia, G. et al., *Bound. Layer Meteorol.*, 113, 187, 2004. With permission.)

Figure 9.8 reports the FSLE measured from our LES data. We can see that the expected behavior of Equation 9.37b occurs from the scale of the spatial resolution to about the size of the domain. There is also a clear region of scaling in which Richardson's law ($R^2(t)=C_2\varepsilon t^3$) holds, so that we can estimate the Richardson constant C_2. To do this, we used a formula that relates C_2 to the FSLE (for details, see Boffetta and Sokolov, 2002), that is,

$$C_2 = \beta \frac{\alpha^3}{\varepsilon} \left(\frac{r^{2/3}-1}{r^{1/3}\ln r} \right)^3 \qquad (9.41)$$

where

    β is a numerical coefficient
    ε is the energy dissipation measured from the LES
    α comes from the best fit of Equation 9.37 (i.e., $\lambda(\delta)=\alpha\delta^{-2/3}$)

In particular, from the fit we extracted the coefficient $\alpha=0.1 \text{ m}^{2/3} \text{ s}^{-1}$, and we know that the mean energy dissipation measured from the LES run is $\varepsilon=6\cdot10^{-4} \text{ m}^2\text{s}^{-3}$. Hence, formula (4.12) gives $C_2\simeq0.5$. This measure is affected at most by a relative error of 0.4, that is, $C_2\simeq0.5 \pm 0.2$, since the quantities ε and α, which are the main source of uncertainty, have been determined within a relative error 0.1. The value of C_2 found is in the range of values obtained from both DNS data and recent experiments (Ott and Mann, 2000). Our analysis is therefore quite satisfactory, especially considering the difficult task of estimating the Richardson constant.

9.4.3 CONCLUSIONS FOR THE LAGRANGIAN EXPERIMENTS

We have investigated the problem of relative dispersion in a neutrally stratified PBL simulated by means of an LES. In particular, we have focused on the celebrated

Richardson's law, which governs relative dispersion theory. When dealing with this problem, the usual difficulty is to generate a PBL with a sufficient extended inertial range of scales; the use of the LES technique allows us to overcome this problem. Furthermore, we applied a nonstandard technique (FSLE) coming from the study of dynamical systems to isolate Richardson's law. Even though this was the first time in which the FSLE has been applied in the context of boundary layer physics, we have obtained good results. In fact, with this new tool of analysis, we have observed a clean region of scaling that shows the occurrence of Richardson's law first in a neutral boundary layer where we did not use any subgrid model. In this preliminary case, it has therefore been possible to estimate the Richardson constant. We have found that for our simulation, its value is $C_2 \simeq 0.5$. This estimate is compatible with recent results that fix its value within the $[0.1 - 1]$ range. In particular, especially the most recent experiments and 3-D DNS (Ott and Mann, 2000; Boffetta and Sokolov, 2002) have found similar values.

9.5 CONCLUSIONS

The dispersion of contaminants in the PBL is commonly investigated in both Eulerian and Lagrangian frameworks. The Lagrangian approach considers the trajectory of marked fluid particles in the flow. Lagrangian particle models are very useful for describing the turbulent dispersion of passive contaminants because they can take into account essential aspects of turbulence, although they are limited to a simplified set of reacting species. The Eulerian approach, on the other hand, is based on the mass conservation equation and can incorporate the various second and high-order chemical kinetic equations necessary to describe photochemical smog generation, which is a challenging open problem. In both approaches, the understanding of the turbulent structure of the PBL is crucial for constructing realistic models. In this context, LES represent a very powerful method for calculating 3-D turbulent structures and are fundamental for describing any dispersion phenomena in the PBL. In this chapter, we used an LES to study dispersion properties of the PBL, considering both classical approaches mentioned above.

From a Eulerian point of view, in order to numerically solve the conservation equation, we used a splitting technique developed in the 1970s by Soviet mathematicians. The advective terms, which present many complications in their numerical discretization, are solved with a cubic-spline technique. Such a scheme can be easily adapted to a domain with an irregular grid-spacing. To test the method, we simulated dispersion from an elevated and low continuous point source in a CBL, and from an elevated continuous source in a neutral boundary layer. In all cases, the nontrivial characteristics of dispersion are adequately captured, and the crosswind integrated concentration distribution closely resembles the well-established numerical and laboratory experiments found in the literature. This means that coupling the LES with a dispersion equation can provide a realistic description of the dispersion processes. This combined approach is very promising and actually represents the state of the art in numerical investigations of PBL dispersion, although it is still affected by the lack of high-performance computations. Our results confirm that the LES constitutes an alternative for field experiments: it can provide databases of dispersion data on which a wide range of dispersion models can be developed and tested, in view of their use in air quality applications.

From a Lagrangian point of view, we investigated the problem of relative dispersion in a neutral boundary layer simulated by means of LES. In particular, we focused our study on relative dispersion and on Richardson's law driving the separation of particle pairs.

In general, there are many difficulties in such analysis, as it is hard to simulate a PBL with a sufficiently extended inertial range of scales. Another difficulty arises from the problem of the overlap between dispersion regimes, which is why we applied a nonstandard technique (FSLE) coming from the study of dynamical systems to better identify Richardson's law. The FSLE is a powerful analysis tool for studying relative dispersion, as shown in many recent works. In this case, the FSLE analysis allowed us to detect a clean scaling range where the expected Richardson's law was observed. This kind of analysis was first carried out in a neutral boundary layer, without any subgrid model. In this case, we found that the value of the Richardson constant is $C_2 \simeq 0.5$. This estimate is compatible with recent results that fix it within the [0.1–1] range.

The new strategy, FSLE, for studying the problem of PBL relative dispersion in LES fields has therefore provided positive results with respect to standard methods. A correct description of the behavior of pair separation provides clear advantages for understanding pollutant dispersion.

Finally, we have confirmed that LES are a powerful instrument to simulate those turbulent structures fundamental to describing dispersion phenomena within the PBL.

REFERENCES

Ahlberg, J.H., E.N. Nilson, and J.L. Walsh. 1967. *The Theory of Splines and Their Applications*. Academic Press, New York.

Andren, A., A.R. Brown, J. Graf, P.J. Mason, C.H. Moeng, F.T.M. Nieuwstadt, and U. Schumann. 1994. Large eddy simulation of a neutrally stratified boundary layer: A comparison of four computer codes. *Q. J. R. Meteorol. Soc.* 120: 1457–1484.

Boffetta, G. and I.M. Sokolov. 2002. Relative dispersion in fully developed turbulence: The Richardson's law and intermittency corrections. *Phys. Rev. Lett.* 88: 494–501

Boffetta, G., A. Celani, M. Cencini, G. Lacorata, and A. Vulpiani. 2000. Non Asymptotic properties of transport and mixing, *Chaos* 10: 1–9.

Deardorff, J.W. 1970. Numerical simulation of turbulent channel flow at large Reynolds number. *J. Fluid Mech.* 41: 452–480.

Deardorff, J.W. 1972. Numerical investigation of neutral and unstable planetary boundary layers. *J. Atmos. Soc.*, 29: 91–115.

Forester, C.K. 1979. Higher order monotonic convective difference schemes. *J. Comput. Phys.* 23: 1–22.

Fox, D.G. and S.A. Orszag. 1973. Pseudo-spectral approximation of two-dimensional turbulence. *J. Comput. Phys.* 11: 612–619.

Gioia, G., G. Lacorata, E.P. Marques Filho, A. Mazzino, and U. Rizza. 2004. The Richardson's law in large-eddy simulations of boundary layer flows. *Bound. Layer Meteorol.* 113: 187–199.

Kemp, J.R. and D.J. Thompson. 1996. Dispersion in stable boundary layers using large eddy simulation. *Atmos. Environ.* 30: 2911–2923.

Lacorata, G., E. Aurell, and A. Vulpiani. 2001. Drifter dispersion in the Adriatic Sea: Lagrangian data and chaotic model. *Ann. Geophys.* 19: 121–129.

Lamb, R.G. 1978. A numerical simulation of dispersion from an elevated point source in a convective planetary boundary layer. *Atmos. Environ.* 12: 1297–1304.

Lamb, R.G. 1979. The effects of release height on material dispersion in the convective planetary boundary layer. Reno NV: *AMS Fourth Symposium on Turbulence, Diffusion and Air Pollution.*

Lamb, R.G. 1981. A numerical investigation of tetroon versus fluid particle dispersion in the convective planetary boundary layer. *J. Appl. Meteorol.* 20: 391–403.

Long, P.E. and D.W. Pepper. 1981. An examination of some simple numerical schemes for calculating scalar advection. *J. Appl. Meteorol.* 20: 146–156.

Marcuk, N.N. 1984. *Metodi del Calcolo Numerico.* Roma: Editori Riuniti.

Mc Rae, G.J., W.r. Goodin, and J.H. Seifeld. 1982. Numerical solution of the atmospheric diffusion equation for chemically reacting flows. *J. Comput. Phys.* 45: 1–42.

Meeder, J.P. and F.T.M. Nieuwstadt. 2000. Large eddy simulation of the turbulent dispersion of a reacting plume from a point source into a neutral boundary layer. *Atmos. Environ.* 34: 3563–3573.

Moeng, C.H. 1984. Large eddy simulation model for the study of planetary boundary layer turbulence. *J. Atmos. Sci.* 41: 2052–2062.

Moeng, C.H. and P.P. Sullivan. 1994. A comparison of shear and buoyancy driven planetary boundary layer flows. *J. Atmos. Sci.* 51: 999–1022.

Moeng, C.H. and J.C. Wyngard. 1988. Spectral analysis of large eddy simulation of the convective boundary layer. *J. Atmos. Sci.* 45: 3573–3587.

Moninger, W.R., Eberhard, W.L., Briggs, G.A., Kropfli, R.A., and Kaimal, J.C. 1983. Simultaneous radar and lidar observations of plumes from continuous point sources, in *21st Conference on Radar Meteorology.* Amer. Meterol. Soc., Boston, pp. 246–250.

Nieuwstadt, F.T.M. 1981. The steady-state height and resistance laws of the nocturnal boundary layer: Theory compared with Cabauw observations. *Bound. Layer Meteorol.* 20: 3–17.

Nieuwstadt, F.T.M. and J.P.M.M. De Valk. 1987. A large eddy simulation of buoyant and non-buoyant plume dispersion in the atmospheric boundary layer. *Atmos. Environ.* 21: 2573–2587.

Ott, S. and J. Mann. 2000. An experimental investigation of the relative diffusion of particle pairs in three-dimensional turbulent flow. *J. Fluid Mech.* 422: 207–223.

Pielke R.A. 1984. *Mesoscale Meteorological Modelling.* San Diego, CA: Academic Press.

Price, G.V. and A.K. Mac Pherson. 1973. A numerical weather forecasting method using cubic splines on a variable mesh. *J. Appl. Meteorol.* 12: 1102–1113.

Reynolds, O. 1895. On the dynamical theory of turbulent incompressible viscous fluids and determination of the criterion. *Phil. Trans. R. Soc. Lond A.* 186: 123–161.

Rizza, U., G. Gioia, C. Mangia, and G.P. Marra. 2003. Development of a grid-dispersion model in a large eddy simulation generated planetary boundary layer. *Nuovo Cimento Sez. C* 26: 297–309.

Rizza, U., C. Mangia, J.C. Carvalho, and D. Anfossi. 2006. Estimation of the Lagrangian velocity structure function constant C_0 by large eddy simulation. *Bound. Layer Meteorol.* 120: 25–37.

Sorbjan, Z. and M. Uliasz. 1999. Large-eddy simulation of air pollution dispersion in the nocturnal cloud-topped atmospheric boundary layer. *Bound. Layer Meteorol.* 91: 145–157.

Spalart, P.R. and R. Moser. 1991. Spectral methods for the Navier–Stokes equations with the one-infinite and two-periodic directions. *J. Comput. Phys.* 96: 297–312.

Sullivan, P.P., J.C. Mc Williams and C.H. Moeng. 1994. A sub-grid scale model for large eddy simulation of planetary boundary layer flows. *Bound. Layer Meteorol.* 71: 247–276.

Van Haren, L.V. and F.T.M. Nieuwstadt. 1989. The behaviour of passive and buoyant plumes in a convective boundary layer, as simulated with a large-eddy model. *J. Appl. Meteorol.* 28: 818–832.

Weil, J.C. 1988. Dispersion in the convective boundary layer. In *Lectures on Air Pollution Modelling*, eds. A. Venkatram and J.C. Wyngaard, pp. 167–221. Boston, MA: American Meteorological Society.

Willis, G.E. and J.W. Deardorff. 1976. A laboratory model of diffusion into the convective planetary boundary layer. *Q. J. R. Meteorol. Soc.* 102: 427–445.

Willis, G.E. and J.W. Deardorff. 1978. A laboratory study of dispersion from an elevated source within a modelled convective planetary boundary layer. *Atmos. Environ.* 12: 1305–1311.

Willis, G.E. and J.W. Deardorff. 1981. A laboratory study of dispersion from a source in the middle of the convective mixed layer. *Atmos. Environ.* 15: 109–117.

Wyngaard, J.C. and L.J. Peltier. 1996. Experimental micrometeorology in an era of turbulence simulations. *Bound. Layer Meteorol.* 78: 71–86.

Yanenko, N.N. 1971. *The Method of Fractional Steps*. Berlin/New York: Springer-Verlag.

10 Photochemical Air Pollution Modeling: Toward Better Air Quality Management

Carlos Borrego, Ana Isabel Miranda, and Joana Ferreira

CONTENTS

10.1 INTRODUCTION

On July 26, 1943, *The Los Angeles Times* reported that a pall of smoke and fumes had descended on Downtown Los Angeles cutting visibility to three blocks. Striking in the midst of a heat wave, it gripped workers and residents who suffered with an eye-stinging, throat-scraping sensation. It also left them with the realization that something had gone terribly wrong in their city, normally praised for its sunny climate.

This episode was the trigger for the first U.S. program for air pollution in Los Angeles in 1947. Since then a long road has been covered and nowadays scientific knowledge about photochemical smog is stable, and it is even possible to forecast photochemical pollutants, such as ozone levels, a day in advance (Vautard et al., 2000; Monteiro et al., 2005a). In simple words, ozone can be considered as a secondary pollutant formed alongside the chemical reactions between precursors such as nitrogen oxides (NO_x) or volatile organic compounds (VOCs), in the presence of solar radiation.

269

This scientific knowledge has been the basis for several air quality management strategies including different international conventions, such as the convention on long-range trans-boundary air pollution, whose original purpose was to deal with acidification and eutrophication, but is currently also clearly related to photochemical pollution implying the reduction of ozone precursors. It is within this scope that the member states of the United Nations Economic Commission for Europe (UNECE, 1999) have included the concepts of critical load and level for planning pollution abatement strategies and as a basis for international agreements concerning limitation of emission of air pollutants, namely, the ozone precursors. Air quality legislation that includes ozone and its main precursors as pollutants, whose ambient air concentration levels have to be controlled within certain air quality thresholds, is also part of the established strategy to manage air quality.

In Europe, the Air Quality Framework Directive (FWD) (96/62/EC) regulates air pollutant effects on both human health and ecosystems, and it includes mechanisms to assure their protection. A fundamental requisite of the FWD is the definition of an evaluation program of ambient air quality in the territories of the member states. This evaluation program should include tropospheric ozone which it should cover via three main components: the monitoring of air quality, emission inventories, and atmospheric modeling.

In the United States, air pollution control strategies date from 1955 with the Air Pollution Control Act followed by the Clean Air Act of 1963. The Clean Air Act, which was last amended in 1990, requires the United States Environmental Protection Agency (USEPA) to set National Ambient Air Quality Standards (NAAQS) for pollutants considered harmful to public health and the environment. Since 2003, there has been the Clear Skies Act—cleaner air, better health, brighter future—which is a mandatory program that will dramatically reduce and cap emissions of sulfur dioxide (SO_2), nitrogen oxides (NO_x), and mercury from electric power generation.

From this increase in knowledge and number of strategies, an improvement in ozone pollution levels would be expected. However, long-term data from the air quality monitoring networks indicate a still increasing trend in ozone background surface levels in the northern hemisphere (Brasseur et al, 2003; UE, 2003). However, analysis of these ozone values is not an easy task, and it is possible to find different trends according to time-averaged values under study (e.g., peak values present a different pattern from median ones), or the type of monitoring station used. This complex behavior only confirms the nonlinearity of photochemical pollution.

Chemical transport models (CTMs) can be important tools to better understand and estimate photochemical air pollution and are quite extensively used to evaluate and forecast air quality.

10.2 PHOTOCHEMICAL AIR POLLUTION MODELING

The spatial and temporal distribution of ozone in the troposphere is controlled by several processes that are responsible for its production and removal. These processes lead to a complex equilibrium dependent upon different factors that, if meteorological conditions are favorable, lead to high levels of ozone. The different sources and

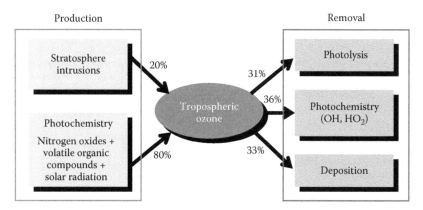

FIGURE 10.1 Global balance of tropospheric ozone. (Adapted from Académie des Sciences. Ozone et propriétés oxydantes de la troposphere, *Académie des Sciences*, rapport n°30, Lavoisier, Tec Doc, London, Paris, New York, 1993.)

sinks of tropospheric ozone, as well as their relative contributions, are represented in a simple and schematic manner in Figure 10.1.

Because of the irreproducibility in time and space of the photochemical pollution processes, either physical or chemical, atmospheric CTMs are very important for the understanding of pollutant dynamics and are the best available way to explain ozone episodes.

A photochemical modeling system should be able to reproduce daily ozone varia-tions due to horizontal turbulence effects, vertical mixing, local removal by nitrogen oxide (NO), and response to fast changing emissions (Hogrefe et al., 2001). In order to do so, meteorological effects, emissions, transport, chemical transformations, and removal processes at the surface of ozone concentration should all be taken into account (Rao et al., 2000).

Historically, air pollution forecasting and numerical weather predictions (NWP) have developed separately (Baklanov et al., 2008). This was unavoidable in previous decades when the resolution of NWP models was too poor for mesoscale air pollution forecasting. Due to modern NWP models that approach meso- and city-scale resolu-tion and the employment of land-use databases with finer resolution, this situation is changing. Most CTMs have embedded meteorological preprocessors/drivers or are coupled to one, and currently two types of photochemical systems are distinguished: online and offline. Figure 10.2 exemplifies the structure of an offline system, where the meteorological and the photochemical models run independently.

However, urban/rural transition processes (e.g., recirculations and feedbacks) are important as is the interaction of these locally forced features with synoptic-scale processes (e.g., fronts and convection). Furthermore, at regional scales, the inter-action of meteorology (e.g., cloud formation) and pollution transport (e.g., cloud nuclei, precipitation) becomes significant. In this case, offline coupling does not allow for the study of feedback of atmospheric pollutants on meteorological pro-cesses, and the access to meteorological fields is limited by the model outputs and the large amount of data exchange (Baklnaov et al., 2008). Online coupling would

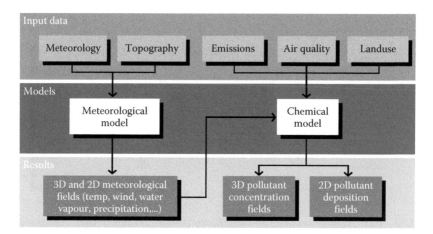

FIGURE 10.2   Structure of an offline system of models, showing the inputs and outputs of a meteorological and a photochemical model.

allow the implementation of "integrated" physical and chemical parameterization schemes.

Moreover, it is also advisable to simulate all the chemical reactions that take place between ozone precursors, namely, NO_x and VOCs. However, an explicit treatment of all these reactions is practically impossible and, for most purposes, the numerical simulations use a condensed kinetic mechanism to avoid excessive numerical costs which means that the chemical scheme is often simplified down to a small number of chemical reactions (Aumont et al., 1997). Therefore, a critical component of a CTM is the chemical mechanism that describes how VOCs and NO_x interact to produce ozone and other oxidants. The daily maximum ozone concentrations generated in mixtures with various initial VOC and NO_x concentrations can be represented by an isopleth diagram using a method called the ozone isopleth plotting method, and can then be used for the assessment of O_3 control strategies. Figure 10.3 gives an example of a set of typical O_3 isopleths. In Figure 10.3, point A represents a base case O_3 concentration and point B represents a control case O_3 concentration as simulated by the OZIPM/EKMA model. In this case, for example, to attain an O_3 level of 0.12 ppm (point B), VOCs need to be reduced from 0.9 ppm (point A) to 0.3 ppm (point B) assuming that NO_x remains the same. The OZIPM/EKMA model was widely used in the 1970s and 1980s for O_3 control strategy assessment before high-performance computer platforms become available.

Hence, chemical mechanisms were first used in models more than 20 years ago. Since then, there has been an enormous growth in our understanding of the chemical processes that lead to oxidant production, especially in that area concerning the role of organic species.

Nowadays, there are several chemical mechanisms included in several different CTMs. The carbon bond mechanism version 4 (CB-IV), the SAPRC99 (Statewide Air Pollution Research Center 99), the KOREM, and the EMEP (cooperative program for monitoring and evaluation of the long-range transmission of air pollutants

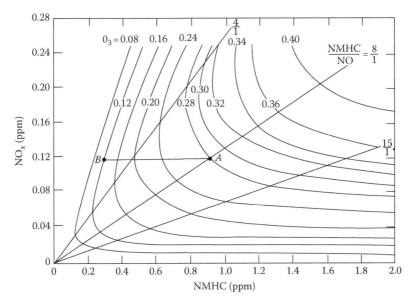

FIGURE 10.3 An example of an O_3 isopleth (ppm) plot produced by the first-generation OZIPM/EKMA model of USEPA. (From Hogo H. and Gery M.W., User's guide for executing OZIPM-4 with CB-IV or optional mechanisms, vol. 1, *Description of the Ozone Isopleth Plotting Package-Version 4*, U.S. EPA/600/8-88/073a, USEPA, RTP, NC, 1988.)

in Europe) are some of the most used mechanisms. CB-IV is a lumped structure mechanism used for both urban and regional-scale modeling. In this lumped structure approach, organics are divided into smaller reaction elements based on the types of carbon bonds existing in each of the 12 VOC species (alkanes, 1-alkenes, ethane, toluene, monoalkybenzene, dialkybenzenes, trialkybenzenes, formaldehyde, other aldehydes isoprene, methylglyixal, and unsaturated dicarbonyls). One hundred and ten chemical reactions and 34 gas species are considered in CB-IV.

The EMEP (Simpson et al., 1993) mechanism describes the tropospheric gas-phase chemistry with 66 species and 139 photochemical reactions, including 34 photolysis reactions. The KOREM, which is simpler, includes 39 chemical reactions and 20 reactive pollutants. KOREM results from the combination of inorganic reactions of the CERT mechanism (Atkinson et al., 1982) and organic reactions of the compact mechanism of Bottenheim and Strausz (1982). Volatile organic compounds are divided into five classes: methane, alkanes, alkenes, aromatics, and aldehydes. On the other hand, the EMEP mechanism considers the following VOC speciation: methane, ethane, *n*-butane, propene, *o*-xylene, formaldehyde, acetaldehyde, methylethylketone, methanol, ethanol, and isoprene (Moussiopoulos, 1992).

The latest version of the SAPRC mechanism, designated SAPRC99 (Carter, 2000), has assignments for 400 types of VOC, and can be used to estimate reactivity for 550 VOC categories. A total of 24 model species are used to represent the reactive organic product species: 11 are explicit and 13 represent groups of similar products using the lumped molecule approach. The peroxyacyl nitrates (PAN) analogue formed from glyoxal is lumped with the rest of the higher PAN analogues. Isoprene photooxidation products in the mechanism are included. SAPRC uses a condensed representation of the reactive organic oxidation products and a highly condensed representation of peroxy reactions.

Jimenez et al. (2003) and Ferreira et al. (2003) have presented comparative studies of different chemical mechanisms in use to simulate photochemical pollution. Their results verify that, even with most chemical schemes yielding similar ozone concentrations, there are still significant discrepancies between the different mechanisms tested. It was not possible to conclude which mechanism is more adequate, but the results did show that a more sophisticated mechanism, with a more complete description of photochemical reactions, does not mean more accurate results.

Moreover, CTM models aiming to quantify the complex nonlinear relationship between the emissions of primary precursors and the maximum ground-level concentrations of O_3 require the input of a number of parameters that are currently either poorly defined and/or contain a high degree of uncertainty (Borrego et al., 2000a). One of the main questions is related to the VOC speciation, meaning that hydrocarbon species have to be treated species by species. There are already some typical VOC class distributions where biogenic hydrocarbon sources should be taken into account. An example of disaggregation factors for VOC species considered in CB-IV chemical mechanism is provided by Zlatev et al. (1993).

10.3 CASE STUDY

High levels of photochemical pollutants, such as ozone, often affect south European countries and concern with this issue has increased substantially during the last decade. The west coast of the Iberian Peninsula, surrounded by the Atlantic Ocean, is characterized by complex topography and some favorable synoptical situations that promote the occurrence of mesoscale circulations, namely, sea/land breezes and anabatic and katabatic flows (Millan et al., 1992; Carvalho et al., 2006). This type of circulation encourages photochemical production of air pollutants leading to smog episodes, which can cause health problems in the population as well as environmental degradation. On the west coast of Portugal, the most densely populated part of the country, several episodes of photochemical pollution have been verified (Borrego et al., 2000b).

The European Directive 2002/3/CE relating to ozone in ambient air has been transposed into Portuguese legislation by *Decreto-Lei 320/2003*. It establishes the public information and alert thresholds for human health as being, respectively, set at 180 and 240 µg m^{-3} as hourly averages. Additionally, and according to the Air Quality Framework Directive ([96/62/CE] transposed by *Decreto-Lei 276/99*), plans and programs should be implemented in zones and agglomerations where the ozone concentrations exceed these thresholds, in order to guarantee that in the future they will be attained.

Notwithstanding the necessary plans and programs, other actions to reduce atmospheric pollutant emissions are foreseen. At the community level, the "National Emission Ceilings" (NEC) Directive was adopted in 2001 in order to limit the negative environmental impacts of acidification, eutrophication, and ground-level ozone

(EC, 2001). This directive obliges each member state to develop and to implement a strategic National Program (PTEN) to comply with the emission ceilings until 2010 for the most critical acidifying air pollutants, namely, sulfur dioxide (SO_2), nitrogen oxides (NO_x), non-methane volatile organic compounds (NMVOC), and ammonia (NH_3), responsible for noxious effects on the environment.

In this context, Portugal has developed technical studies aimed at setting up a reference scenario by 2010 and at evaluating the compliance with the emission ceilings established for this target year. In addition to this reference scenario, high and low emission reduction scenarios have also been defined (IA, 2004a,b,c). Within this perspective, it is important to evaluate the trends in atmospheric ozone precursor emissions, and consequently predict ozone levels in the near future, taking into account this kind of national reduction strategy. The objective of this case study was to evaluate the impact of these national emission reduction scenarios on ozone levels in continental Portugal and in the main agglomerations of Lisbon and Porto, verifying the fulfillment of the air quality thresholds for 2010, using atmospheric modeling techniques.

10.3.1 Methodology

Photochemical simulations for continental Portugal (not considering Madeira and the Azores) and for two urban areas were carried out using the comprehensive air quality model (CAMx) regional CTM with the meteorological forcing coming from the meteorological model MM5. The analysis of typical synoptical and meteorological conditions associated with ozone exceeding episodes in Portugal for 2001 led to the selection of the period May 27–29 as the study period.

The air quality photochemical model CAMx (ENVIRON, 2004) is an offline CTM based on the integration of the continuity equation for the concentrations of several chemical species in each cell of a given 3D grid domain. This model has been used for several research applications covering short- and long-term photochemical and aerosol simulations (Soong et al., 2005; Morris et al., 1997). CAMx has also been applied in sensitivity analysis to grid resolution, chemical mechanisms, and emissions including source apportionment techniques (Bedogni et al., 2005; Ferreira et al., 2003; Morris et al., 1998).

The meteorological input variables come from MM5 (Dudhia, 1993), which was initialized with reanalysis data from the European Centre for Medium Range Weather Forecasting (ECMWF) with 2.5° of resolution and applied to five domains that make up a large part of Europe, the Iberian Peninsula, Portugal, Lisbon, and Porto with a spatial resolution of 90, 30, 10, and 2 km, respectively (see Figure 10.4).

Besides the meteorological input, the CAMx model needs initial and boundary conditions, emission data, and land-use and topography characterization. In order to evaluate the impact of the emission reduction scenarios on air quality, numerical simulations with the CAMx model were performed for continental Portugal and the urban areas of Porto and Lisbon, first for the 2001 baseline period and then compared with the 2010 scenarios, using the same 2001 meteorological conditions. The national total emission data for each simulated year (2001 and 2010) were disaggregated according to a top-down methodology (Monteiro et al., 2005b). Therefore,

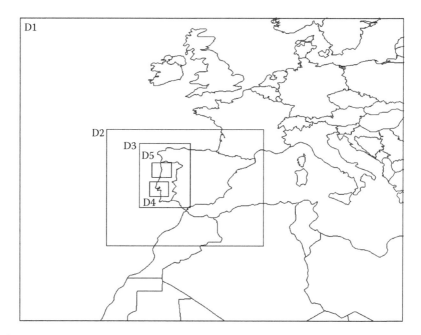

FIGURE 10.4   Meteorological simulation domains.

CAMx was applied to a domain of $430 \times 740\,km$ with $10\,km^2$ resolution, for Portugal and, using its nesting capabilities, to two smaller domains covering Lisbon and Porto, having $134 \times 134\,km$ and $124 \times 124\,km$, respectively, both with $2 \times 2\,km$ resolution. The comparison between model results for the study period of 2001 and measured data allowed the baseline simulation to be evaluated.

10.3.2   Emission Ceiling Scenarios

The numerical values for the 2010 emission ceilings for the individual member states were based on the findings of extensive analysis using the "regional air pollution information and simulation" (RAINS) model developed by the International Institute for Applied Systems Analysis (IIASA) (Amann, 2004). This model was applied to find the internationally least costly allocation of emission control measures. Concerning Portugal, the National Research Centre developed a reference scenario for the economy (IA, 2004a), which is based on a "business as usual" macro-scenario. Two different scenarios (high and low) were established according to macroeconomic and sector-based indicators. The low scenario corresponds to a low growth of the gross domestic product (GDP), in opposition to the high scenario that reflects a faster economic evolution, with a higher investment rate.

Table 10.1 presents a comparison between total SO_2, NO_x, NMVOC, and NH_3 emissions estimated for 2001 baseline year and the 2010 scenarios. The most recent revision of the NEC values is also presented (IA, 2004b).

The reduction defined for the 2010 scenarios is significantly large for all the pollutants (~40%–60%). All pollutant emissions, with the exception of NH_3, exceeded in 2001 the

TABLE 10.1

Comparison between the Total Emissions Estimated for 2001 Baseline Year and 2010 Scenarios, and the National Emissions Ceilings for 2010

Emissions (kt/Year)	SO_2	NO_x	NMVOC	NH_3
Baseline year (2001)	242	383	468	89
Low scenario (2010)	139	220	200	85
High scenario (2010)	145	232	222	85
Emission ceilings (2010)	160	250	180	90

national ceilings for 2010 (see Table 10.1). Nevertheless, the emission reductions projected for 2010 will be enough for both scenarios (low and high) to accomplish the target values for SO_2 and NO_x. This is not the case for the NMVOC, for which the reduction measures seem not to be sufficient to fulfill the NEC requirement for this pollutant. An additional effort to reduce (by 10%–20%) these emissions is being planned, affecting mainly the domestic sector, public construction, and the storage and distribution of fossil fuels (IA, 2004c). Table 10.1 also shows that the difference between low and high scenarios is not significant, with a maximum variance of 10%.

The most important emission reduction measures planned for future scenarios are focused on road transport activities, rather than industrial and residential combustion processes. Finally, it should be considered (and not forgotten) that there is a high level of uncertainty affecting all these future emissions estimations, deriving from the following:

- Uncertainties about the 2001 baseline case and consequently also the projections for 2010
- Uncertainties associated with the political instruments in force and the implementation of new ones
- Uncertainties associated with the estimation of the potential emission reduction of each measure

10.3.3 EVALUATION OF THE MODEL PERFORMANCE

The model performance was evaluated using the baseline simulation results for Portugal and the urban areas of Porto and Lisbon. 3D ozone concentration fields and hourly temporal series were obtained. Figure 10.5 shows, as an example, the ozone hourly averaged surface concentration fields obtained for the May 29, 2001, at 10:00, 14:00, and 18:00h.

With this still too coarse resolution, ozone values estimated for the Lisbon and Porto regions at 10:00h are quite low never exceeding 180 µg m^{-3}. However, results from the nested simulations over these same areas give different concentration values. Figure 10.6 presents the hourly averaged ozone concentration fields for the smallest domains of 2×2 km resolution for Lisbon and Porto, for May 29, 2001, at 10:00h.

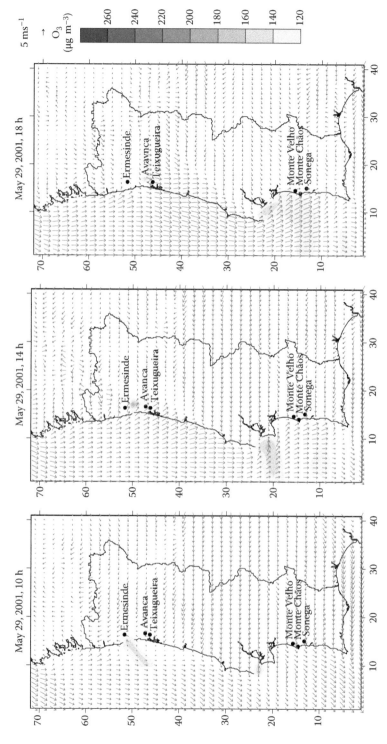

FIGURE 10.5   Hourly averaged wind and ozone surface concentration fields over Portugal, simulated by CAMx, with 10×10 km spatial resolution.

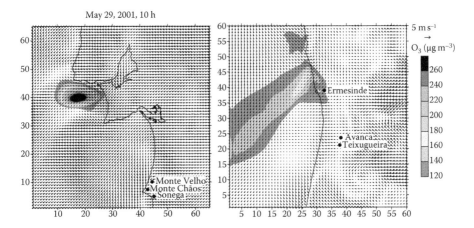

FIGURE 10.6 (See color insert following page 234.) Hourly averaged wind and ozone surface concentration fields, simulated by CAMx, with 2×2 km resolution for the domains Lisbon and Porto.

Six air quality monitoring stations from the national air quality network, registered values that exceed the established thresholds throughout the study period (two rural stations at Avanca and Monte Velho; three industrial stations at Teixugueira/Estarreja, Monte Chãos, and Sonega and one suburban station, Ermesinde). The location of these air quality monitoring stations is indicated in Figure 10.6. A careful analysis of the concentration fields led to the conclusion that these high levels of ozone result from the production and transport of primary pollutants from the urban areas of Porto and Lisbon to both coastal and inland areas of the country. The results of the higher resolution (2×2 km) of the Porto and Lisbon domains confirm these facts.

Figure 10.7 shows the temporal evolution of hourly averaged concentrations of ozone simulated for the three days, at four of the six monitoring stations where excessive values were registered.

CAMx simulates the overall tendency of the ozone episode for the course of the three days of simulation (May 27–29, 2001) reasonably well. The results are significantly better when analyzing the stations located in the North of Portugal (Avanca and Ermesinde). This can probably be put down to an overestimation of the wind speed observed in the south of the country (Figure 10.5), that pushes the plume out over the Atlantic Ocean. In all the stations, except the Ermesinde suburban station, CAMx underestimates the observed concentrations of ozone; this fact is quite evident in Monte Velho.

With the aim of evaluating quantitatively the results obtained, a statistical analysis was performed considering some adequate indicators, namely, root mean square error (RMSE), estimated bias (BIAS), correlation coefficient (R), fraction of predictions within a factor of two observations (FAC2) (Chang and Hanna, 2004), and the normalized accuracy of the domain-wide maximum 1 h concentration unpaired in space and time (Au) (Canepa et al., 2001). Table 10.2 summarizes these parameters for each monitoring station.

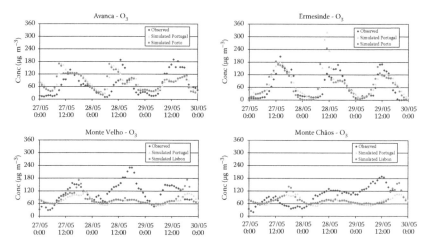

FIGURE 10.7 (See color insert following page 234.) Temporal evolution of hourly averaged concentrations of ozone (μg m^{-3}), simulated for Portugal, Porto, and Lisbon domains, and comparison with measured data.

TABLE 10.2
Statistical Evaluation of Model Performance

Station	Parameter				
	RMSE	BIAS	R	Au	FAC2
Avanca	46.52	8.17	0.52	12.6	1.7
Ermesinde	61.70	13.30	0.66	−55.6	2.0
Teixugueira	76.07	−20.12	0.46	122.6	1.2
Monte Velho	56.87	−29.64	0.19	32.9	0.9
Monte Chãos	56.28	−22.29	−0.39	19.8	1.1
Sonega	60.53	−53.60	0.18	38.9	0.6

This quantitative approach confirms the analysis of the temporal series (Figure 10.7) regarding the good performance of the model in respect of the Avanca and Ermesinde locations and the higher errors in Monte Chãos and Monte Velho. The negative BIAS for all the stations, except Avanca and Ermesinde, accords with the observed trend of the model to underestimate the measured concentrations of ozone. In spite of these verified errors, the model presents good correlation coefficients, except in the three industrial stations located in the southern part of Portugal: Monte Velho, Monte Chãos, and Sonega. This fact indicates that the model simulates well the physics and chemistry of the atmosphere and that the deviations found reflect mistakes in input data (meteorology or emissions). More details about this evaluation exercise can be found in Salmim et al. (2005) and Miranda et al. (2006).

10.3.4 PTEN Scenario Results

Concerning the National Emission Ceiling Scenarios (Borrego et al., 2003), Figure 10.8 shows the differential of ozone concentrations between the reference scenario (2001) and both 2010 scenarios, high and low, for the first day of the simulation at 12:00, 14:00, and 18:00 h.

From the analysis of Figure 10.8, it is possible to conclude that by 2010, considering the application and fulfillment of the National Emission Ceiling Program, there will be a significant decrease in ozone concentrations, especially near the urban areas of Lisbon and Porto where, in 2001, the ozone concentrations exceeded the thresholds of both information and alert. There are no significant divergences between the results of high and low scenarios, probably due to the small differences between the two emissions scenarios.

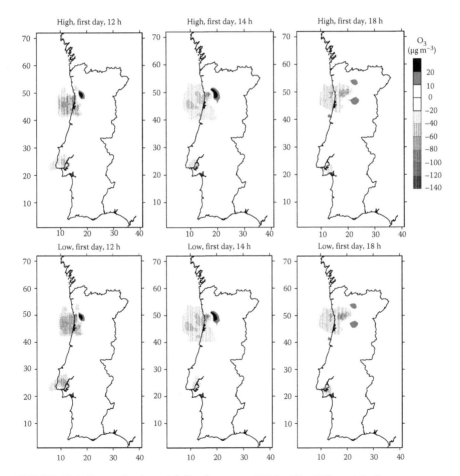

FIGURE 10.8 (See color insert following page 234.) The differential of ozone concentration between the reference scenario (2001) and both, high and low, scenarios for 2010 given by the National Emission Ceiling Program.

10.4 FINAL COMMENTS

One of the main aims of European environmental policy is to improve air quality in Europe. The Framework Directive on air quality assessment and management addresses air quality near major sources and in cities. Other instruments, such as the National Ceiling Directive, the Large Combustion Plant Directive, and the Gothenburg Protocol, address long-range trans-boundary issues as well as local ones. The effective implementation of European legislation requires a strategy for policy decision support based on modeling applications that are powerful tools for air quality management.

The key aspect is the use of models to attribute measured concentrations to the sources from which the pollution may have been emitted, and to assess the environmental impact of pollutants or pollution reduction strategies. The presented case study is an example of the application of modeling tools to the assessment and evaluation of air quality management strategies. The analysis of the modeling results reveals a reasonable simulation of ozone levels. The statistical evaluation allowed the conclusion that despite the relative deviation from the observed concentrations, the modeling system presents good correlation coefficients in most of the air quality stations. The simulation of the Porto and Lisbon domains indicates the importance of refining the grid to achieve better detail and definition of results. Concerning the 2010 scenario simulation of the National Emission Ceiling Program, the significant decrease in ozone concentration is noticeable when the limits are compared with the baseline year of 2001. That decrease is more significant in the urban areas of Lisbon and Porto. However, the need is well recognized to deepen the study in order to improve results which means, in practice, updating the emission inventory for instance and its spatiotemporal disaggregation, and refining the initial and boundary conditions.

Modeling results should be carefully analyzed and interpreted. The uncertainties associated with the input data, and with the model itself, need to be properly evaluated before their predictions can be used with confidence. Dispersion is primarily controlled by turbulence, which is random by nature, and thus cannot be precisely described or predicted by means of basic statistical properties. As a result, there is spatial and temporal variability that occurs naturally in the observed concentration field. On the other hand, uncertainty in the model results could also be due to factors such as errors in the input data and model formulation. Because of the effects of uncertainty and its inherent randomness, it is not possible for an air quality model ever to be "perfect," and there is always a base amount of scatter that cannot be removed (Chang and Hanna, 2004).

Notwithstanding these limitations, air quality models should be considered as powerful tools to predict the fate of pollutant gases or aerosols upon their release into the atmosphere, and to influence decisions with significant public health and economic consequences.

ACKNOWLEDGMENTS

The authors are grateful for the collaboration of L. Salmim and would like to thank her for her participation in the case study modeling work. Also the collaboration of the Portuguese Institute for the Environment and the Northern Regional Commission for Coordination and Development are acknowledged.

REFERENCES

Académie des Sciences. 1993. Ozone et propriétés oxydantes de la troposphere. *Académie des Sciences*, rapport n°30. Lavoisier, Tec Doc, London, Paris, New York.

Amann M. 2004. *The RAINS Model.* Documentation of the model approach prepared for the RAINS peer review. February 2004, www.iiasa.ac.at/rains/.

Atkinson R., Lloyd A.C., and Winges L. 1982. An updated chemical mechanism for hydrocarbon/NO_x/SO_2 photo-oxidations suitable for inclusion in atmospheric simulation models. *Atmospheric Environment* 16: 1341–1355.

Aumont B., Jaecker-Voirol A., Martin B., and Toupance G. 1997. Tests of some reduction hypotheses made in photochemical mechanisms. *Atmospheric Environment* 30: 2061–2077.

Baklanov A., Fay B., Kaminsky J., and Sokhi R. 2008. Overview of existing integrated (off-line and on-line) mesoscale meteorological and chemical transport modelling systems in Europe. Joint Report of COST Action 728 and GURME. GAW Report No. 177. World Meteorological Organization.

Bedogni M., Gabusi V., Finzi G., Minguzzi E., and Pirovano G. 2005. Sensitivity analysis of ozone long term simulations to grid resolution. *International Journal of Environment and Pollution* 24(1–4): 51–63.

Borrego C., Gomes P., Barros N., and Miranda A.I. 2000a. Importance of handling organic atmospheric pollutants for assessing air quality. *Journal of Chromatography A* 889: 271–279.

Borrego C., Barros N., and Tchepel O. 2000b. An assessment of modelling ozone control abatement strategies in Portugal: The Lisbon urban area. *International Technical Meeting of NATO/CCMS on "Air Pollution Modelling and its Application."* Boulder, CO, May 15–19, 2006, 439–446.

Borrego C., Miranda A.I., Monteiro A., Salmim L., Ferreira J., Coutinho M., and Ribeiro C. 2003. *Avaliação e Previsão da Qualidade do Ar em Portugal—Definição de Cenários.* Relatório R2, Universidade de Aveiro, Aveiro, Portugal: AMB-QA-13/03. November 2003.

Bottenheim J.W. and Strausz O.P. 1982. Modelling study of a chemically reactive power plant plume. *Atmospheric Environment* 16: 85–106.

Brasseur G., Prinn R., and Pszenny A. 2003. *Atmospheric Chemistry in a Changing World. An Integration and Synthesis of a Decade of Tropospheric Chemistry Research.* The IGBP Series. Springer, the Netherlands, ISBN: 3-540-43050-4.

Canepa E., Builtjes P.J.H., and Van Loon M. 2001. An outline of photochemical model evaluation techniques. *Fifth GLOREAM Workshop.* Wengen, Switzerland, September 24–26. Conference Proceedings on CD-ROM.

Carter W.P.L. 2000. Documentation on the SAPRC-99 chemical mechanism for VOC reactivity assessment. Final report to California Air Resources Board Contract No. 92–329 and 95–308. May 2000.

Carvalho A.C., Carvalho A., Gelpi I., Barreiro M., Borrego C., Miranda A.I., and Pérez-Muñuzun V. 2006. Influence of topography and land use on pollutants dispersion in the Atlantic coast of Iberian Peninsula. *Atmospheric Environment* 40: 3969–3982.

Chang J.C. and Hanna S.R. 2004. Air quality model performance evaluation. *Meteorology and Atmospheric Physics* (June) Springer-Verlag, online issue.

Dudhia J. 1993. A nonhydrostatic version of the Penn State/NCAR mesoscale model: Validation tests and simulation of an Atlantic cyclone and clod front. *Monthly Weather Review* 121: 1493–1513.

ENVIRON. 2004. *Comprehensive Air Quality Model with Extensions (CAMx), version 4.10s.* User's guide, ENVIRON International Corporation, Novato, CA, 120 pp.

European Commission. Directive 2001/81/EC on national emissions ceilings for certain atmospheric pollutants.

Ferreira J., Carvalho A., Carvalho A.C., Monteiro A., Martins H., Miranda A.I., and Borrego C. 2003. Chemical mechanisms in two photochemical modelling systems: a comparison procedure. In *International Technical Meeting of NATO-CCMS on "Air Pollution Modelling and its Application,"* vol. 26, Istanbul, Turkey, May 26–30, 87–96.

Hogo H. and Gery M.W. 1988. User's guide for executing OZIPM-4 with CB-IV or optional mechanisms. *Description of the Ozone Isopleth Plotting Package-Version 4*, vol. 1. U.S. EPA/600/8–88/073a. USEPA, RTP, NC.

Hogrefe C., Rao S., Trivikrama Kasibhatla P., Hao W., Sistla G., Mathur R., and McHenry J. 2001. Evaluating the performance of regional-scale photochemical modelling systems: Part II—ozone predictions, *Atmospheric Environment* 35: 4175–4188.

IA–Instituto do Ambiente. *Programa para os Tectos de Emissão Nacional–Estudos de base: Cenário de cumprimento*, 2004a: www.iambiente.pt.

IA–Instituto do Ambiente. *Programa para os Tectos de Emissão Nacional–Estudos de base: Cenário de referência*, 2004b: www.iambiente.pt.

IA–Instituto do Ambiente. *Programa para os Tectos de Emissão Nacional—Estudos de Base: Medidas para o controlo das emissões nacionais de COVNM*, 2004c: www.iambiente.pt.

Jimenez P., Baldasano J.M., and Dabdub D. 2003. Comparison of photochemical mechanisms for air quality modeling. *Atmospheric Environment* 37: 4179–4194.

Millan M., Artinano B., Alonso L., Castro M., Fernandez-Patier R., and Goberna J. 1992. *Mesometeorological Cycles of Air Pollution in the Iberian Peninsula.* Air Pollution Research Report 44. European Community Commission. Brussels, Belgium.

Miranda A.I., Monteiro A., Lopes M., and Borrego C. 2006. National emission ceilings for 2005 and 2010 and their impact on Portuguese air quality. In *Air Pollution.* New Forest, U.K., May 22–24, pp. 693–712.

Monteiro A., Vautard R., Lopes M., Miranda A.I., and Borrego C. 2005a. Air pollution forecast in Portugal: A demand from the new Air Quality Framework Directive. *International Journal of Environment and Pollution* 25(2): 4–15.

Monteiro A., Vautard R., Borrego C., and Miranda A.I. 2005b. Long-term simulations of photo oxidant pollution over Portugal using the CHIMERE model. *Atmospheric Environment* 39: 3089–3101.

Morris R.E., Yarwood G., and Yocke M. 1997. Recent advances in regional photochemical grid modeling and application using the OTAG modeling databases. *90th Annual Meeting of the Air and Waste Management Association.* New Orleans, LA, June 1997.

Morris R.E., Tai E., Wilson G., and Weiss D. 1998. Use of advanced ozone source apportionment techniques to estimate area of influence (AOI) of emissions contributions to elevated ozone concentrations. *91st Annual Meeting of the Air and Waste Management Association.* San Diego, CA, June 14–18, 1998.

Moussiopoulus N. 1992. *MARS (Model for Atmospheric Dispersion of Reactive Species): Technical Reference.* Aristotle University, Thessaloniki, Greece.

Rao S.T., Hogrefe C., Mao H., Biswas J., Zurbenko I., Porter P.S., Kasibhatla P., and Hansen D.A. 2000. *24th International Technical Meeting of NATO/CCMS on Air Pollution Modelling and Its Application*, Istanbul, Turkey, ed. Gryning and Schiermeier. Kluwer Academic/Plenum Press, New York, pp. 25–34.

Salmim L., Ferreira J., Monteiro A., Miranda A.I., and Borrego C. 2005. Emissions reduction scenarios for 2010: Impact on air quality in Portuguese urban areas. In *Fifth International Conference of Urban Air Quality 2005* (UAQ), March 29–31, Valencia, Spain, ed. Ranjeet Sokhi and Nicolas Moussiopoulos, p. 172.

Simpson D., Andersson-Sköld Y.A., and Jenkin M.E., 1993. *Updating the Chemical Scheme for the EMEP MSC-W Note 2/93*. The Norwegian Meteorological Institute, Oslo, Norway.

Soong S.-T., Martien P.T., Archer C., Tanrikulu S., Bao J.-W., Wilczak. J.M., Michelson S.A., Jia Y., and Emery C. 2005. *Comparison of WRF/CAMx and MM5/CAMx Simulations for an Ozone Episode in California*. 14th Joint Conference on the Applications of Air Pollution Meteorology with the Air and Waste Management Assoc, Amer. Meteor. Soc. J1.4.

UE. 2003. Ozone-climate interactions. Air pollution research report n° 81. Directorate-General for Environment and sustainable development programme. ISBN: 92-894-5619-1.

UNECE-United Nations Economic Commission for Europe. 1999. *Protocol to Abate Acidification, Eutrophication and Ground-level Ozone*. Available at www.unece.org/env/lrtap/multi_h1.htm.

Vautard R., Beekmann M., Roux J., and Gombert D. 2000. Validation of a deterministic forecasting system for the ozone concentrations over the Paris area. *Atmospheric Environment* 35: 2449–2461.

Zlatev Z., Christensen J., and Eliassen A. 1993. Studying high ozone concentrations by using the Danish Eulerian model. *Atmospheric Environment—Part A General Topics* 27A(6): 845–865.

11 Inversion of Atmospheric CO_2 Concentrations

Ian G. Enting

CONTENTS

The space–time distribution of concentrations of long-lived tracers, most notably CO_2, contains considerable information about sources and sinks. However, atmospheric transport and mixing makes the estimation of source–sink strengths a challenging mathematical problem. This chapter reviews the main inversion techniques and the results that have been obtained.

11.1 INTRODUCTION

Carbon dioxide (CO_2) is the most important of the so-called anthropogenic greenhouse gases. The complexities of the carbon cycle (Falkowski et al., 2000), along with the perceived difficulties in achieving reductions of fossil carbon emissions, have prompted intensive study of the global carbon cycle. Direct observations of the global increase in CO_2 provide important information about how the carbon cycle responds to steadily growing inputs, but only limited information about how the cycle would behave under other conditions (Enting, 2007). In order to obtain additional information about the carbon cycle, extensive investigations of indirect data have been undertaken.

One widely-applied indirect technique involves the deduction of surface CO_2 fluxes from data on the spatial distribution of concentrations. This type of inverse calculation goes back to Bolin and Keeling (1963), with little further study until the late 1980s, and then widespread application since then. Important milestones were the development of techniques including formal analysis of uncertainties arising from data error (Enting et al., 1993, 1995) and the establishment of the TransCom working group to study the role of model error in CO_2 inversions.

The principle of trace gas inversions is that the observed spatial distribution of concentrations reflects the combined effect of the spatial distribution of sources and sinks and of atmospheric transport. Therefore, in principle, if the contribution of transport can be calculated, then the net result of sources and sinks can be estimated. In practice, the ability to resolve details of sources is severely limited by the mathematical characteristics—technically termed ill-conditioning—of the estimation problem.

While much of the research effort in global-scale trace gas inversions has concentrated on CO_2, inversion studies have been undertaken for other gases such as methane (CH_4) and various halogenated compounds. There have been a number of reviews of trace gas inversions (Enting, 2000a,b; Prinn, 2000). One of my books (Enting, 2002) gives an extensive discussion of many aspects of trace gas inversion. More recently, Heimann et al. (2004) have reviewed the contribution of space–time inversions to our understanding of the carbon cycle.

This chapter reviews inversion techniques, places them in the context of statistical estimation, and considers the limits to resolution imposed by the ill-conditioning of the inverse problem. The results of inversions of CO_2 data are considered in some detail, noting applications to other long-lived tracers. Many of these topics are addressed in greater detail in my book (Enting, 2002), but the field is undergoing continuous rapid development. Consequently, in Section 11.5, the present chapter tries to identify new emerging areas of development.

11.2 TECHNIQUES

11.2.1 ESTIMATION AND STATISTICS

For a quantitative uncertainty analysis, an inversion should be treated as a process of statistical estimation. The statistical formulation of inverse problems is espoused by Tarantola (1987) as a general principle, and forms the basis of the analysis of meteorological data assimilation by Kalnay (2003). For trace gas inversions, the statistical formulation of inversions is the theme of my book (Enting, 2002). The statistical approach to inverse problems has recently been advocated by Evans and Stark (2002). In particular, they suggest that nonparametric statistical approaches are most appropriate.

In discussing the question of what has been learned from CO$_2$ inversions, we adopt a number of principles identified by Enting (2002):

- Whatever cannot be modeled deterministically should be modeled statistically.
- Any statistical analysis is assuming (either explicitly or implicitly) some statistical model. Clearly, explicit identification of the assumptions is preferable.

Statistical formalisms can be used in various ways. Tarantola (1987) considers that *the* answer to any inversion problem is given by Bayesian posterior distribution:

$$\text{Pr}_{\text{posterior}}(\mathbf{x} \mid \mathbf{z}) \propto \text{Pr}(\mathbf{z} \mid \mathbf{x})\,\text{Pr}_{\text{prior}}(\mathbf{x}) \tag{11.1}$$

for parameters **x** given observations **z**.

A more usual approach is to give an estimate with uncertainty, often derived using maximum likelihood (or maximum of posterior distribution in the Bayesian case). The likelihood for parameters **x** given **z** is

$$L(\mathbf{x} \mid \mathbf{z}) = \text{Pr}(\mathbf{z} \mid \mathbf{x}) \tag{11.2}$$

regarded as a function of **x** with **z** fixed.

Maximum likelihood can be treated as a special case of minimization of a "cost" (or "penalty") function, Θ, by putting

$$\Theta(\mathbf{x}) = -\ln(L(\mathbf{x} \mid \mathbf{z})) \tag{11.3a}$$

or

$$\Theta(\mathbf{x}) = -\ln(\text{Pr}_{\text{posterior}}(\mathbf{x} \mid \mathbf{z})) \tag{11.3b}$$

in the Bayesian case. Taking negatives of logarithms converts multivariate normal distributions into quadratic cost functions.

A linear relation, defined by a matrix **G**, between parameters and the "true" value of noisy data leads to a generic cost function:

$$\Theta = [\mathbf{Gx} - \mathbf{z}]^{\text{T}} \mathbf{X}[\mathbf{Gx} - \mathbf{z}] \tag{11.4a}$$

where **X** is the inverse covariance matrix of the observations. The Bayesian case, with a multivariate normal distribution (with inverse covariance matrix **Y**) for the priors, has

$$\Theta = [\mathbf{Gx} - \mathbf{z}]^{\text{T}} \mathbf{X}[\mathbf{Gx} - \mathbf{z}] + [\mathbf{x} - \mathbf{x}_{\text{prior}}]^{\text{T}} \mathbf{Y}[\mathbf{x} - \mathbf{x}_{\text{prior}}] \tag{11.4b}$$

11.2.2 EARLY INVERSION TECHNIQUES

The initial CO_2 inversion study (Bolin and Keeling, 1963) modeled atmospheric transport as a one-dimensional diffusion process, by taking averages of over height and longitude. Diffusive models have been used in studies of restricted domains (Heimann and Keeling, 1986; Keeling and Heimann, 1986) and for analyzing ill-conditioning and error statistics (Newsam and Enting, 1988; Enting and Newsam, 1990; Enting, 1993). However, in most contexts, explicit consideration of the advective nature of atmospheric transport is essential. A class of low-resolution models, used for inversion of halogenated gases, are the box models used in the AGAGE study (and its predecessors) (Prinn et al., 2000). Early mass-balance inversions used two-dimensional models (averaged over longitude). Most recent inversions, particularly for CO_2, have used full three-dimensional transport models.

Very early in the development of trace gas inversions, two main types of calculations emerged:

Mass-balance inversion: In this technique, an atmospheric transport model is run with boundary conditions specified in terms of concentrations, and surface fluxes are deduced from the requirement of local conservation of mass. The technique was developed using two-dimensional models (Enting and Mansbridge, 1989; Tans et al., 1989; Enting and Mansbridge, 1991; Conway et al., 1994), but there has been some later work with 3-D models (Law, 1999) requiring extensive interpolation of concentration data. Dargaville and Simmonds (2000) developed a related "data-assimilation" technique, using multiple model runs to interpolate the concentration data.

Synthesis inversion: In this technique, defined by Equations 11.5a and b, the source is broken into blocks with specified space–time distributions of fluxes multiplied by unknown scale factors. For each block, concentrations resulting from unit sources are calculated using a transport model. The fluxes are estimated by the process of estimating the scale factors. This was done by fitting observed concentrations with a linear combination of the calculated concentrations. Early applications (Keeling et al., 1989; Tans et al., 1990) used ad hoc fits. The process was formalized as one of (Bayesian) statistical estimation by Enting et al. (1993, 1995). The term "synthesis" was apparently first used for this type of inversion in the study of methane by Fung et al. (1991).

The mass-balance technique takes the equations for a vector of concentrations $\mathbf{c}(t)$ with elements $c_j(t)$ and transport $T_j(\mathbf{c}(t))$:

$$\frac{\mathrm{d}}{\mathrm{dt}} c_j(t) = T_j(\mathbf{c}(t)) + S_j(t) \quad \text{for all sites} \tag{11.5a}$$

and, for a set of surface sites, rewrites them as

$$S_j(t) = \frac{\mathrm{d}}{\mathrm{dt}} c_j(t) - T_j(\mathbf{c}(t)) \quad \text{for surface sites} \tag{11.5b}$$

This means that at all the surface sites, one needs to know the $c_j(t)$ and their rates of change. Given the tendency of numerical differentiation to amplify noise, this requires smoothing of the data.

In contrast, the synthesis techniques construct a source representation as a sum of specified basis functions, σ_α:

$$S_j(t) = \sum_\alpha a_\alpha \sigma_{j\alpha} \tag{11.6a}$$

and calculate the atmospheric responses by integrating the equations:

$$\frac{d}{dt} c_j^{[\alpha]} = T_j(\mathbf{c}^{[\alpha]}) + \sigma_{j\alpha}(t) \tag{11.6b}$$

to give a set of responses $c_j^{[\alpha]}(t)$. Linear combinations of these responses are fitted to observations as

$$z_k(t_m) \approx \sum_\alpha a_\alpha c_k^{[\alpha]}(t_m) \tag{11.6c}$$

in order to obtain estimates, $\hat{a}_\alpha$, of the coefficients a_α. The $c_j^{[\alpha]}(t)$ are a discretization of Green's function of the set of equations 11.5a. The source is reconstructed as an estimate:

$$\hat{S}_j(t) = \sum_\alpha \hat{a}_\alpha \sigma_{j\alpha}(t) \tag{11.6d}$$

Some of the important characteristics of this approach are

- The data set in Equation 11.6c does not have to span the whole globe.
- One can take linear combinations of the fluxes and/or data, for example, those defining Fourier components of cycle. This case gives the so-called "cyclo-stationary" inversions of the earliest studies. Fully time-dependent inversions came later (Rayner et al., 1999a).

One structural way in which these synthesis and mass-balance methods differ is in the type of interpolation that is occurring. Mass-balance inversions need to interpolate observations of concentrations in order to define the boundary condition that specifies concentrations at all points on the surface. Synthesis inversion interpolates fluxes to estimate a flux distribution as a linear combination of specified distributions. In the forms that they were originally developed, neither form of interpolation is likely to be optimal. The approach described in a recent high-resolution inversion study (Rödenbeck et al., 2003) (using the gradient techniques described below) derives the effective interpolation from the data statistics (of both observations and priors) and should be much closer to optimal.

However, one of the most important characteristics of synthesis inversion is that the fit can be, and since 1995 usually is, performed as a statistical fit and weighted to take into account the uncertainties in the data. Using the generic notation of Section 11.2.2 (as opposed to the specific discretization of (Equation 11.4a through 11.4d)), with $\mathbf{z}$ as a data vector, $\mathbf{x}$ as a generic discretization of fluxes, and $\mathbf{G}$ as the corresponding Green's functions ($c_k^{[\alpha]}$), leads to the cost functions 11.4a or 11.4b. In the absence of a Bayesian constraint, Equation 11.4a must be supplemented by some other form of regularization.

Minimizing the quadratic cost function leads to linear relations for the estimates, the Bayesian case being given by

$$\hat{\mathbf{x}} = [\mathbf{G}^T\mathbf{X} + \mathbf{Y}]^{-1}[\mathbf{G}^T\mathbf{X}\mathbf{z} + \mathbf{Y}\mathbf{x}_{\text{prior}}] \qquad (11.7a)$$

with a covariance

$$E[(\hat{\mathbf{x}} - \mathbf{x}) \otimes (\hat{\mathbf{x}} - \mathbf{x})] = [\mathbf{G}^T\mathbf{X}\mathbf{G} + \mathbf{Y}]^{-1} \qquad (11.7b)$$

Various modifications to the basic synthesis approach have been explored, including

- Estimation of the data covariance matrix $\mathbf{X}^{-1}$, in the sense of assuming $\mathbf{X}$ is diagonal with variances equal for sites within classes defined by site-specific conditions—these variances are then estimated by maximum likelihood (Michalak et al., 2005).
- Replacement of the Bayesian constraint by regularization based on a geostatistics approach (Michalak et al., 2004).
- Shrinkage estimators to reduce variance, at the expense of additional bias, leading to an overall reduction in mean-square-error (Shaby and Field, 2006).

11.2.3 GRADIENT METHODS

Two difficulties are concealed by the mathematical elegance of the linear estimation equations that come from multivariate normal distributions:

- The formalism does not apply if either the linear relation, $\mathbf{z} \approx \mathbf{G}\mathbf{x}$, or the normality assumption is invalid.
- Even if these assumptions apply, using the linear equations may not be the best way to find the minimum of the cost function.

An alternative approach to data fitting is to use gradient techniques that aim to minimize the cost function directly, using generic minimization techniques, based on the gradient. For the linear case derived from Equation 11.4:

$$\frac{1}{2}\nabla_x\Theta = \mathbf{G}^T\mathbf{X}[\mathbf{G}\mathbf{x} - \mathbf{z}] + \mathbf{Y}[\mathbf{x} - \mathbf{x}_{\text{prior}}] \qquad (11.8)$$

In terms of computational efficiency, $\mathbf{Gx} - \mathbf{z}$ is easy to evaluate, since $\mathbf{Gx}$ is obtained by a single model integration with sources $\mathbf{x}$. Multiplying this vector by $\mathbf{X}$ is a simple matrix-times-vector operation in general, and specific cases are often much simpler if $\mathbf{X}$ has a block-diagonal (or even diagonal) structure due to independence of various subsets of data. The difficulty comes from multiplying this vector by $\mathbf{G}^{\mathsf{T}}$. The direct approach, used in synthesis inversion, is to calculate the full matrix $\mathbf{G}$ by integrating the model with a set of basis functions. The number of such integrations is the dimensionality of $\mathbf{x}$. For high-resolution inversions, this becomes computationally infeasible.

The alternative to such "brute-force" is to use what is known as an adjoint model. This is a model whose operation corresponds to the effect of $\mathbf{G}^{\mathsf{T}}$. This adjoint model is then run with $\mathbf{X}[\mathbf{Gx} - \mathbf{z}]$ as its input. There are software tools (Giering, 2000; Griewank, 2000) that take the computer code (that formally computes $\mathbf{Gx}$ for arbitrary $\mathbf{x}$) and automatically generates "adjoint" code that calculates $\mathbf{G}^{\mathsf{T}}\mathbf{z}$ for arbitrary $\mathbf{z}$. Adjoint techniques apply to linearized calculations as well as the fully-linear case described here.

These adjoint techniques have been the basis for obtaining high-resolution inversions, regularized, at least in part, by smoothness constraints rather than relying on fixing the spatial structure of low-resolution basis functions (Rödenbeck et al., 2003).

11.2.4 Process Inversions

The inversions described in Section 2.2 obtain estimates of fluxes, given observations of concentrations. Much of the interest in such flux estimates comes from trying to identify the processes responsible for the spatial and temporal variability in these fluxes. This immediately suggests the desirability of inversions that directly estimate characteristics of processes. Some aspects of process inversion have been reviewed (Enting, 2002, Section 12.4), mainly in the context of the likely need for nonlinear estimation in such studies. This review noted some "precursor" studies that involved aspects of process inversion including (a) the estimation of hydroxyl (OH) concentrations from studies of halogenated compounds and (b) the use of satellite data for prescribing the timing of seasonal vegetation fluxes in a study of the seasonal cycle of CO$_2$ (Fung et al., 1987).

In the last few years, there have been several studies that could be classed as "genuine" process inversions. These have taken the form of estimating parameters in terrestrial models (Knorr and Heimann, 1995; Kaminski et al., 2002; Rayner et al., 2005). If one has fluxes as functions $\mathbf{x}(\mathbf{p})$ of parameters, $\mathbf{p}$, then the gradient of the cost function is given by

$$\frac{1}{2}\nabla_{\mathrm{p}}\Theta = (\nabla_{\mathrm{p}}\mathbf{x})\mathbf{G}^{\mathsf{T}}\mathbf{X}[\mathbf{Gx}(\mathbf{p}) - \mathbf{z}] \tag{11.9}$$

These calculations have been performed using the adjoint of the combined terrestrial + transport model, that is $(\nabla_{\mathrm{p}}\mathbf{x})\mathbf{G}^{\mathsf{T}}$, to obtain the gradients used in the iterative minimization of the cost function. (In this paper, the superscript T denotes the matrix transpose (or adjoint in the case of operators) and the notation does not distinguish between row and column vectors, relying on the context to indicate the appropriate case.)

However, given the linearity of the transport, if only a small number of parameters, p_α, are involved, an alternative way of calculating gradients would be to obtain the sensitivities of the terrestrial model (i.e., from $\partial x/\partial p_\alpha$, the tangent linear model for the terrestrial component) and use these sensitivities as basis functions in an integration of the transport model. Equation 11.9 becomes

$$\frac{1}{2}\frac{\partial \Theta}{\partial p_\alpha} = \frac{\partial \mathbf{x}}{\partial p_\alpha}\mathbf{G}^\mathrm{T}\mathbf{X}[\mathbf{Gx(p)}-\mathbf{z}] \qquad (11.10)$$

which can be evaluated by calculating $\mathbf{G}\partial x/\partial p_\alpha$, that is, running the forward transport model with surface forcing taken as the tangent linear sensitivities of the flux model and then taking the scalar product of this vector with the vector $\mathbf{X}[\mathbf{Gx(p)} - \mathbf{z}]$. While Equation 11.10 is formally less efficient than Equation 11.9, since the evaluation of Equation 11.10 needs to be repeated for each of the parameters p_α, it avoids the need to have adjoints of the transport model, the terrestrial model, or the combined terrestrial + transport model.

This simplification may be useful in some cases, but extensions to more complex problems may require the full adjoint:

- Consideration of spatially varying uncertainties in the functional relationships in the terrestrial model means that the sensitivities will require full spatial resolution rather than being represented by a small set of basis functions.
- Cases where the terrestrial model is "forced" by functions obtained by the assimilation of proxy data (e.g., NPP derived from the model with satellite data as a constraint) may again lead to a gradient relation that requires the full adjoint $(\nabla_p x)\mathbf{G}^\mathrm{T}$ in order to achieve acceptable computational efficiency.

In looking at future directions, just as flux estimates are a step toward understanding processes, process estimates are a step toward understanding the role of these processes in the coupled earth system. In this context, inversion studies of CO_2 are being subsumed into a more general framework termed "model-data fusion" (Raupach et al., 2005). CO_2 inversions have recently been reviewed from this perspective by Wang et al. (2009).

11.2.5 Other Tracers

While much of the development of trace gas inversions has been in the context of CO_2, other compounds have been analyzed in this way. A number of studies are reviewed in Enting (2002). The discussion in the present section is mainly focussed on methodology, identifying aspects where the techniques differ from the practice of CO_2 inversions. Such generalizations may be useful in future CO_2 studies, particularly as carbon cycle studies become increasingly involved with the type of process inversion described in Section 11.2.4.

Carbon isotopes: The use of various carbon isotopes to help elucidate the behavior of the carbon cycle provides an excellent example of the way in which different types of data can reveal different aspects of the system. Radiocarbon (^{14}C) has long been used to characterize the global-scale carbon cycle responses (Oeschger et al., 1975; Enting and Pearman, 1987, for example). Atmospheric nuclear testing, particularly over 1961–1962 produced a large spike of ^{14}C in the atmosphere—tracking this perturbation through the carbon cycle has provided important information about the decadal-scale carbon cycle response. The natural ^{14}C distribution, derived from "pre-bomb" data, provides information about the behavior of the carbon cycle on time scales of centuries or longer, reflecting the ^{14}C half-life of 5730 years. Spatial distributions of the ^{14}C "spike" from nuclear testing have been analyzed to identify the seasonal modulation caused by biospheric exchange in the immediate posttesting period when there was a large isotopic disequilibrium between atmosphere and biosphere (Randerson et al., 2002).

The minority (ca. 1%) stable carbon isotope, ^{13}C, provides, in principle, a way of separating terrestrial from oceanic carbon exchanges. This reflects the differences in carbon fractionation between photosynthesis and air–sea exchange. The quantitative interpretation of ^{13}C data requires incorporation of additional effects, particularly the so-called "isofluxes" that arise from isotopic disequilibrium, even in the absence of net carbon exchange, and also the differences in fractionation between the C-3 and C-4 mechanisms of photosynthesis. A discussion of isotopic budgeting is given in Enting (2002, Section 10.5).

Some ^{13}C data were used in initial synthesis inversions, both as a trend (giving a constraint on the oceanic–terrestrial partitioning of net CO$_2$ uptake) and as sparse spatially-distributed data. Inversions making use of larger ^{13}C data sets followed Ciais et al. (1995a, b).

Methane: Fung et al. (1991) undertook a study of methane (CH$_4$), apparently coining the term "synthesis" for this type of study. For CH$_4$, the main difference from CO$_2$ inversions is the presence of a sink process (oxidation by hydroxyl radicals) in the free atmosphere. If the sink rate were known, and strictly proportional to methane concentrations, then it would be possible to define Green's functions that described the surface sources and the resulting sink, and combine these in a synthesis (Enting, 2002, Section 15.2). However, most, if not all, studies have treated the sink as an unknown that must be estimated and performed iterative calculations to obtain a consistent budget.

Mass-balance inversions of CH$_4$ have been obtained by Law and Vohralik (2001), with a prescribed atmospheric sink, and by Butler et al. (2004), using a chemical transport model. The use of a full chemical model can take account of the effect of methane concentration changes affecting the strength of the methane sink, through reduction of hydroxyl concentrations.

Halogenated compounds: Various halogenated compounds, originally the chlorofluorocarbons and more recently their "ozone-friendly" replacements, have been studied within the AGAGE program (Prinn et al., 2000; see also Enting, 2002, chapter 16). The main aim of the initial inversions was to estimate a destruction rate (or equivalently, its inverse, termed an "atmospheric lifetime"). The calculations were performed at low resolution with the atmosphere divided into a small number of boxes. The estimation technique was based on the Kalman filter (Gelb, 1974). This approach was later extended to studies with full 3-D atmospheric transport models (Hartley and Prinn, 1993; Haas-Laursen et al., 1996; Huang, 2000). As discussed in Enting (2002), when using the Kalman filter formalism it is important to distinguish the case of estimating a fixed source (with the estimate evolving in time as additional data are obtained) from estimating a changing source. These cases will differ in the model that is used in the Kalman filter, the former corresponding to a fixed state (so that the estimation takes the form of recursive regression) while the latter requires a prior stochastic representation of the evolving state.

11.3   STATISTICAL CHARACTERISTICS

11.3.1   Ill-Conditioning

Enting (2002) defined an inverse problem as one in which the direction of mathematical inference was opposite to that of real-world causality for the system in question. Since many real-world processes are dissipative, information is lost or severely attenuated, particularly on small-scales. Consequently, the associated inverse problems, in attempting to reconstruct causes from attenuated signals, are subject to severe amplification of both model error and data error. This characteristic is known as ill-conditioning. Indeed the standard usage is to take such ill-conditioning as the defining characteristic of inverse problems (i.e., the mathematical study of inverse problems is effectively defined as the mathematical study of these computationally challenging inverse problems.)

The ill-conditioning in trace gas inversions has been recognized from the very first of such calculations. Bolin and Keeling (1963) noted the problem and stated that "no details of the sources and sinks are reliable." Similarly, in reviewing the work of Bolin and Keeling, Junge and Czeplak (1968) concluded that "It seems hardly likely that detailed information could be obtained on the latitudinal dependence of K and the CO_2 source function from atmospheric CO_2 observations, even if the number and quality of the data were very considerably increased." (K was a "diffusion coefficient" characterizing atmospheric transport). Less pessimistically, a 1980 WMO report (Pearman, 1980) estimated the observational requirements for various objectives as

- About three stations: Determine global inventories and trends
- Additional 5–10 stations: Determine meridional transport of CO_2
- Over 100 stations: Determine air-surface exchange within regions with significant anthropogenic influence

The difficulty of the inversion can be quantified, in part, by determining how rapidly the error amplification grows as the resolution increases. This has been analyzed for the dependence on latitudinal wave-number n. If the forward problem attenuates fluxes of wave number n with an $n^{-\alpha}$ decay, then the inverse problem involves an error amplification that grows as n^{α} In their 2D numerical modeling,

Enting and Mansbridge found a n^1 growth in the extent to which errors with lati-tudinal wave-number n were amplified. In contrast, the Bolin and Keeling analysis implied n^2 growth, fully justifying the pessimistic assessments quoted above. Enting and Mansbridge (incorrectly) attributed the difference to their use of advective trans-port as opposed to the diffusive transport approximation used by Bolin and Keeling. The real reason for the difference between the n^2 and n^1 growth was identified by Newsam and Enting (1988) as having the n^2 growth appear as an artifact of approxi-mating the boundary-value problem as a vertically-averaged problem. Numerical studies with the GISS 3D transport model (Enting, 2000a) showed that this case also exhibits the n^1 growth in error amplification. Thus, although the concerns of Bolin and Keeling were overstated, the degree of ill-conditioning is an important limitation on the global-scale trace gas inversion problem.

11.3.2 CHARACTERIZING RESOLUTION

Since ill-conditioned inverse problems are typically characterized by increasing difficulty in estimating successively smaller scale detail, an important theme of inver-sion studies, across all areas of science, has been the analysis of what features can be validly estimated (Backus and Gilbert, 1968; Jackson, 1972). The main trade-off is between resolution and variance in the estimates. Below some limiting resolution, there can be an explosive growth in variance.

The principle, identified by Wunsch and Minster (1982), is that only a small number of modes can be resolved and that these specific modes will be defined by the data. Failure to recognize this, with a small number of modes selected for estima-tion without reference to the resolution implied by the data, was one of the failings in the analysis by Fan et al. (1998) discussed in Section 11.4.1.

While the resolution of inverse problems requires a careful analysis for each particular case, qualitative characterizations may be possible. A set of characteristic numbers has been defined (Enting, 2002, Section 8.1), defining such things as the number of signal components about the noise level, the number of effectively inde-pendent data values, and the computational resolution. This can lead to significant "truncation error," which is often neglected. An appropriate formulation is given by Trampert and Sneider (1996) in the context of seismology. The applicability to CO_2 inversions was confirmed by Kaminski et al. (2001). A schematic of the formalism for assessing truncation error has been given in Figure 8.3 of Enting (2002), and Enting (2008) has used examples from digital filtering of time series to illustrate how various forms of noise are affected by truncation.

The conditions for working with reduced computational resolution are either

A The small-scale variability is known to be small **or both of**
B1 The small-scale variability is not of interest, **and**
B2 The small-scale variability can be separated from the larger-scale variability.

A generic mathematical technique for tackling inverse problems is known as regularization (Tikhonov, 1963). Various regularizations have been applied to CO_2

inversions (Fan et al., 1999). However, in order to obtain valid estimates from regularization, one needs to have one or other of the alternatives above applying. A number of workers have identified regularization as an implicit Bayesian constraint, in that it is making prior assumptions about the small-scale behavior having low variability.

A high degree of negative autocorrelation occurs between successive zonal flux estimates. This reflects the ill-conditioning in that the smallest scales are the most-poorly resolved. It was suggested (Enting, 2002, Section 8.2) that it may be appropriate to communicate results in terms of integrated sources. As well as having plots that may be easier to interpret, the use of integrated fluxes facilitates comparison of results that have been calculated using different discretizations.

A further complication occurs in multi-tracer inversions where the signal-to-noise characteristics of the different tracers may lead to differences in resolution. For that example, in the study by Ciais et al. (1995b), the ^{13}C resolution, and the consequent land-ocean partitioning, is poorer than the resolution of the total flux that derives from the denser CO_2 data set.

11.3.3 ERRORS IN TRANSPORT MODELS

Section 9.2 of Enting (2002) addresses the issue of transport model error. Model error is a troublesome issue across a wide range of inverse problems. One initiative for trace gas inversions has been the TransCom model intercomparison, set up to parallel other earth system intercomparisons such as AMIP and OCMIP. In earth system science, the term "intercomparison" has come to mean a comparison between models run under equivalent conditions, often with idealized cases designed to reveal (and hopefully categorize) model differences.

There have been several phases of TransCom, targeted at assessing the effect of model error on CO_2 inversions:

Phase 1: CO_2 transport This study compared "forward" calculations for the two main source components: fossil emissions and the seasonal biotic exchange. The results were published by Law et al. (1996) with further detail in a technical paper (Rayner and Law, 1995). An important issue was the so-called rectifier effect (see Section 11.4.3).

Phase 2: SF_6 transport This was undertaken as a transport experiment using a compound for which there were good observations and well-known sources (Denning et al., 1999).

Phase 3: CO_2 inversions This is the "core" TransCom experiment. Gurney et al. (2002) gave initial summary. This was followed by more detailed studies of annual means by Gurney et al. (2003), seasonal cycles by Gurney et al. (2004), and a study of interannual variability by Baker et al. (2006).

Other studies of interannual variability have been reported by Patra et al. (2005) and Gurney et al. (2008). A consistent result is that interannual variability is determined more precisely than long-term means.

At the time of writing, ongoing TransCom activities are

Non-surface data looking at the role of aircraft data and similar data sets.

TCCON The continuous data experiment, looking at the role of high-frequency (typically hourly) data, as opposed to the monthly mean data used in most inversions.

TransCom exists primarily to study the impact of model error in CO$_2$ inversions (although inversion needs to be regarded more broadly than the Bayesian synthesis that is the "standard" TransCom case). However, one of the consequences of working on a template negotiated well in advance of the calculations is that as inversions, TransCom calculations will not be state-of-the-art, but rather closer to a lowest-common-denominator of what can be achieved by a number of groups.

Some of the features where recent TransCom presentations lag current-best-practice inversions are

- Omission of carbon fluxes into the atmosphere in forms other than CO$_2$ (see Section 11.4.2).
- Omission of any uncertainty in the fossil component.
- The calculations have low spatial resolution without any specification of the truncation error discussed above.

Indeed, the first two omissions are specific simplifications for TransCom since a fossil fuel uncertainty and a CO flux were included in the first published Bayesian inversions of CO$_2$ (Enting et al., 1993, 1995), and increasing resolution in inversion calculations is reducing the discretization problems.

11.3.4 EXPERIMENTAL DESIGN

Inversion studies can also be used for purposes of experimental design. The idea is that an inversion technique that includes a systematic assessment of uncertainty can be applied to assess the utility (as measured by reduction in uncertainty) of putative new data.

This is particularly simple in the case of a linear model with multivariate normal error because in this case the posterior uncertainty does not depend on the values of the putative data, but only on its uncertainty (as indicated by Equation 11.7b). Examples using Bayesian synthesis inversion of CO$_2$ were given in the initial paper by Enting et al. (1995), considering aspects such as the utility of ^{13}C data and improved precision in CO$_2$ data. My book described joint reductions in uncertainties of the land-ocean partitioning from various forms of ^{13}C data (Enting, 2002, Fig. 13.1).

A more extensive application of "experimental design" was the evaluation of the CO$_2$ sampling network, initially as optimal location of additional stations and also as reconfiguration of the entire network (Rayner et al., 1996). The latter case involved a complex optimization with multiple local minima, and a simulated annealing technique was used, drawing on applications in seismology (Hardt and Scherbaum, 1994). A similar approach to network design, using simulated annealing as the optimization technique, has also been applied in oceanography. Further network design studies were reported by Patra and Maksyutov (2002) and Patra et al. (2003). Rayner (2004) has extended network design calculations through optimization in the presence of model error, this being based on results from TransCom.

Two areas of ongoing research where trace inversions are being used for "experimental design" described in the following sections are

- Analyzing the utility of continuous data (hourly or better time-resolution)
- Analyzing the utility of satellite data

Most global-scale inversions of CO_2 have used monthly mean data, obtained mainly from flask sampling. However, there are long-term continuous measurements from the NOAA CMDL (NOAA GMCC) program at the South Pole, Samoa, Mauna Loa, and Point Barrow. Other national programs also maintain continuous analyzers.

The development of the LoFlo analyzer (Francey and Steele, 2003) potentially allows wider deployment of continuous analyzers by reducing the demands for calibration gases. Law and collaborators have undertaken a series of experimental design studies, exploring the utility of various modes of deployment (Law et al., 2002, 2003b, 2004).

Inversions using continuous data (often inverting hourly values) place severe demands on the modeling:

1. The transport modeling needs to be based on analyzed winds in order to match specific events, in contrast to low-resolution inversions that often use transport field generated by GCMs.
2. The statistical modeling needs to capture a complicated space–time structure of the transport fields.

For experimental design studies, point (1) is not a problem, since the analysis can be done using synthetic data. Point (2) remains a challenge to researchers. Additional research is being conducted through the TransCom continuous data activity.

The potential for measuring CO_2 concentrations from satellites brings the prospect of global-scale coverage at high spatial resolution and regular time sampling. Against this stand the technical difficulties of achieving reliable measurements from space. As well as future missions targeted at CO_2 (OCO and GOSAT) there have also been efforts to obtain CO_2 information from existing instruments (e.g., AIRS and SCIAMACHY) that were designed for other purposes.

This creates a valuable role for "experimental design" studies that assess the potential utility of satellite data. Initial studies (Rayner and O'Brien, 2001a,b; O'Brien and Rayner, 2002) suggested that a useful reduction in uncertainty in estimated fluxes could be achieved with column-integrated values with an $8° \times 10°$ footprint and 2.5 ppm uncertainty on monthly mean values. Pak and Prather (2001) compared the relative utility of various cases such as tropospheric versus whole column and spatial resolution versus coverage. They noted in particular that the most useful data would be for tropical regions, that is, those that currently have least surface data. An overview of the OCO (Orbiting Carbon Observatory) is given by Crisp et al. (2004). Houweling et al. (2004) compared OCO and SCIAMACHY. In particular they found that techniques based on near-infrared (rather than thermal infrared) were preferable because of better performance near the surface. The OCO launch failure in February 2009 has been a major setback to carbon cycle studies.

11.4 CO_2

11.4.1 Context

Section 14.1 of Enting (2002) gives an overview of the progression of key ideas about the carbon cycle (based on pers. comm. from Roger Francey). This can be summarized (and updated) to give

- Before 1958: the growth in CO_2 was not well established.
- Post-1958: about half of the fossil CO_2 was found to remain in the atmosphere.
- Circa 1975: models calibrated using ^{14}C indicated an imbalance that became known as the "missing sink."
- Circa 1980: estimates of large emissions from deforestation (subsequently revised downwards) exacerbated these discrepancies.
- 1983: Response function analysis (Oeschger and Heimann, 1983) pointed out that present atmospheric budget depends, in part, on past imbalances— the budget for a particular time cannot be considered in isolation from previous changes.
- 1985 Onwards: ice-core data provide the requisite information about past changes, allowing a deconvolution of past emissions (Siegenthaler and Oeschger, 1987).
- Circa 1990: conflicting views of the carbon budget: low ocean uptake versus high ocean uptake from both inversion (Tans et al., 1990; Keeling et al., 1989) and isotopic (Tans et al., 1993; Quay et al., 1992) studies.
- 1992: Much of the discrepancy resolved by Sarmiento and Sundquist (1992) by clarifying distinctions between different type of atmospheric budget as described below—a consistent analysis of the $CO_2/^{13}CO_2$ budget was produced by Heimann and Maier-Reimer (1996).
- Circa 1995: ^{13}C records indicated large interannual variability (Francey et al., 1995; Keeling et al., 1995).
- Mid-1990s: additional constraints were provided by measurements of trends in O_2/N_2 obtained by a variety of techniques (Keeling et al., 1993; Bender et al., 1994; Manning et al., 1999).
- Circa 1995: concern about the "rectifier effect" that is, mean gradients induced by diurnal and seasonal covariance between fluxes and transport, and the extent to which inversions were biased due to models underestimating the effect (Denning et al., 1995).

- Circa 2000: increasing focus on processes rather than budgets.
- Concern about feedbacks arising from disruption of the carbon cycle by anthropogenic climate change (Friedlingstein, 2004; Friedlingstein et al., 2006).

There was also a burst of scientific activity (and a large degree of politicized misrepresentation) following the estimate by Fan et al. (1998) that the U.S. carbon sink (actually for the region south of 50°N) exceeded the U.S. fossil carbon emissions. Some of the points to note concerning this study are

- The model transport was at an extreme of the range in the TransCom intercomparison.
- The statistical analysis of uncertainty was compromised by performing calculations in terms of offsets from the South Pole, thereby violating the underlying independence assumptions.
- As noted above, the calculation attempted to achieve high spatial resolution in regions where such resolution was not supported by the data.
- The inversion involved CO_2, thereby neglecting carbon emitted as CO and other reduced forms of carbon from both natural and anthropogenic sources.
- In addition, the period studied included a time of lower natural emissions globally, apparently arising from reduced respiration in the lower temperatures that followed the Pinatubo eruption.

Further details are given in TransCom studies such as Law et al. (2003a).

11.4.2 BUDGETS

Interpretation of inversion results needs to be undertaken in terms of a consistent budget. Failure to achieve this has been the reason for a number of apparent discrepancies in past analyses. The main issues are

- On a regional scale, the net CO_2 flux into the atmosphere differs from the net carbon flux because some carbon enters the atmosphere in reduced forms such as carbon monoxide (CO), methane (CH_4), and other organic compounds. An initial study (Enting and Mansbridge, 1991) using a two-dimensional model quantified the corrections that would be required for inversions that ignored these fluxes. Recently, a study with a full three-dimensional model (Suntharalingam et al., 2005) has largely confirmed these early results. Note that although the earliest synthesis inversions did include the role of CO (Enting et al., 1993, 1995), this was not followed in the TransCom intercomparison. Similarly, the role of CO and its precursors has generally been neglected in the CO_2 inversion literature.
- The respective carbon fluxes between the atmosphere and land and ocean systems do not represent the rate of change of carbon content of these reservoirs because of the transfer of carbon from terrestrial systems to the oceans via rivers (Sarmiento and Sundquist, 1992).

The effects of neglecting CO and its precursors depend on what is required:

1. Neglect of a free atmosphere source in the set of basis functions means that other sources will correspondingly increase. Other things being equal, this error will have a fairly uniform distribution, somewhat peaked toward tropics. However, a Bayesian inversion will preferentially distribute the error into the regions with the loosest priors. A correction for this term will lead to a surface CO$_2$ inversion that is formally correct.
2. More seriously, many CO$_2$ inversions are presented in terms of nonfossil fluxes by subtracting the full fossil carbon emissions, rather than fossil CO$_2$ emissions. To correct such a budget to produce nonfossil CO$_2$ emissions requires adding back on the non-CO$_2$ fossil emissions. This produces a surface CO$_2$ flux budget for nonfossil CO$_2$ fluxes.
3. If a surface carbon flux budget is required, then one needs to add the surface biotic emissions of non-CO$_2$ carbon (excluding components that are rapidly oxidized to products that return to the ecosystem via rain-out).
4. A carbon storage budget would require an additional correction for carbon cycled from biosphere to oceans via rivers before returning to the atmosphere.

There are three ways of evaluating these corrections:

- From the 2-D analysis by Enting and Mansbridge (1991)
- From the 3-D study by Suntharalingam et al. (2005)
- By running synthesis inversions with and without proper treatment of the various effects

Note that all but item (1) are, in principle, off-line corrections, and really constitute a change in what is being estimated from the inversion. The point where failure to be consistent in the treatment of (2), (3), and (4) will affect the inversion calculation (as opposed to its interpretation) is if priors (especially tight priors) are based on the wrong sort of budget.

As noted above, the analysis by Sarmiento and Sundquist (1992) showed that a number of the discrepancies in published results arose from the use of different forms of atmospheric budget. They also noted the effect on inversion estimates of a skin effect (Robertson and Watson, 1992) that implied a correction to flux estimates derived from ocean carbon measurements.

Around 1990, there were conflicting inversion estimates (Keeling et al., 1989; Tans et al., 1990) about the size of the ocean sink, with a relatively sharp distinction between estimates of 1 GtC y^{-1} or less and those that were nearer to 2 GtC y^{-1}. Each of these inversions was based on ad hoc fits, chosen to be consistent with various other data—they were undertaken prior to the introduction of systematic Bayesian statistical inversions into CO$_2$ studies.

The inversions by Keeling et al. (1989) indicated a large northern ocean sink. In part, this was forced by the application of a constraint of a large total ocean sink, a result derived from double-deconvolution of the long-term CO$_2$ and ^{13}CO$_2$ records.

(The advantage of such a double deconvolution calculation is that it automatically incorporates consistent estimates of the so-called isoflux contributions.) In contrast, the inversion by Tans et al. (1990) used ocean CO_2 partial pressure data in northern and tropical regions with the inversion mainly determining southern ocean sinks and northern and tropical land fluxes.

Although the various contending calculations were performed before the advent of systematic uncertainty analysis from the Bayesian synthesis approach, it is possible to revisit the various cases, cast them in Bayesian synthesis form, and compare the uncertainty ranges. The results are shown in Table 11.1. Since this calculation is intended to illustrate the difference, each of these calculations follows the original form, neglecting the role of CO. Additional details are given by Wang et al. (2009).

The various cases are

1. *Constrained ocean* This applies an ocean constraint derived from multi-decadal box modeling. The initial estimate is from Keeling et al. (1989). In a Bayesian inversion, applying an ocean constraint (in this case applied as a prior with $\pm 0.1\,\text{GtCy}^{-1}$ uncertainty) collapses the initial inversion, although not greatly reducing the range of estimated northern ocean fluxes.

2. *pCO₂* This case shows the effect of adding p_{CO2} data for the northern and tropical oceans. This has greatly constrained the northern uptake, while only slightly constraining the range of possible total uptake. The first four cases are values found in ad hoc fits by Tans et al. (1990). Thus the

---

TABLE 11.1

Comparison of Carbon Budget Estimates from ca. 1990, Showing Discrepancies and Resolution

Case	Northern	Global	Reference
Constrained ocean	−2.5	−1.6	Keeling et al. (1989)
Constrained ocean	−2.2 ± 0.4	−1.8 ± 0.1	
pCO₂	−0.7	−0.3	Tans et al. (1990)
pCO₂	−0.7	−0.5	Tans et al. (1990)
pCO₂	−0.7	−0.6	Tans et al. (1990)
pCO₂	−0.7	−0.8	Tans et al. (1990)
pCO₂	−0.8 ± 0.1	−0.6 ± 0.5	—
pCO₂ (corrected)	−0.8 ± 0.1	−1.4 ± 0.5	
Inverse	−2.2 ± 0.4	−1.8 ± 0.8	

Note: Original estimates from Tans et al. (1990) and Keeling et al. (1989) are listed and compared to Bayesian synthesis estimates (those showing uncertainties) based on essentially the same data. A synthesis estimate is also given using pCO₂ data with a skin temperature correction applied. The final line gives the Bayesian estimate that is obtained without using the ocean constraint and using pCO₂ as a weak constraint.

statistical fit spans a somewhat wider range than the cases considered by Tans et al. (1990).
3. *Skin correction* This repeats case 2, with a correction for the skin effect identified by Robertson and Watson (1992) to the priors for northern and tropical oceans.
4. *Inverse* The final line represents an inversion with very loose priors, and using neither the total ocean constraint nor the pCO$_2$ data. As indicated by the following two cases, additional constraints can greatly influence the fit. A wide range of ocean total uptake is consistent with the data. (The original Bayesian synthesis calculations (Enting et al., 1995) used ^{13}C data to constrain the land–ocean partitioning.)

11.4.3   Results from Inversions

Heimann et al. (2004) reviewed CO$_2$ inversions and noted that much of the small-scale information is really coming from the priors. Conversely, on the very largest scales (i.e., global totals) inversions contribute relatively little to land–ocean partitioning (see Enting, 2002, Fig. 13.1). On the continental scale, inversions can produce useful information. A summary of results is given below, including results from the TransCom control experiment (Gurney et al., 2003), which was a 22-region synthesis inversion of monthly mean data covering 1992–1996.

Ocean sink Section 11.4.2 describes discrepancies in estimates of ocean carbon uptake produced circa 1990, and the resolution through consistent budgeting.

There were also, slightly later, two different analyses based on ^{13}C that also differed between high ocean uptake (Quay et al., 1992) and low ocean uptake (Tans et al., 1993). A consistent ^{13}C-based budget was presented by Heimann and Maier-Reimer (1996). For an overview of isotopic budgeting calculations, see Enting (2002, Section 10.4), which identifies the Quay et al. budget as a "storage" budget and the Tans et al. budget as a flux budget. The TransCom result is a sink of -1.5 ± 1.1 GtC y^{-1}.

The overall result is that inversions have played only a limited role in determining the net ocean uptake. In spite of the concept that interhemispheric differences are strongly correlated with land–ocean differences, the spatial gradient is only a poor proxy for process differences. However, as discussed below, inversion studies have put useful constraints on southern hemisphere ocean fluxes. To a large extent, some of the strongest constraints on estimates of ocean carbon uptake come from measurements of O$_2$:N$_2$ ratios. What these determine, when combined with CO$_2$ trends, is the changes of oxidized carbon (essentially storage in the ocean) versus reduced carbon (primarily in terrestrial biota). As such it is a "storage" budget rather than a flux budget.

Tropical biota: The appreciation of the large carbon releases from land-use change in tropical regions brought the carbon budget discrepancies (the issue of the so-called missing sink) into particular prominence. The ability to estimate net tropical fluxes from global inversions is severely hampered by (a) the shortage of sampling sites in the relevant regions and (b) the nature of the atmospheric circulation where tropical air is transported vertically, so that any anomalous concentration signal makes little contribution to concentration gradients at tropical latitudes. Early mass-balance

inversions (Tans et al., 1989; Enting and Mansbridge, 1991) found a small low latitude (12°S–12°N) source, comparable to (or slightly less than) the expected oceanic
source, implying a near-zero net flux from terrestrial biota. The TransCom estimate
of 1.1 ± 1.3 GtC y^{-1} suggest less of an apparent discrepancy, but insufficient data to
resolve the fluxes.

Interannual variability: Two studies (Francey et al., 1995; Keeling et al., 1995)
used joint globally aggregated $CO_2/^{13}CO_2$ budgets to partition the interannual variability between atmosphere and oceans. Bousquet et al. (2000) described a time-
dependent synthesis of the period 1980–1998. They found that in the 1980s, most
interannual variability came from tropical land regions while in the 1990s northern
land ecosystems contributed most variability. In particular, they found a large but
short-lived carbon uptake over North America over the period 1992–1993, associated with colder than average temperatures. Various studies, including Law and
Rayner (1999) described below, and most recently from TransCom (Baker et al.,
2006; Gurney et al., 2008), have noted that estimates of interannual variability are
more robust than estimates of annual means.

Rayner et al. (1999b) compared CO_2 inversion results to the Southern Oscillation
Index (SOI), as a characterization of the ENSO phenomenon. The SOI-CO_2 connection has been known since the 1976 study by Bacastow (1976) and has been studied
by many other workers (Thompson et al., 1986; Elliott and Angell, 1987; Elliott
et al., 1991). In order to help ensure robustness of the results, Rayner et al. used three
different inversion techniques: data assimilation (Dargaville and Simmonds, 2000),
mass balance, and synthesis. They found an apparent close correlation when the CO_2
flux led the SOI by some months, in spite of a lack of any physical mechanism. Since
the ESNO is quasi-periodic, with preferential phasing relative to the annual cycle,
such correlation techniques, designed for stationary time series, will not necessarily be appropriate. A more detailed analysis of the results showed an early-ENSO
CO_2 decline closely synchronized with the SOI (and apparently affecting the tropical
oceans), while the CO_2 increase in the later ENSO phase seemed to lag the SOI and
(in two of the three analyses) be associated with land biota.

NH sink: As noted above, an early inversion (Keeling et al., 1989) indicated a
large northern ocean sink. In part, this was forced by the application of the constraint
of a large total ocean sink—a biased constraint due to the failure to distinguish
between flux budgets and storage budgets (as described in Section 11.4.2). Ciais et al.
(1995b) identified the northern sink as terrestrial on the basis of ^{13}C data. Bousquet
et al. (1999) found, for the 1985–1995 average, northern sinks (in GtC y^{-1}) of $0.7 \pm
0.7$ for North America, 0.2 ± 0.3 for North Pacific, 0.5 ± 0.8 for Europe, 0.7 ± 0.3 for
North Atlantic, and 1.2 ± 0.8 for North Asia. Note that most of these estimates seem
to have been produced by subtracting fossil carbon sources, rather than fossil CO_2
sources and so are likely to estimate too large a sink, especially in North America.

Rectifier: It has long been appreciated that covariance between seasonal variations
in fluxes and transport would create mean spatial gradients. In particular, it is expected
that there was a mean interhemispheric CO_2 gradient in preindustrial times. The effect
was apparent in Phase 1 of TransCom with a "split" into two groups, depending on how
models treated the boundary layer. The issue was emphasized by Denning et al. (1995)
who also noted the effect of covariance and concentration over the diurnal cycle.

A set of comparisons by Law and Rayner (1999) suggested that the modeling of the rectifier could have a significant effect on estimates of land–ocean partitioning of fluxes within zones, but a lesser effect on larger (semihemispheric) scales. They also suggested that estimates of interannual variability were more robust (with respect to rectifier uncertainty) than estimates of long-term means. The TransCom control partitions the northern sink as -2.3 ± 0.6 GtC y^{-1} to land and -1.1 ± 0.5 GtC y^{-1} to oceans.

SH sink: A relatively low southern hemisphere uptake has been a feature of most inversion studies. One area of difference is whether there is a net source at latitudes south of about 50°S. This has been further studied by Roy et al. (2003) based on additional *in situ* ocean measurements. One of the difficulties in extracting an overview is the small latitude range involved, given the differences in spatial discretization of various modeling groups. One reason for believing the sink to be stronger than previous estimates is that, with only spatially-sparse sampling, Cape Grim, Tasmania had the lowest annual mean CO$_2$ concentration. More recently, Easter Island has been found to have lower concentrations. The TransCom estimate for the southern hemisphere is -0.8 ± 0.7 GtC y^{-1} to oceans (and -0.2 ± 1.1 GtC y^{-1} to land). This is an area of ongoing scientific debate.

11.5   EMERGING TRENDS

In looking to the future, one can foresee changes with analysis of smaller scales, the incorporation of new data streams, the development of new inversion techniques, and changing objectives. Some emerging trends are

Smaller spatial scales: The majority of the inversions described in this chapter apply on the global scale, aiming to resolve fluxes representative of large regions. In principle, similar approaches can be applied on smaller scales although details of the computational techniques change (Enting, 2002, Chapter 17). In analyzing smaller spatial scales, much of the relevant information is in the concentration variations on small timescales, the so-called continuous data (to be contrasted with discrete flask sampling and intervals of about a week). A number of "experimental design" studies (Law et al., 2002, 2003b, 2004) have analyzed the requirements and potential gains in using high-resolution CO$_2$ data in global (and regional) inversions. At the time of writing, this work is being extended as a TransCom project.

Additional/enhanced data streams: There is increasing recognition that improved understanding of the carbon cycle will involve the interpretation of new types of data. This expanded framework is sometimes referred to as multiple constraints (Kruijt et al., 2001; Wang and Barrett, 2003). Of course, the basic approach of the Bayesian synthesis inversion is built on the use of data other than atmospheric concentrations as providing the constraints needed to regularize the ill-conditioned inverse problem. The range of carbon cycle data has been reviewed by Canadell et al. (2000) and the statistical characteristics of these various data sets discussed (qualitatively) by Raupach et al. (2005). Of the emerging data streams, satellite data will provide formidable computational challenges. The launch failure of the Orbiting Carbon Observatory (OCO) represents a major setback. In the process of developing applications involving these new data streams, the use of inversions studies as a tool for experimental design can be expected to continue.

Inversion techniques: Further development of inversion techniques can be expected, both to refine the type of calculations reviewed here and to accommodate the new types of data. A few new techniques are noted in Section 11.2.4. Some other likely future directions include the increased use of nonlinear estimation (Enting, 2002, Chapter 12). Linear estimation is appropriate when the model relations are linear and the statistics are multivariate Gaussian, as in the classic "synthesis" approach to estimating fluxes. As more complex systems are analyzed, these simplifying conditions are likely to prove less and less adequate as working approximations. Applications involving varying forms of data assimilation are also likely to form an increasing part of carbon cycle research (Wang et al., 2009). In these various developments, the use of adjoint modeling (Giering, 2000) is likely to become increasingly important.

Evolving objectives: As noted in Section 11.4.1, much of the current interest in the carbon cycle involves the coupling between the carbon cycle and the physical climate system. Such coupling implies feedbacks that could exacerbate the ongoing anthropogenic warming. A range of initial studies are being undertaken in the context of the C4MIP intercomparison (Fung et al., 2000; Rayner, 2001; Friedlingstein et al., 2006). Carbon cycle inversions may have a future role in detecting the "fingerprints" of such feedbacks when and if they occur. In the more immediate future, carbon cycle inversions are likely to focus on the calibration of the models that are being used to project climate change in the presence of such feedbacks. As discussed in Section 11.2.4, such "process inversion" may often involve "data assimilation" using proxy data to estimate, via model relations, the forcing of the carbon system. It is also possible that the links between photosynthesis and plant water use may mean that carbon data provide information about terrestrial water balance, leading to possible applications in synoptic and seasonal forecasting. In these various inversion problems, the multiple interacting scales, a characteristic of complex systems, (c.f., Falkowski et al., 2000), lead to a new class of mathematically challenging inverse problems. As a further step, the Global Carbon Project plan (Global Carbon Project, 2003) envisages modeling of the coupled interactions of the climate system with human systems. This will require a major realignment of modeling approaches.

11.6 CONCLUDING REMARKS

In the concluding chapter of my book, I noted the rapid growth in the field of trace gas inversions. This rapid growth has continued and the pace is accelerating. New data streams will stimulate further growth, particularly once satellite data become available.

An analysis of the science of ozone depletion and the ozone hole (Christie, 2000) notes that for both the British Halley Bay observations and the NASA satellite retrievals, limitations in data processing and analysis led to a delay in recognition of the ozone hole for a number of years after it appeared. It was suggested that this reflected the less-developed state of computer systems in the early 1980s and that a similar situation is now less likely to occur. Compared to the 1980s, raw computing power is indeed less of a problem now, although the projected volume of raw satellite data is formidable. There would seem to be a high risk of a gap between observational capability and scientific understanding in the development of techniques that extract the significance of these data. Understanding the carbon cycle, with its characteristic "complex systems" features of feedbacks, nonlinearity, and multiple timescales, creates a need to go beyond "business-as-usual" analyses.

ACKNOWLEDGMENTS

The Centre of Excellence for Mathematics and Statistics of Complex Systems (MASCOS) is funded by the Australian Research Council. The author's fellowship at MASCOS is supported in part by CSIRO. Nathan Clisby of MASCOS provided helpful comments on the manuscript. As with my book, the present account draws on the work of many collaborators, both from CSIRO (and a number formerly with CSIRO) and beyond. Particular thanks are due to Prof. Inez Fung. My recent thinking on trace gas inversions has been greatly stimulated by the lecturers and students at the 2006 MSRI-NCAR workshop on Carbon Data Assimilation.

REFERENCES

R. B. Bacastow. Modulation of atmospheric carbon dioxide by the Southern Oscillation. *Nature*, 261:116–118, 1976.

G. Backus and F. Gilbert. The resolving power of gross earth data. *Geophys. J. R. Astr. Soc.*, 13:247–276, 1968.

D. F. Baker, R. M. Law, K. R. Gurney, P. Rayner, P. Peylin, A. S. Denning, P. Bousquet et al. TransCom 3 inversion intercomparison: Impact of transport model errors on the interannual variability of regional CO_2 fluxes, 1988–2003. *Glob. Biogeochem. Cycles*, 20:GB1002, doi:10.1029/2004GB002439, 2006.

M. L. Bender, P. P. Tans, J. T. Ellis, J. Orchado, and K. Habfast. A high precision isotope ratio mass spectrometry method for measuring O_2/N_2 ratio of air. *Geochim. et Cosmochim. Acta*, 58:4751–4758, 1994.

G. Boer. Climate Model Intercomparisons, In P. Mote and A. O'Neill, eds., *Numerical Modelling of the Global Atmosphere*, Kluwer, Dordrecht, the Netherlands, 2000, pp. 443–464.

B. Bolin and C. D. Keeling. Large-scale atmospheric mixing as deduced from the seasonal and meridional variations of carbon dioxide. *J. Geophys. Res.*, 68:3899–3920, 1963.

P. Bousquet, P. Ciais, P. Peylin, and P. Monfray. Inverse modelling of annual atmospheric CO_2 sources and sinks 2. Sensitivity study. *J. Geophys. Res.*, 104D:26179–26193, 1999.

P. Bousquet, P. Peylin, P. Ciais, C. Le Quéré, P. Friedlingstein, and P. P Tans. Regional changes in carbon dioxide fluxes of land and oceans since 1980. *Science*, 290:1342–1346, 2000.

T. M. Butler, I. Simmonds, and P. J. Rayner. Mass balance inverse modelling of methane in the 1990s using a chemistry transport model. *Atmos. Chem. Phys.*, 4:2561–2580, 2004.

J. P. Canadell, H. A. Mooney, D. D. Baldocchi, J. A. Berry, J. R. Ehleringer, C. B. Field, S. T. Gower et al. Carbon metabolism of the terrestrial biosphere: A multitechnique approach for improved understanding. *Ecosystems*, 3:115–130, doi: 10:1007/s100210000014, 2000.

M. Christie. *The Ozone Layer: A Philosophy of Science Perspective*. CUP, Cambridge, U.K., 2000.

P. Ciais, P. P. Tans, J. W. C. White, M. Trolier, R. J. Francey, J. A. Berry, D. R. Randall, P. J. Sellers, J. G. Collatz, and D. S. Schimel. Partitioning of ocean and land uptake of CO_2 as inferred by $\delta^{13}C$ measurements from the NOAA climate monitoring and diagnostics laboratory global air sampling network. *J. Geophys. Res.*, 100D:5051–5070, 1995a.

P. P. Ciais, P. Tans, M. Trolier, J. W. C. White, and R. J. Francey. A large northern hemisphere terrestrial CO_2 sink indicated by the C^{13}/C^{12} ratio of atmospheric CO_2. *Science*, 269:1098–1102, 1995b.

T. J. Conway, P. P. Tans, L. S. Waterman, K. W. Thoning, D. R. Kitzis, K. A. Masarie, and N. Zhang. Evidence for interannual variability of the carbon cycle from the National Oceanic and Atmospheric Administration/Climate Monitoring and Diagnostics Laboratory global air sampling network. *J. Geophys. Res.*, 99D:22831–22855, 1994.

D. Crisp, R. M. Atlas, F. M. Breon, L. R. Brown, J. P. Burrows, P. Ciais, B. J. Connor et al. The Orbiting Carbon Observatory (OCO) mission. *Adv. Space Res.*, 34:700–709, 2004.

R. J. Dargaville and I. Simmonds. Calculating CO_2 fluxes by data assimilation coupled to a three dimensional mass balance inversion. In Kasibhatla et al. (2000), pp. 255–264.

A. S. Denning, I. Y. Fung, and D. Randall. Latitudinal gradient of atmospheric CO_2 due to seasonal exchange with the land biota. *Nature*, 376:240–243, 1995.

A. S. Denning, M. Holzer, K. R. Gurney, M. Heimann, R. M. Law, P. J. Rayner, I. Y. Fung et al. Three-dimensional transport and concentration of SF_6: A model intercomparison study (TransCom 2). *Tellus*, 51B:266–297, 1999.

W. P. Elliott and J. K. Angell. On the relation between atmospheric CO_2 and equatorial sea-surface temperature. *Tellus*, 39B:171–183, 1987.

W. P. Elliott, J. K. Angell, and K. W. Thoning. Relation of atmospheric CO_2 to tropical sea-surface temperatures and precipitation. *Tellus*, 43B:144–155, 1991.

I. G. Enting. Inverse problems in atmospheric constituent studies: III. Estimating errors in surface sources. *Inverse Problems*, 9:649–665, 1993.

I. G. Enting. Constraints on the atmospheric carbon budget from spatial distributions of CO_2. In T. Wigley and D. Schimel, eds., *The Carbon Cycle*, Chapter 8. CUP, Cambridge, U.K., 2000a.

I. G. Enting. Green's function methods of tracer inversion. In Kasibhatla et al. (2000), pp. 19–31.

I. G. Enting. *Inverse Problems in Atmospheric Constituent Transport*. CUP, Cambridge, U.K., 2002.

I. G. Enting. Laplace transform analysis of the carbon cycle. *Environ. Model. Softw.*, 22: 1488–1497, 2007.

I. G. Enting. Assessing the information content in environmental modelling: A carbon cycle perspective. *Entropy*, 10:556–575, 2008.

I. G. Enting and J. V. Mansbridge. Seasonal sources and sinks of atmospheric CO_2: Direct inversion of filtered data. *Tellus*, 41B:111–126, 1989.

I. G. Enting and J. V. Mansbridge. Latitudinal distribution of sources and sinks of CO_2: Results of an inversion study. *Tellus*, 43B:156–170, 1991.

I. G. Enting and G. N. Newsam. Inverse problems in atmospheric constituent studies: II. Sources in the free atmosphere. *Inverse Problems*, 6:349–362, 1990.

I. G. Enting and G. I. Pearman. Description of a one-dimensional carbon cycle model calibrated using techniques of constrained inversion. *Tellus*, 39B:459–476, 1987.

I. G. Enting, R. J. Francey, C. M. Trudinger, and H. Granek. Synthesis inversion of atmospheric CO_2 using the GISS tracer transport model. Technical Paper no. 29, CSIRO Division of Atmospheric Research, 1993.

I. G. Enting, C. M. Trudinger, and R. J. Francey. A synthesis inversion of the concentration and $\delta^{13}C$ of atmospheric CO_2. *Tellus*, 47B:35–52, 1995.

S. N. Evans and P. B. Stark. Inverse problems as statistics. *Inverse Problems*, 18:R55–R97, 2002.

P. Falkowski, R. J. Scholes, E. Boyle, J. Canadell, D. Canfield, J. Elser, N. Gruber et al. The global carbon cycle: A test of our knowledge of the earth as a system. *Science*, 290:291–296, 2000.

S.-M. Fan, M. Gloor, J. Mahlman, S. Pacala, J. Sarmiento, T. Takahashi, and P. Tans. A large terrestrial carbon sink in North America implied by atmospheric and oceanic carbon dioxide data and models. *Science*, 282:442–446, 1998.

S.-M. Fan, J. L. Sarmiento, M. Gloor, and S. W. Pacala. On the use of regularization techniques in the inverse modeling of atmospheric carbon dioxide. *J. Geophys. Res.*, 104D:21503–21512, 1999.

R. J. Francey and L. P. Steele. Measuring atmospheric carbon dioxide—the calibration challenge. *Accred. Qual. Assur.*, 8:200–204, doi: 10.1007/s00769–003–0620–1, 2003.

R. J. Francey, P. P. Tans, C. E. Allison, I. G. Enting, J. W. C. White, and M. Trolier. Changes in oceanic and terrestrial carbon uptake since 1982. *Nature*, 373:326–330, 1995.

P. Friedlingstein. Climate-carbon cycle interactions. In C. B. Field and M. R. Raupach, editors, *The Global Carbon Cycle: Integrating Humans, Climate and the Natural World*, Chapter 10, pp. 217–224. Island Press, Washington, DC, 2004.

P. Friedlingstein, P. Cox, R. Betts, L. Bopp, W. von Bloh, V. Brovkin, P. Cadul et al. Climate-carbon cycle feedback analysis: Results from the C4MIP model intercomparison. *J. Clim.*, 19:3337–3353, 2006.

I. Fung, P. Rayner, P. Friedlingstein, and D. Sahagian. Full-form earth-system models: Coupled carbon-climate interaction experiment (the 'flying leap'). *Glob. Change Newslett.*, 41:7–8, 2000.

I. Y. Fung, C. J. Tucker, and K. C. Prentice. Application of advanced very high resolution radiometer vegetation index to study atmosphere–biosphere exchange of CO$_2$. *J. Geophys. Res.*, 92D:2999–3015, 1987.

I. Y. Fung, J. John, J. Lerner, E. Matthews, M. Prather, L. P. Steele, and P. J. Fraser. Three-dimensional model synthesis of the global methane cycle. *J. Geophys. Res.*, 96D:13033–13065, 1991.

A. Gelb, editor. *Applied Optimal Estimation*. MIT Press, Cambridge, MA, 1974.

R. Giering. Tangent linear and adjoint biogeochemical models. In Kasibhatla et al. (2000), pp. 33–48.

Global Carbon Project. Science framework and implementation, 2003. GCP Report no. 1.

A. Griewank. *Evaluating Derivatives: Principles and Techniques of Algorithmic Differentiation*. SIAM, Philadelphia, PA, 2000.

K. R. Gurney, R. M. Law, A.S. Denning, P. J. Rayner, D. Baker, P. Bousquet, L. Bruhwiler et al. Towards robust regional estimates of CO$_2$ sources and sinks using atmospheric transport models. *Nature*, 415:626–630, 2002.

K. R. Gurney, R. M. Law, A. S. Denning, P. J. Rayner, D. Baker, P. Bousquet, L. Bruhwiler et al. TransCom 3 CO$_2$ inversion intercomparison: 1. Annual mean control results and sensitivity to transport and prior flux information. *Tellus*, 55B:555–579, 2003.

K. R. Gurney, R. M. Law, A. S. Denning, P. J. Rayner, B. C. Pak, D. Baker, P. Bousquet et al. TransCom 3 inversion intercomparison: Model mean results for the estimation of seasonal carbon sources and sinks. *Glob. Biogeochem. Cycles*, 18:1010, 2004, doi:10.1029/2003GB002111.

K. R. Gurney, D. Baker, P. Rayner, and S. Denning. Interannual variations in continental-scale net carbon exchange and sensitivity to observing networks estmated from atmospheric CO$_2$ inversions for the period 1980–2005. *Glob. Biogeochem. Cycles*, 22:GB3025, 2008 doi: 10,1029/2007/GB003082.

D. E. Haas-Laursen, D. E. Hartley, and R. G. Prinn. Optimizing an inverse method to deduce time-varying emissions of trace gases. *J. Geophys. Res.*, 101D:22823–22831, 1996.

M. Hardt and F. Scherbaum. The design of optimum networks for aftershock recordings. *Geophys. J. Int.*, 117:716–726, 1994.

D. Hartley and R. Prinn. Feasibility of determining surface emissions of trace gases using an inverse method in a three-dimensional chemical transport model. *J. Geophys. Res.*, 98D:5183–5197, 1993.

M. Heimann and C. D. Keeling. Meridional eddy diffusion model of the transport of atmospheric carbon dioxide. 1. seasonal carbon cycle over the tropical Pacific Ocean. *J. Geophys. Res.*, 91D:7765–7781, 1986.

M. Heimann and E. Maier-Reimer. On the relations between the oceanic uptake of CO_2 and its isotopes. *Glob. Biogeochem. Cycles*, 10:89–110, 1996.

M. Heimann, C. Rödenbeck, and M. Gloor. Spatial and temporal distribution of sources and sinks of carbon dioxide. In C. B. Field and M. R. Raupach, eds., *The Global Carbon Cycle: Integrating Humans, Climate and the Natural World*, Chapter 8, pp. 187–204. Island Press, Washington, DC, 2004.

S. Houweling, F.-M. Breon, I. Aben, C. Rödenbeck, M. Heimann, and P. Ciais. Inverse modelling of CO_2 sources and sinks using satellite data: A synthetic inter-comparison of measurement techniques and their performance as a function of space and time. *Atmos. Chem. Phys.*, 4:523–538, 2004.

J. Huang. *Optimal Determination of Global Tropospheric OH Concentrations Using Multiple Trace Gases*. PhD thesis, MIT, Cambridge, MA, 2000. Report no. 65 of Center for Global Change Science.

D. D. Jackson. Interpretation of inaccurate, insufficient and inconsistent data. *Geophys. J. R. Astr. Soc.*, 28:97–109, 1972.

C. E. Junge and G. Czeplak. Some aspects of the seasonal variation of carbon dioxide and ozone. *Tellus*, 20:422–434, 1968.

E. Kalnay. *Atmospheric Modeling, Data Assimilation and Predictability*. CUP, Cambridge, U.K., 2003.

T. Kaminski, P. J. Rayner, M. Heimann, and I. G. Enting. On aggregation errors in atmospheric transport inversions. *J. Geophys. Res.*, 106:4703–4715, 2001.

T. Kaminski, W. Knorr, P. J. Rayner, and M. Heimann. Assimilating atmospheric data into a terrestrial biosphere model: A case study of the seasonal cycle. *Glob. Biogeochem. Cycles*, 16:1066, 2002. doi:10.1029/2001GB001463.

P. Kasibhatla, M. Heimann, P. Rayner, N. Mahowald, R. G. Prinn, and D. E. Hartley, eds. *Inverse Methods in Global Biogeochemical Cycles (Geophysical Monograph no. 114)*. American Geophysical Union, Washington, DC, 2000.

C. D. Keeling and M. Heimann. Meridional eddy diffusion model of the transport of atmospheric carbon dioxide. 2. Mean annual carbon cycle. *J. Geophys. Res.*, 91D:7782–7796, 1986.

C. D. Keeling, S. C. Piper, and M. Heimann. A three-dimensional model of atmospheric CO_2 transport based on observed winds. 4: Mean annual gradients and interannual variations. In D. H. Peterson, ed., *Aspects of Climate Variability of the Pacific and Western Americas. Geophysical Monograph 55*. American Geophysical Union, Washington, DC, 1989.

C. D. Keeling, T. P. Whorf, M. Wahlen, and J. van der Plicht. Interannual extremes in the rate of rise of atmospheric carbon dioxide since 1980. *Nature*, 375:666–670, 1995.

R. F. Keeling, R. P. Najjar, M. L. Bender, and P. P. Tans. What atmospheric oxygen measurements can tell us about the global carbon cycle. *Glob. Biogeochem. Cycles*, 7:37–67, 1993.

W. Knorr and M. Heimann. Impact of drought stress and other factors on seasonal land biosphere CO_2 exchange studied through an atmospheric tracer transport model. *Tellus*, 47B:471–489, 1995.

B. Kruijt, A. J. Dolman, J. Lloyd, J. Ehleringer, M. Raupach, and J. Finnigan. Assessing the regional carbon balance: Towards an integrated, multiple constraints approach. *Change*, 56, (March–April 2001):9–12, 2001.

R. M. Law. CO_2 sources from a mass-balance inversion: Sensitivity to the surface constraint. *Tellus*, 51B:254–265, 1999.

R. M. Law and P. J. Rayner. Impacts of seasonal covariance on CO_2 inversions. *Glob. Biogeochem. Cycles*, 13:845–856, 1999.

R. M. Law and P. Vohralik. Methane sources from mass-balance inversions: Sensitivity to transport. CSIRO Atmospheric Research Technical Paper no. 50, 2001. Electronic edition at: *http://www.dar.csiro.au/publications/Law_2001a.pdf*.

R. M. Law, P. J. Rayner, A. S. Denning, D. Erickson, I. Y. Fung, M. Heimann, S. C. Piper et al. Variations in modeled atmospheric transport of carbon dioxide and the consequences for CO$_2$ inversions. *Glob. Biogeochem. Cycles*, 10:783–796, 1996.

R. M. Law, P. J. Rayner, L. P. Steele, and I. G. Enting. Using high temporal frequency data for CO$_2$ inversions. *Glob. Biogeochem. Cycles*, 16, 2002, doi:10.1029/2001GB001593.

R. M. Law, Y.-H. Chen, K. R. Gurney, and TransCom3 modellers. Transcom 3 CO$_2$ inversion inter-comparison: 2. Sensitivities of annual mean results to data choices. *Tellus*, 55B:580–595, 2003a.

R. M. Law, P. J. Rayner, L. P. Steele, and I. G. Enting. Data and modelling requirements for CO$_2$ inversions using high-frequency data. *Tellus*, 55B:512–521, 2003b.

R. M. Law, P. J. Rayner, and Y. P. Wang. Inversion of diurnally varying synthetic CO$_2$: Network optimization for an Australian test case. *Glob. Biogeochem. Cycles*, 18:1044, 2004, doi:10.1029/2003GB002136.

A. C. Manning, R. F. Keeling, and J. P. Severinghaus. Precise atmospheric oxygen measurements with a paramagnetic oxygen analyzer. *Glob. Biogeochem. Cycles*, 13:1107–1115, 1999.

A. Michalak, L. Bruhwiler, and P. P. Tans. A geostatistical approach to surface flux estimation of atmospheric trace gases. *J. Geophys. Res.*, 109:D14109, 2004, doi:10.1029/2003/JD004422.

A. Michalak, A. Hirsch, L. Bruhwiler, K. R. Gurney, and P. P. Tans. Maximum likelihood estimation of covariance parameters for Bayesian atmospheric trace gas surface flux inversions. *J. Geophys. Res.*, 110:D24107, 2005, doi:10.1029/205/JD005970.

G. N. Newsam and I. G. Enting. Inverse problems in atmospheric constituent studies: I. Determination of surface sources under a diffusive transport approximation. *Inverse Problems*, 4:1037–1054, 1988.

D. M. O'Brien and P. J. Rayner. Global observations of the carbon budget. 2. CO$_2$ column from differential absorption of reflected sunlight in the 1.61 µm band of CO$_2$. *J. Geophy. Res.*, 107D, 2002, doi:10.1029/2001JD000617.

H. Oeschger and M. Heimann. Uncertainties of predictions of future atmospheric CO$_2$ concentrations. *J. Geophys. Res.*, 88C:1258–1262, 1983.

H. Oeschger, U. Siegenthaler, U. Schotterer, and A. Gugelmann. A box diffusion model to study the carbon dioxide exchange in nature. *Tellus*, 27:168–192, 1975.

B. C. Pak and M. J. Prather. CO$_2$ source inversions using satellite observations of the upper troposphere. *Geophys. Res. Lett.*, 28:4571–4574, 2001.

P. K. Patra and S. Maksyutov. Incremental approach to the optimal network design for CO$_2$ surface source inversion. *Geophys. Res. Lett.*, 29:1459, 2002. doi:10.1029/2001/GL013943.

P. K. Patra, S. Maksyutov, and TransCom 3 Modellers. Optimal network design for improved CO$_2$ source inversion. *Tellus*, 55B:498–511, 2003.

P. K. Patra, S. Maksyutov, M. Ishizawa, T. Nakazawa, and J. Ukita. Interannual and decadal changes in the sea-air CO$_2$ flux from atmospheric CO$_2$ inverse modelling. *Glob. Biogeochem. Cycles*, 19:GBC4013, 2005, doi:10/1029/2004GB002257.

G. I. Pearman. Atmospheric CO$_2$ concentration measurements. A review of methodologies, existing programmes and available data. Technical Report Report no. 3, WMO Project on Research and Monitoring of Atmospheric CO$_2$, Geneva, 1980.

R. G. Prinn. Measurement equation for trace chemicals in fluids and solution of its inverse. In Kasibhatla et al. (2000), pp. 3–18.

R. G. Prinn, R. F. Weiss, P. J. Fraser, P. G. Simmonds, D. M. Cunnold, F. N. Alyea, S. O'Doherty et al. A history of chemically and radiatively important gases in air deduced from ALE/GAGE/AGAGE. *J. Geophys. Res.*, 105D:17751–17792, 2000.

P. D. Quay, B. Tilbrook, and C. S. Wong. Oceanic uptake of fossil fuel CO_2: Carbon-13 evidence. *Science*, 256:74–79, 1992.

J. T. Randerson, I. G. Enting, E.A.G. Schuur, K. Caldeira, and I. Y. Fung. Seasonal and latitudinal variability of tropospheric $\Delta^{14}CO_2$: Post bomb contributions from fossil fuels, oceans and the stratosphere, and the terrestrial biosphere. *Glob. Biogeochem. Cycles*, 16:1112, doi:10:1029/2002GB001876, 2002.

M. R. Raupach, P. J. Rayner, D. J. Barrett, R. S. DeFries, M. Heimann, D. S. Ojima, S. Quegan, and C. C. Schmullius. Model-data synthesis in terrestrial carbon observation: Methods, data requirements and data uncertainty specifications. *Glob. Change Biol.*, 11:378–397, 2005, doi:10.1111/j.1365–2486.2005.00917.x.

P. J. Rayner. Flying leap becomes C4MIP. *Res. GAIM*, 4:2 (winter 2001):8, 2001.

P. J. Rayner. Optimizing CO_2 observing networks in the presence of model error: Results from TransCom 3. *Atmos. Chem. Phys.*, 4:413–421, 2004.

P. J. Rayner and R. M. Law. A comparison of modelled responses to prescribed CO_2 sources. CRC-SHM Technical Paper no. 1, 1995 (and CSIRO Division of Atmospheric Research Technical Paper no. 36) (CSIRO: Australia).

P. J. Rayner and D. O'Brien. The utility of remotely sensed CO_2 concentration data in surface source inversions. *Geophys. Res. Lett.*, 28:175–178, 2001a.

P. J. Rayner and D. M. O'Brien. Correction to 'The utility of remotely sensed CO_2 concentration data in surface source inversions'. *Geophys. Res. Lett.*, 28:2429, 2001b.

P. J. Rayner, I. G. Enting, and C. M. Trudinger. Optimizing the CO_2 observing network for constraining sources and sinks. *Tellus*, 48B:433–444, 1996.

P. J. Rayner, I. G. Enting, R. J. Francey, and R. Langenfelds. Reconstructing the recent carbon cycle from atmospheric CO_2, $\delta^{13}C$ and O_2/N_2 observations. *Tellus*, 51B:213–232, 1999a.

P. J. Rayner, R. M. Law, and R. Dargaville. The relationship between tropical CO_2 fluxes and the El Niño-Southern Oscillation. *Geophys. Res. Lett.*, 26:493–496, 1999b.

P. J. Rayner, M. Scholze, W. Knorr, T. Kaminski, R. Giering, and H. Widmann. Two decades of terrestrial carbon fluxes from a carbon cycle data assimilation system (CCDAS). *Glob. Biogeochem. Cycles*, 19:GB2026, 2005, doi: 10.1029/2004GB002254.

J. E. Robertson and A. J. Watson. Thermal skin effect on the surface ocean and its implications for CO_2 uptake. *Nature*, 358:738–740, 1992.

C. Rödenbeck, S. Houwerling, M. Gloor, and M. Heimann. CO_2 flux history 1982–2001 inferred from atmospheric data using a global inversion of atmospheric transport. *Atmos Chem. Phys. Discuss.*, 3:2575–2659, 2003.

T. Roy, P. Rayner, R. Matear, and R. Francey. Southern hemisphere ocean CO_2 uptake: Reconciling atmospheric and oceanic estimates. *Tellus*, 55B:701–710, 2003.

J. L. Sarmiento and E. T. Sundquist. Revised budget for the oceanic uptake of anthropogenic carbon dioxide. *Nature*, 356:589–593, 1992.

B. A. Shaby and C. B. Field. Regression tools for CO_2 inversions: Application of a shrinkage estimator to process attribution. *Tellus*, 58B:279–292, 2006.

U. Siegenthaler and H. Oeschger. Biospheric CO_2 emissions during the past 200 years reconstructed by deconvolution of ice core data. *Tellus*, 39B:140–154, 1987.

P. Suntharalingam, J. T. Randerson, N. Krakauer, D. J. Jacob, and J. A. Logan. Influence of reduced carbon emissions and oxidation on the distribution of atmospheric CO_2: Implications for inversion analyses. *Glob. Biogeochem. Cycles*, 19:GB4003, 2005, doi:10.1029/2005GB002493.

P. P. Tans, T. J. Conway, and T. Nakazawa. Latitudinal distribution of the sources and sinks of atmospheric carbon dioxide derived from surface observations and an atmospheric transport model. *J. Geophys. Res.*, 94D:5151–5172, 1989.

P. P. Tans, I. Y. Fung, and T. Takahashi. Observational constraints on the global atmospheric CO_2 budget. *Science*, 247:1431–1438, 1990.

P. P. Tans, J. A. Berry, and R. F. Keeling. Oceanic C^{13}/C^{12} observations: A new window on oceanic CO_2 uptake. *Glob. Biogeochem. Cycles*, 7:353–368, 1993.

A. Tarantola. *Inverse Problem Theory: Methods for Data Fitting and Model Parameter Estimation*. Elsevier, Amsterdam, the Netherlands, 1987.

M. L. Thompson, I. G. Enting, G. I. Pearman, and P. Hyson. Interannual variation of atmospheric CO_2 concentrations. *J. Atmos. Chem.*, 4:125–155, 1986.

A. N. Tikhonov. On the solution of incorrectly posed problems. *Sov. Math. Dokl.*, 4:1035–1042, 1963.

J. Trampert and R. Sneider. Model estimation biased by truncated expansions: Possible artifacts in seismic tomography. *Science*, 271:1257–1260, 1996.

Y.-P. Wang and D. J. Barrett. Estimating regional terrestrial carbon fluxes for the Australian continent using a multiple-constraint approach: I. Using remotely sensed data and ecological observations of net primary production. *Tellus*, 55B:270–289, 2003.

Y.-P. Wang, C.M. Trudinger, and I. G. Enting. Applications of model-data fusion to studies of terrestrial carbon fluxes at different scales. *Agric. Forest Meteorol.* (in press) 2009.

C. Wunsch and J.-F. Minster. Methods for box models and ocean circulation tracers: Mathematical programming and non-linear inverse theory. *J. Geophys. Res.*, 87C:5647–5662, 1982.

APPENDIX A: NOTATION

c	Vector of calculated trace gas concentrations, **c**, especially for CO_2
G	Green's function describing source-to-concentration relation
n	Latitudinal wave number
p	Generic parameter vector, elements denoted p_α
$S_j(t)$	CO_2 flux for region j
$T_j(\mathbf{c})$	Transport operator, giving contribution to rate of change of c_j due to transport, given a concentration distribution, **c**
t	Time
x	Generic parameter vector in linear model
X	Generic data covariance matrix for observations, **z**
Y	Generic data covariance matrix for priors, $\mathbf{x}_{prior}$
z	Generic data vector
Θ	Objective function, minimized in inversion process. Often derived from a log-likelihood expression

APPENDIX B: ACRONYMS AND ABBREVIATIONS

AGAGE	Advanced Global Atmospheric Gases Experiment (Prinn et al., 2000). A successor to GAGE and (ALE Atmospheric Lifetime Experiment)
AMIP	Atmospheric Model Intercomparison Project (Boer, 2000)
CCDAS	Carbon Cycle Data Assimilation System (Rayner et al., 2005)
CSIRO	Commonwealth Scientific and Industrial Research Organisation (Australia)
C4MIP	Coupled-Carbon-Cycle-Climate Intercomparison Program (Rayner, 2001)

ENSO	El Niño/Southern Oscillation; A large-scale multi-year fluctuation in the coupled atmosphere-ocean system in the southern and equatorial Pacific region
GOSAT	Greenhouse gas Observing Satellite; Japanese mission, scheduled for 2009
NOAA	National Oceanic and Atmospheric Administration (United States)
OCMIP	Ocean Carbon Modeling Intercomparison Project
OCO	Orbiting Carbon Observatory (Crisp et al., 2004); Proposed satellite mission, failed on launch, February 2009
SCIAMACHY	Scanning Imaging Absorbtion spectroMeter for Atmospheric ChartograpHY. (European Space Agency)
SOI	Southern Oscillation Index; A measure, defined in terms of monthly-mean pressure differences across the Pacific, of the atmospheric component of ENSO
WMO	World Meteorological Organization

Index

T - #0368 - 071024 - C356 - 234/156/16 - PB - 9780367384814 - Gloss Lamination